Series in BioEngineering

The Series in Bioengineering serves as an information source for a professional audience in science and technology as well as for advanced students. It covers all applications of the physical sciences and technology to medicine and the life sciences. Its scope ranges from bioengineering, biomedical and clinical engineering to biophysics, biomechanics, biomaterials, and bioinformatics.

Indexed by WTI Frankfurt eG, zbMATH.

Kamarul Arifin Ahmad ·
Nor Asilah Wati Abdul Hamid ·
Mohammad Jawaid · Tabrej Khan · Balbir Singh
Editors

High Performance Computing in Biomimetics

Modeling, Architecture and Applications

Springer

Editors
Kamarul Arifin Ahmad
Department of Aerospace Engineering
Faculty of Engineering
Universiti Putra Malaysia
Serdang, Selangor, Malaysia

Mohammad Jawaid
Department of Chemical and Petroleum Engineering
College of Engineering
United Arab Emirates University
Al Ain, Abu Dhabi, United Arab Emirates

Balbir Singh
Department of Aeronautical and Automobile Engineering
Manipal Institute of Technology
Manipal Academy of Higher Education (MAHE)
Manipal, Karnataka, India

Nor Asilah Wati Abdul Hamid
Institute of Mathematical Research
Universiti Putra Malaysia
Serdang, Selangor, Malaysia

Tabrej Khan
Department of Engineering Management
College of Engineering
Prince Sultan University
Riyadh, Saudi Arabia

ISSN 2196-8861 ISSN 2196-887X (electronic)
Series in BioEngineering
ISBN 978-981-97-1016-4 ISBN 978-981-97-1017-1 (eBook)
https://doi.org/10.1007/978-981-97-1017-1

This Springer imprint is published by the registered company Springer Nature Singapore Pte Ltd.
The registered company address is: 152 Beach Road, #21-01/04 Gateway East, Singapore 189721, Singapore

Paper in this product is recyclable.

Preface

In recent years, the field of biomimetics has emerged as a fascinating and rapidly evolving area of research, combining the principles of biology, engineering, and computer science to create innovative solutions inspired by nature. The remarkable ability of living organisms to adapt, optimize, and perform complex tasks has inspired scientists and engineers to develop computational models and architectures that mimic biological systems. At the heart of these endeavors lies high-performance computing, a field that has revolutionized the way we approach scientific and engineering challenges. "High-Performance Computing in Biomimetics: Modeling, Architecture, and Applications" explore the intersection of these two dynamic fields, exploring the vast potential of high-performance computing in advancing biomimetic research and development. This book aims to provide a comprehensive overview of the theoretical foundations, cutting-edge methodologies, and practical applications of high-performance computing in biomimetics.

The journey begins with an introduction to biomimetics, laying the groundwork for understanding the underlying principles of nature-inspired design and engineering. We explore the rich diversity of biological systems, ranging from the intricate neural networks of the brain to the efficient locomotion mechanisms of animals, drawing inspiration for computational models and architectures. By emulating the inherent intelligence and adaptability found in nature, we can unlock new avenues for innovation across various domains. Next, we enter into the world of high-performance computing, looking into the technologies, architectures, and algorithms that enable the rapid processing and analysis of large-scale biomimetic models. We explore the power of parallel and distributed computing, GPU acceleration, and emerging technologies such as quantum computing, showcasing how these advancements have revolutionized the computational capabilities of researchers and engineers. With a solid foundation in place, we embark on a captivating journey through the realm of biomimetic modeling, bio-inspired computing, and its associated algorithms. We explore various methodologies for creating computational models inspired by biological systems, including neural networks, genetic algorithms, swarm intelligence, and cellular automata. By combining these models with high-performance computing techniques, we can simulate and optimize complex biological phenomena, paving the way for groundbreaking advancements in diverse fields such as robotics, materials science, drug discovery, and environmental sustainability.

Throughout the book, we feature some real-world applications of high-performance computing in biomimetics. From designing autonomous robots capable of navigating complex terrains to the impact of bio-inspired computing algorithms on the available high-performance computation facilities, these applications showcase the transformative potential of this interdisciplinary field. We also highlight the challenges and future directions in high-performance computing and biomimetics, shedding light on the possibilities and limitations that lie ahead. It is our hope that this book will serve as a valuable resource for researchers, engineers, and students interested in the exciting field of biomimetics and its integration with high-performance computing. By bridging the gap between biology, computer science, and engineering, we can unlock novel solutions to some of the most pressing challenges of our time. The future is full of possibilities, and we invite you to join us on this extraordinary journey into the realm of high-performance computing in biomimetics. We will keep updating this book in the subsequent editions. We thank all the authors who contributed to this book and who made our proposal come true. We also thank the Springer Nature, Singapore, for their help and support.

Serdang, Malaysia — Kamarul Arifin Ahmad
Serdang, Malaysia — Nor Asilah Wati Abdul Hamid
Al Ain, United Arab Emirates — Mohammad Jawaid
Riyadh, Saudi Arabia — Tabrej Khan
Manipal, India — Balbir Singh

Acknowledgements

We would like to express our deepest gratitude and appreciation to all the individuals who have contributed to the creation of this book, "High-Performance Computing in Biomimetics: Modeling, Architecture, and Applications." First and foremost, we would like to thank the authors of this book for their tireless efforts and dedication in researching, writing, and editing this comprehensive guide on nanomaterials. Their expertise and insights have helped to make this book an invaluable resource for anyone seeking to learn more about this rapidly evolving field. We would also like to extend our thanks to the publishers and all editorial members who have played a crucial role in bringing this book to fruition. Their guidance, support, and attention to detail have been instrumental in ensuring that the book meets the highest standards of quality and accuracy.

In addition, we are grateful to the reviewers who provided valuable feedback and constructive criticism, which helped to improve the content and readability of the book. Last but not least, we would like to acknowledge the contributions of all the researchers, scientists, and engineers who have dedicated their careers to advancing the field of computational biology and bio-inspired computing. Their groundbreaking work has paved the way for many of the technological innovations we enjoy today.

Thank you to all who have contributed to the creation of this book, and we hope it serves as a valuable resource for students, researchers, and practitioners in the field of high-performance computing and biomimetics.

Contents

Introduction to Biomimetics, Modelling and Analysis 1
Balbir Singh, Adi Azriff Basri, Noorfaizal Yidris, Raghuvir Pai, and Kamarul Arifin Ahmad

High Performance Computing and Its Application in Computational Biomimetics 21
Mohd. Firdaus bin Abas, Balbir Singh, and Kamarul Arifin Ahmad

Bio-inspired Computing and Associated Algorithms 47
Balbir Singh and Manikandan Murugaiah

Cloud Computing Infrastructure, Platforms, and Software for Scientific Research 89
Prateek Mathur

Expansion of AI and ML Breakthroughs in HPC with Shift to Edge Computing in Remote Environments 129
Kumud Darshan Yadav

Role of Distributed Computing in Biology Research Field and Its Challenges 147
Bahiyah Azli and Nurulfiza Mat Isa

HPC Based High-Speed Networks, ARM Processor Architecture and Their Configurations 163
Srikanth Prabhu, Richa Vishwanath Hinde, and Balbir Singh

High-Performance Computing Based Operating Systems, Software Dependencies and IoT Integration 175
Nor Asilah Wati Abdul Hamid and Balbir Singh

GPU and ASIC as a Boost for High Performance Computing 205
Rajkumar Sampathkumar

Biomimetic Modeling and Analysis Using Modern Architecture Frameworks like CUDA 223
Balbir Singh, Kamarul Arifin Ahmad, and Raghuvir Pai

Unsteady Flow Topology Around an Insect-Inspired Flapping Wing Pico Aerial Vehicle 241
Balbir Singh, Adi Azriff basri, Noorfaizal Yidris, Raghuvir Pai, and Kamarul Arifin Ahmad

Machine Learning Based Dynamic Mode Decomposition of Vector Flow Field Around Mosquito-Inspired Flapping Wing 251
Balbir Singh, Adi Azriff basri, Noorfaizal Yidris, Raghuvir Pai, and Kamarul Arifin Ahmad

Application of Cuckoo Search Algorithm in Bio-inspired Computing Using HPC Platform 259
Tabrej Khan

Application of Machine Learning and Deep Learning in High Performance Computing 271
Manikandan Murugaiah

The Future of High Performance Computing in Biomimetics and Some Challenges 287
Lanston Pramith Fernandes, Palash Kharate, and Balbir Singh

About the Editors

Prof. Ir. Dr. Hj. Kamarul Arifin Ahmad is currently Director of the Putra Science Park UPM and Professor in Aerospace Engineering Department in UPM. He has bachelor degree from Universiti Sains Malaysia, M.Sc. in Aerodynamics (Aerospace Dynamics) from Cranfield University, and Doctorate from Queens University of Belfast. In 2015, he obtained his Professional Engineer (PE) status from Board of Engineers Malaysia (BEM). Between 2016 and 2018, he was seconded to King Saud University, Saudi Arabia, in the Department of Mechanical Engineering. He is also Former Head of the Aerospace Malaysia Research Centre (AMRC) UPM and Head of the Aerospace Engineering Department in UPM. He has vast experiences in the Aerospace Engineering field and has been teaching in the same area for more than 15 years. He also taught aerospace engineering related subjects namely introduction to aerospace engineering, aerodynamics, gas dynamics, and computational fluid dynamics.

Dr. Nor Asilah Wati Abdul Hamid is currently Lab Head of Laboratory of Computational Sciences and Mathematical Physics, Institute for Mathematical Research, Universiti Putra Malaysia, where she is also Associate Professor in the Department of Communication Technology and Network, Faculty of Computer Science and Information Technology. She has received her Ph.D. in Computer Science from University of Adelaide in 2008. She has been Visiting Scholar with the High Performance Computing Lab, the George Washington University, USA, from 2013 until 2015. She is Receiver of the CUDA Teaching Centre award from NVIDIA in 2015 and established the CUDA lab in her faculty. She has authored or co-authored more than 80 journals and conference proceedings in her area, and her work was funded by government and some by industry. Her research interests are in parallel and distributed high performance computing, cloud computing and data intensive computing.

Prof. Dr. Mohammad Jawaid is currently working as Distinguished Professor at Chemical and Petroleum Engineering Department, United Arab Emirates University, UAE. Earlier he worked as Senior Fellow (Professor) at Biocomposite Technology Laboratory, Institute of Tropical Forestry and Forest Products (INTROP), Universiti

Putra Malaysia, Serdang, Selangor, Malaysia. He also worked as Visiting Professor at King Saud University and Distinguished Visiting Professor at Malaysian Japan International Institute of Technology (MJIIT), Malaysia. He received his Ph.D. from Universiti Sains Malaysia, Malaysia. He has more than 20 years of experience in teaching, research, and industries. His area of research interests includes hybrid reinforced/filled polymer composites and advanced materials. So far, he has published 75 books, 90 book chapters, more than 450 peer-reviewed international journal papers and several published review papers under top 25 hot articles in science direct during 2013–2020. He is also International Advisory Board member of Springer Series on Polymer and Composite Materials. He is Reviewer of several high impact ISI journals (200 journals). He is also Fellow and Chartered Scientist of IOM, UK.

Dr. Tabrej Khan is a material scientist with Ph.D. in Aerospace Engineering (Specialization in materials) from University Putra Malaysia in 2020. He is currently working as a postdoctoral fellow in the field of material science and engineering at Department of Engineering Management, College of Engineering Prince Sultan University, Riyadh, Saudi Arabia. He has a pool of good publications in reputed journals, patents, chapters in books and has lot of experience in the field of Hybrid Composites, Advanced Materials, Structural Health Monitoring and Impact Studies, Damage Detections and Repairs, Impact Studies, Signal Processing and Instrumentations, Non-destructive Testing, and Destructive Testing.

Dr. Balbir Singh is currently working as a senior assistant professor in Aerospace Engineering at the Department of Aeronautical and Automobile Engineering, Manipal Institute of Technology, Manipal Academy of Higher Education, India. He has a doctoral degree in the field of Aerospace Engineering from Department of Aerospace Engineering, Faculty of Engineering. Universiti Putra Malaysia. He has good number of publications and grants in his favour. His research focuses on insect inspired miniaturized robots for planetary studies, advanced aerospace materials (bio, natural, or nano) at all the scales, high performance computing and modelling, unsteady aerodynamics, space debris removal and traffic management, space systems engineering, and clean energy.

Introduction to Biomimetics, Modelling and Analysis

Balbir Singh, Adi Azriff Basri, Noorfaizal Yidris, Raghuvir Pai, and Kamarul Arifin Ahmad

Abstract Biomimetics, also known as biomimicry or biologically inspired design, is a multidisciplinary field that draws inspiration from nature to develop innovative solutions for complex engineering challenges. This chapter introduces the fundamental concepts of biomimetics, highlighting its significance and potential applications. The chapter also explores the role of modeling and analysis in biomimetic research, emphasizing their importance in understanding and replicating the intricate designs and functions found in biological systems. The chapter begins by discussing the motivation behind biomimetics and its overarching goal of emulating nature's strategies and principles to solve engineering problems. It explores the benefits of biomimetic approaches, including enhanced efficiency, adaptability, and sustainability. Examples of successful biomimetic designs and their impact across various fields are presented to illustrate the practical applications of biomimetics. The chapter then looks into the crucial role of modeling and analysis in biomimetic research. It explores different modeling techniques, such as computational modeling, mathematical modeling, and computer simulations, enabling researchers to understand biological systems' complex behavior and functionality. The use of advanced analytical tools and techniques to analyze biological structures, functions, and processes is also discussed, highlighting their role in extracting design principles for biomimetic applications. Moreover, the chapter emphasizes the importance of interdisciplinary collaboration between biologists, engineers, material scientists, and other experts

B. Singh · A. A. Basri · N. Yidris · K. A. Ahmad (✉)
Department of Aerospace Engineering, Faculty of Engineering, Universiti Putra Malaysia, Serdang 43400, Selangor, Malaysia
e-mail: aekamarul@upm.edu.my

B. Singh
Department of Aeronautical and Automobile Engineering, Manipal Institute of Technology, Manipal Academy of Higher Education, Manipal, Karnataka 576104, India

A. A. Basri · N. Yidris · K. A. Ahmad
Aerospace Malaysia Research Centre, Faculty of Engineering,, Universiti Putra Malaysia, Serdang 43400, Selangor, Malaysia

R. Pai
Department of Mechanical and Industrial Engineering, Manipal Institute of Technology, Manipal Academy of Higher Education, Manipal, Karnataka 576104, India

K. A. Ahmad et al. (eds.), *High Performance Computing in Biomimetics*, Series in BioEngineering, https://doi.org/10.1007/978-981-97-1017-1_1

in biomimetic research. It highlights the need for a comprehensive understanding of biological and engineering principles to successfully bridge the gap between nature and technology. The chapter concludes by emphasizing the promising future of biomimetics, modeling, and analysis. It underscores the potential for biomimetic designs to drive technological advancements and sustainability and the importance of continued research and development in this field. This chapter provides an introductory overview of biomimetics, modeling, and analysis. It sets the stage for subsequent chapters that look into specific aspects of biomimetics, exploring in-depth the principles, approaches, challenges, and applications within this exciting and rapidly evolving field.

Keywords Biomimetics · Modeling · Biomimetic optimization · Structural materials · Adhesives and coatings · Bio-inspired robotics

1 Definition and Scope of Biomimetics

Biomimetics, also known as biomimicry or bioinspiration, is the field of study that draws inspiration from nature's design principles, structures, and processes to develop innovative solutions for human problems. It involves imitating, analyzing, and applying biological systems, materials, and functions to create new technologies, materials, and designs. Biomimetics, also known as biomimicry or biologically inspired design, is a multidisciplinary field that seeks to emulate and apply principles, processes, and systems found in nature to solve human problems and develop innovative technologies. It involves studying and understanding biological structures, functions, and strategies and then using that knowledge to design and engineer new materials, devices, and systems.

The term "biomimetics" is derived from the Greek words "bios" (life) and "mimesis" (imitation). It emphasizes the core idea of imitating or emulating nature to create efficient, sustainable, and well-adapted solutions to their intended purpose. Biomimetics draws inspiration from various biological domains, including organisms, ecosystems, and evolutionary processes. The scope of biomimetics is broad, covering various disciplines and applications. As far as the scope of biomimetics is concerned, it offers insights into the unique properties and structures of biological materials, such as bones, shells, and spider silk, which have evolved to exhibit exceptional strength, flexibility, and resilience. Researchers in this field aim to replicate these properties in synthetic materials for engineering, construction, aerospace, and medicine applications. Robotics and Artificial Intelligence is another area of scope in which biomimetics inspires the design and control of robots and autonomous systems. By studying animal locomotion, sensing mechanisms, and cognitive abilities, researchers develop robots that can navigate complex environments, adapt to changing conditions, and interact with humans more effectively.

Biomimetics has revolutionized the field of medicine by drawing inspiration from biological systems. Examples include the development of prosthetic limbs and implants that mimic natural movement and integrate seamlessly with the body and the design of drug delivery systems based on the structure and function of biological cells. Nature has evolved efficient strategies for energy capture, storage, and utilization. Biomimetics offers insights into photosynthesis, energy conversion in animals, and energy-efficient structures like termite mounds, which can inform the development of renewable energy technologies, smart buildings, and sustainable urban planning. Biomimetics plays a crucial role in nanotechnology by studying biological structures and processes at the nanoscale. Researchers aim to replicate natural nanoscale structures, such as lotus leaves' self-cleaning properties or gecko feet's adhesive abilities, to develop advanced materials, coatings, and adhesives with unique properties. Biomimetics has influenced the design of transportation systems, such as streamlined vehicles inspired by fish or birds' aerodynamics or the development of novel materials for lightweight aircraft construction based on the structure of bird bones.

It is important to note that biomimetics is not limited to the direct replication of biological structures or processes. It also encompasses extracting fundamental principles from nature and their creative application to solve engineering challenges. The field requires collaboration between biologists, engineers, material scientists, physicists, and other specialists to translate biological knowledge into technological innovations [1–12].

2 Historical Background of Biomimetics

Biomimetics is rooted in ancient civilizations, where humans observed and imitated nature to solve practical challenges. However, the formalization of biomimetics as a scientific discipline began in the twentieth century with notable contributions from scientists such as Otto Schmitt and Janine Benyus. Today, biomimetics has gained significant attention across various disciplines, including engineering, materials science, robotics, and medicine. The concept of biomimetics, or looking to nature for inspiration and solutions, has deep historical roots. Humans have long observed and imitated nature to enhance their own capabilities and solve problems. The history of biomimetics can be traced back to ancient civilizations, where early inventors and scientists drew inspiration from natural phenomena.

One of the earliest examples of biomimetics can be seen in ancient Egyptian and Greek architecture. The Egyptians observed the strength and stability of natural structures like the lotus flower and incorporated these principles into their own architectural designs. The Greeks were inspired by the flight of birds, leading to the development of early flying machines like Daedalus and Icarus. During the Renaissance, Leonardo da Vinci exemplified the spirit of biomimetics through his detailed observations of nature. He studied birds' flight, fish swimming, and water flow, among other phenomena. Da Vinci's sketches and designs laid the foundation for future engineers and inventors to explore biomimetics.

In the eighteenth and nineteenth centuries, biomimetic principles gained more attention. The Industrial Revolution sparked interest in harnessing natural processes for technological advancements. Jan Swammerdam, a Dutch biologist, made significant contributions by studying the intricate structures of insects and illustrating them with precision. His work on the compound eye of insects laid the groundwork for developing lenses and optics. In the twentieth century, biomimetics emerged as a formal field of study. In the 1950s, Swiss engineer Georges de Mestral invented Velcro after observing the mechanism by which burrs stuck to his dog's fur. This inspired him to create a fastening system based on tiny hooks and loops. Velcro became a prime example of biomimetics and its practical applications.

The term "biomimetics" was coined in 1960 by American biophysicist and polymath Otto Schmitt, who recognized the need for a unified field that combined biology, engineering, and design. Schmitt's vision paved the way for establishing biomimetics as an interdisciplinary scientific discipline. Since then, biomimetics has gained increasing attention and recognition. The development of advanced imaging techniques and computational tools has allowed researchers to look deeper into the structures and functions of biological systems, enabling more accurate replication and utilization.

Biomimetics is a rapidly growing field with wide-ranging applications in various domains. Researchers draw inspiration from biological systems, such as the self-cleaning mechanism of lotus leaves, the incredible strength of spider silk, or the energy conversion processes in plants, to develop innovative technologies, materials, and designs [1, 4, 8, 11–15].

3 Key Principles and Approaches in Biomimetics

Biomimetics, also known as biomimicry or biologically inspired design, is guided by key principles and approaches that enable researchers and engineers to draw inspiration from nature and apply it to technological advancements. These principles provide a framework for understanding and emulating biological systems' efficiency, adaptability, and sustainability. Here are some key principles and approaches in biomimetics:

Emulating Form and Function: Biomimetics focuses on understanding the form and function of biological structures and processes. This involves studying organisms' anatomy, morphology, and behavior to identify key design features and functionalities. By replicating these features, engineers can develop innovative solutions in various fields, ranging from materials science to robotics.

Systems Thinking: Biomimetics takes a holistic approach by considering the interconnectedness and interdependencies within natural systems. Researchers can gain insights into how biological systems achieve efficiency and sustainability by studying the relationships between organisms, their environment, and the larger ecosystem.

This system-thinking approach informs the design of integrated and synergistic solutions.

Hierarchical Design: Biological systems often exhibit hierarchical organization, where complex functionalities emerge from the interaction of simpler components. Biomimetics leverages this principle by adopting a hierarchical design approach. Engineers can develop scalable and adaptable solutions by breaking down complex systems into smaller components and understanding their interplay.

Material and Energy Efficiency: Nature is known for its resource-efficient strategies. Biomimetics aims to replicate and improve upon these strategies to develop sustainable technologies. For example, studying the self-cleaning properties of lotus leaves has led to the development of superhydrophobic coatings that reduce water consumption and maintenance in various applications.

Adaptation and Resilience: Biological systems possess remarkable adaptability and resilience, allowing them to thrive in changing and harsh environments. Biomimetics draws inspiration from these adaptive mechanisms to create solutions responding to dynamic conditions. Examples include self-healing materials inspired by the regenerative abilities of living organisms and adaptive control systems in robotics.

Bio-inspired Materials: Biomimetics involves the development of new materials that mimic the properties and structures found in nature. This includes bio-inspired composites, self-assembling materials, and nanomaterials with enhanced functionalities. By emulating the properties of biological materials, such as strength, flexibility, and self-repair, engineers can create innovative materials for various applications.

Learning from Evolution: Biomimetics recognizes the power of evolution as a design process. Nature has undergone billions of years of optimization and selection, resulting in highly efficient and effective solutions. By studying the principles of evolution, such as variation, selection, and reproduction, biomimetics can uncover design strategies and develop innovative solutions.

These key principles and approaches in biomimetics provide a framework for researchers and engineers to understand and harness the principles of nature in the design of new technologies and systems. The framework of the biomimetics process is shown in Fig. 1.

4 Modeling and Analysis in Biomimetics

Modeling and analysis play a crucial role in biomimetics by providing a systematic framework to understand, replicate, and optimize nature's complex structures and processes. These tools allow researchers to simulate and evaluate the performance of biomimetic designs, assess their feasibility, and guide the development of innovative technologies. Here is a detailed description of modeling and analysis in biomimetics:

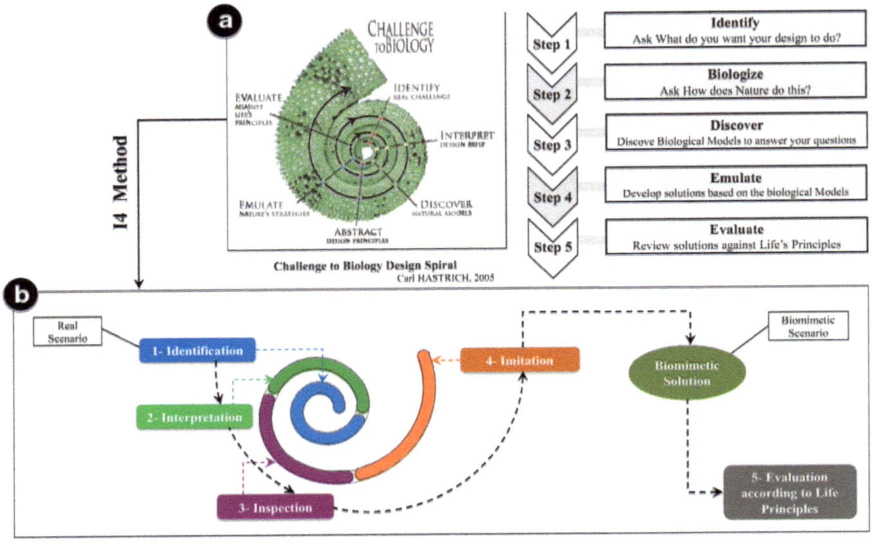

Fig. 1 **a** Hastrich's direct approach (Challenge to Biology); **b** process of the Biomimetic methodological approach "Challenge to Biology Design Spiral". Adapted from ref [75], copyright (2022), with permission from Springer

Computational Modeling: Computational modeling involves using computer simulations and mathematical algorithms to mimic and analyze biological systems. It allows researchers to explore the behavior and performance of biomimetic designs under different conditions. Finite element analysis (FEA), computational fluid dynamics (CFD), and multi-body dynamics are common techniques used to simulate biomimetic systems' mechanical, fluidic, and structural properties.

Biomechanical Modeling: Biomechanical modeling focuses on understanding and replicating the mechanical behavior of biological structures. It involves analyzing the forces, stresses, strains, and motions experienced by living organisms. By studying the biomechanics of natural systems, researchers can develop accurate mathematical models that capture the underlying principles and guide the design of biomimetic structures, such as artificial limbs or exoskeletons as shown in Fig. 2.

Structural Analysis: Structural analysis is crucial in biomimetics to ensure biomimetic designs' integrity, strength, and stability. Finite element analysis (FEA) and other structural analysis techniques are employed to evaluate the mechanical properties and performance of materials and structures inspired by biological counterparts. These analyses help optimize the design for efficiency, weight reduction, and resistance to external loads.

Fluid Dynamics Analysis: Fluid dynamics analysis is essential for understanding biological systems' flow patterns, efficiency, and drag reduction. Computational fluid dynamics (CFD) models and simulates the fluid flow around biomimetic designs,

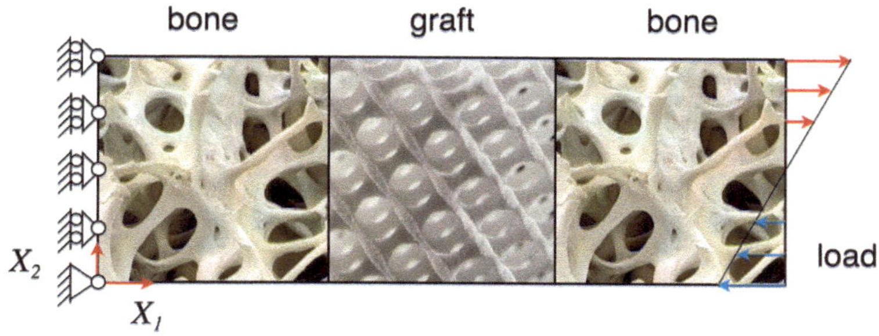

Fig. 2 Modeling of bone tissues joined by biomaterial graft (*Source* https://doi.org/10.3390/biomimetics6010018)

such as streamlined vehicle shapes inspired by fish or bird wings. This analysis aids in improving the aerodynamic performance of vehicles, turbines, and other fluid-based systems.

Kinematics and Dynamics Analysis: Kinematics and dynamics analysis focus on studying motion and the forces involved in biological systems. By applying principles of kinematics and dynamics, researchers can analyze the movement and interactions of organisms or develop biomimetic robots and mechanisms that replicate or enhance natural motion. This analysis helps optimize efficiency, stability, and adaptability in biomimetic designs.

Biomimetic Optimization: Optimization techniques are employed to enhance biomimetic designs and systems. These techniques, such as genetic algorithms, evolutionary algorithms, or machine learning, help researchers identify optimal parameters, configurations, or materials based on desired performance criteria. Optimization can be used to refine biomimetic structures, such as improving the efficiency of solar panels inspired by plant photosynthesis or optimizing the wing design of an aircraft inspired by bird flight. Some of these methods are depicted in Fig. 3.

Multi-disciplinary Analysis: Biomimetics often requires a multi-disciplinary approach integrating knowledge from various fields such as biology, engineering, materials science, and physics. The multi-disciplinary analysis brings together experts from different domains to address the complex challenges of translating biological principles into functional technological solutions. It enables the synthesis of knowledge, collaboration, and cross-pollination of ideas.

Biomimetic Prototyping and Testing: Modeling and analysis are complemented by prototyping and testing to validate and refine biomimetic designs. Prototyping involves the fabrication of physical models or prototypes based on computational models, allowing researchers to evaluate the functionality and performance of the biomimetic system. Testing helps verify the models' accuracy, identify design improvements, and assess the system's viability.

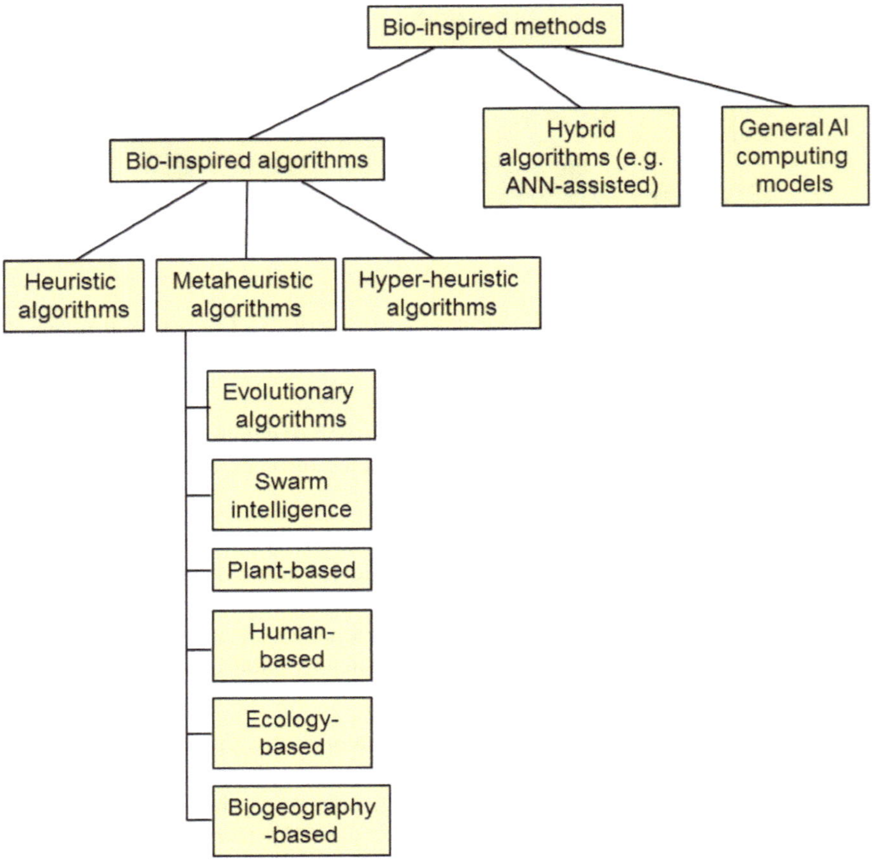

Fig. 3 Possible classification of bio-inspired optimization methods [76]

These references provide insights into the modeling and analysis techniques used in biomimetics and offer a starting point for further exploration of the subject. They highlight the importance of computational modeling, biomechanics, structural analysis, fluid dynamics analysis, optimization, and multi-disciplinary analysis in developing and refining biomimetic designs [1, 2, 4, 8, 11, 13, 16, 17].

5 Applications of Biomimetics

Biomimetics, also known as biomimicry or biologically inspired design, has various applications across various fields. By drawing inspiration from nature and emulating biological principles, biomimetics offers innovative solutions to address complex challenges. Here is a detailed description of the various applications of biomimetics.

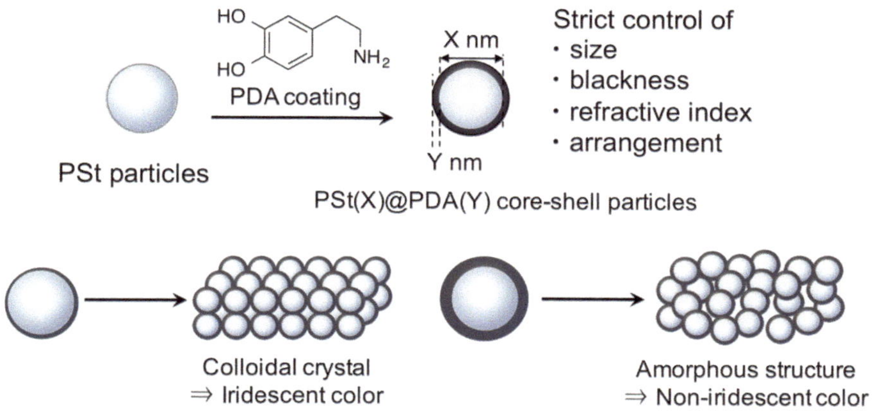

Fig. 4 Schematic representation of the fabrication of core–shell particles with biomimetic melanin-like PDA shell layers [77]

5.1 Materials Science and Engineering

a. **Structural Materials**: Biomimetics provides insights into the design and fabrication of advanced structural materials. Examples include the development of bio-inspired composites that mimic the hierarchical structure and mechanical properties of natural materials like bone and seashells [18, 19].
b. **Adhesives and Coatings**: Biomimetics offers inspiration for developing bio-inspired adhesives and coatings. Examples include gecko-inspired adhesives that replicate the unique sticking properties of gecko feet, leading to applications in robotics, manufacturing, and space exploration [20, 21].
c. **Self-Healing Materials**: Biomimetic self-healing materials aim to replicate the regenerative abilities found in living organisms. These materials have the ability to autonomously repair damage autonomously, leading to increased durability and reliability [22, 23].
d. **Photonic Materials**: Biomimetics offers insights into the intricate structures and optical properties of natural photonic systems, leading to the development of biomimetic photonic materials with enhanced light manipulation capabilities [24, 25]. Figure 4 depicts the schematics of fabrication of core–shell particles with biomimetic melanin-like PDA shell layers.

5.2 Robotics and Automation

a. **Locomotion and Mobility**: Biomimetics inspires the design and control of robotic systems capable of efficient locomotion and navigation in various environments. Examples include the development of robotic fish, insect-inspired flying robots, and snake-like robots [26, 27].

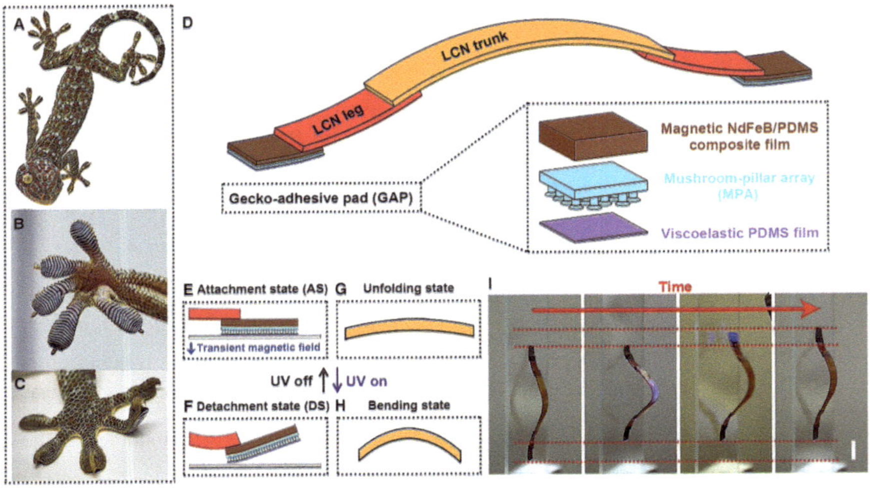

Fig. 5 Design principles of a holistic untethered soft robot, GeiwBot [78]

b. **Sensing and Perception**: Biomimetics enables the development of sensors and perception systems that replicate the capabilities of biological organisms. Examples include vision systems inspired by the compound eyes of insects and tactile sensors inspired by human skin [28, 29].
c. **Soft Robotics**: Biomimetics contributes to soft robotics by emulating living organisms' compliant and flexible characteristics. Soft robotic systems inspired by biological structures, such as octopus arms and elephant trunks, offer enhanced dexterity and adaptability [30, 31]. For example Fig. 5 shows the design and development procedure of GeiwBot using biomimetic principles.

5.3 *Medicine and Healthcare*

a. **Prosthetics and Implants**: Biomimetics is vital in developing advanced prosthetic devices and implants. By mimicking the form and function of natural limbs and organs, biomimetic prosthetics offer improved mobility, dexterity, and integration with the human body [32, 33] (see Fig. 6).
b. **Drug Delivery**: Biomimetics inspires the design of drug delivery systems that mimic biological mechanisms for targeted and controlled drug release. Examples include liposomal drug carriers and micro/nanoparticles designed to replicate the functionality of biological cells [34, 35].
c. **Tissue Engineering**: Biomimetics contributes to tissue engineering by replicating the structure and properties of natural tissues to create biomimetic scaffolds and engineered organs for regenerative medicine [36, 37].

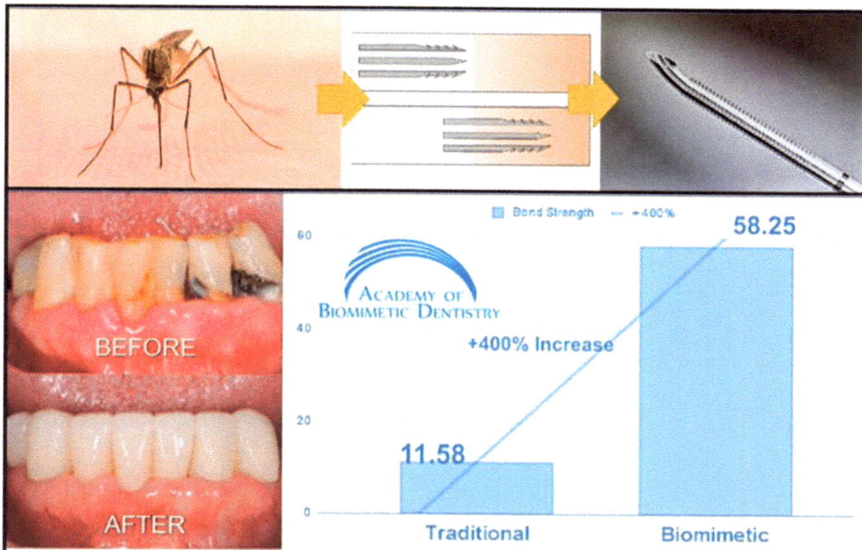

ABOVE: Pain-less Needle for Injections inspired by Maxillae of Mosquitos.
BELOW: bioMIMETIC DENTISTRY and the adhesive (restorative material) dental bond achieved with 400% increase in strength (equivalent to strength of bond between natural enamel and dentin).

Fig. 6 Two applications of biomimetics in the field of surgery and dentistry [79]

5.4 Energy and Sustainable Design

a. **Solar Energy**: Biomimetics offers insights into the efficient capture and utilization of solar energy. Researchers draw inspiration from photosynthesis and the light-harvesting capabilities of plants to develop biomimetic solar cells and energy conversion systems [38, 39].
b. **Energy-Efficient Buildings**: Biomimetics informs the design of energy-efficient buildings by studying natural systems' thermal regulation, ventilation, and structural strategies. Examples include biomimetic building materials and designs inspired by termite mounds or bird nests [40, 41]. Research is continuous in constructing sustainable buildings by 2050, enhanced by biomimetics as shown in Fig. 7.

c. **Sustainable Manufacturing**: Biomimetics contributes to sustainable manufacturing practices by developing eco-friendly processes inspired by natural systems. This includes bio-inspired manufacturing techniques, waste reduction strategies, and resource-efficient production methods [42, 43].

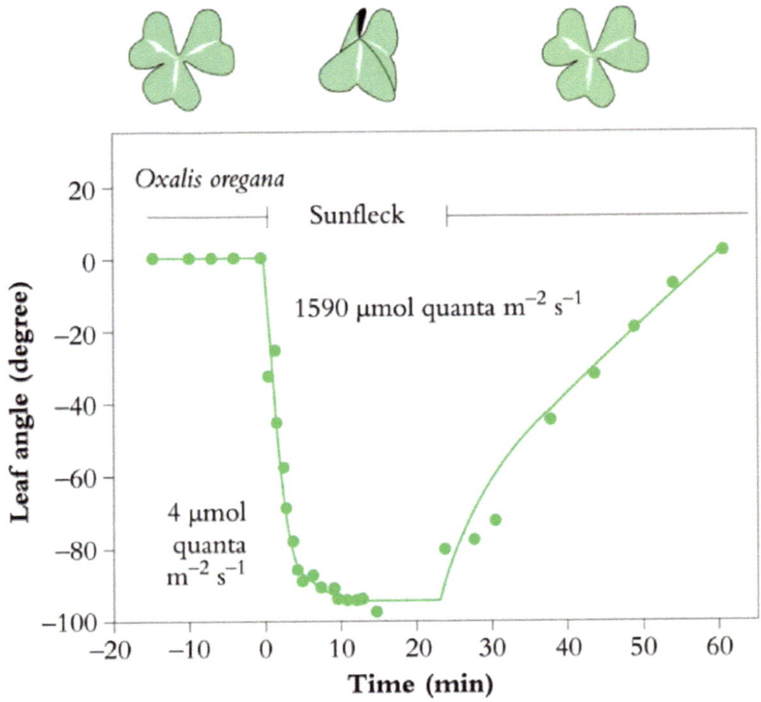

Fig. 7 Time course of leaf unfolding in Oxalis oregano used as an adaptive biomimetic façade that helps reduce the energy consumption of high glazed building significantly [80]

6 Challenges and Future Directions

As a rapidly evolving field, biomimetics faces several challenges and holds great potential for future advancements. Overcoming these challenges and exploring new directions is crucial for unlocking the full potential of biomimetics. Here is a detailed description of the challenges and future directions of biomimetics:

Understanding Biological Complexity: Biological systems are incredibly complex, with intricate structures and intricate interactions. Replicating this complexity in biomimetic designs remains a challenge. Advancements in high-resolution imaging, computational modeling, and systems biology approaches are needed to better understand biological systems and translate that knowledge into practical biomimetic applications [44, 45].

Bridging the Gap between Biology and Engineering: Effective collaboration between biologists and engineers is essential to bridge the gap between understanding biological principles and translating them into engineering solutions. Interdisciplinary education, collaborative research platforms, and shared vocabulary must facilitate effective communication and collaboration between the two disciplines [1, 46].

Scaling Biomimetic Solutions: Scaling up biomimetic solutions from laboratory prototypes to real-world applications poses challenges. Factors such as manufacturing scalability, cost-effectiveness, and regulatory requirements must be considered to successfully translate biomimetic designs into practical products [8, 47].

Ethical and Sustainability Considerations: As biomimetics advances, ethical and sustainability considerations become increasingly important. Ensuring biomimetic solutions are socially responsible, environmentally sustainable, and adhere to ethical standards. Assessing biomimetic technologies' long-term environmental impacts and ethical implications is essential for responsible development [48, 49].

Integration with Emerging Technologies: Future directions of biomimetics involve integrating it with emerging technologies such as nanotechnology, artificial intelligence, and additive manufacturing. Harnessing the power of these technologies in combination with biomimetic approaches can lead to breakthrough innovations in areas such as nanobiotechnology, biomimetic robots, and advanced materials with unprecedented properties [1, 50].

Bio-inspired Artificial Intelligence: Integrating biomimetic principles with artificial intelligence (AI) is promising. By mimicking biological systems' information processing, learning, and decision-making capabilities, bio-inspired AI can lead to the development of more efficient and adaptable intelligent systems [51, 52].

Exploration of Unexplored Biological Phenomena: Nature is full of fascinating and unexplored phenomena. Future directions of biomimetics involve delving into these unexplored areas of biology to uncover new principles and mechanisms that can inspire innovative technological solutions [1, 53].

Education and Awareness: Educating the next generation of scientists, engineers, and designers about biomimetics is crucial for its future growth. Integrating biomimetic principles into educational curricula and fostering awareness about the potential of biomimetics can inspire innovative thinking and contribute to sustainable solutions [54, 55].

These challenges provide insights into the challenges and future directions of biomimetics. They highlight the need to understand biological complexity, bridge interdisciplinary gaps, scale biomimetic solutions, address ethical and sustainability considerations, integrate with emerging technologies, explore unexplored biological phenomena, promote education and awareness, and harness the potential of bio-inspired artificial intelligence.

6.1 Challenges Related to Biomimetics

Biomimetics, also known as biomimicry or biologically inspired design, faces several challenges that must be addressed to unlock its full potential. Overcoming these challenges is crucial for successfully translating biological principles into practical applications. Here is a detailed description of the challenges related to biomimetics:

Complexity of Biological Systems: Biological systems are incredibly complex, often involving intricate structures, processes, and interactions. Replicating this complexity in biomimetic designs remains a challenge. Understanding and mimicking nature's multifaceted behaviors, adaptability, and self-regulation require interdisciplinary collaboration, advanced modeling techniques, and a deeper understanding of biological systems [1, 56].

Translation from Biology to Engineering: Translating the knowledge gained from biological systems into practical engineering solutions presents a significant challenge. Bridging the gap between biological understanding and engineering implementation requires effective collaboration between biologists, engineers, and designers. Understanding the limitations and constraints of engineering materials, manufacturing processes, and scalability is essential for successful translation [4, 57].

Scalability and Manufacturing Processes: Scaling up biomimetic solutions from laboratory prototypes to real-world applications poses challenges. Manufacturing biomimetic structures and materials at larger scales while maintaining their desired properties and functionality requires advanced manufacturing techniques, such as additive manufacturing and nanostructuring. Developing cost-effective and sustainable manufacturing processes for biomimetic designs is a key challenge [58, 59].

Ethical and Sustainability Considerations: As biomimetics progresses, it is essential to consider biomimetic solutions' ethical and sustainability aspects. Ensuring responsible innovation and addressing potential ethical implications, such as intellectual property rights, biosecurity concerns, and the impact on ecosystems, is crucial for developing and implementing biomimetic technologies [12, 48].

Lack of Standardization and Best Practices: Biomimetics is a multidisciplinary field, and the lack of standardization and best practices poses a challenge. Establishing standardized protocols, methodologies, and design guidelines for biomimetic research and development can improve results' reproducibility, comparability, and reliability. Collaboration among researchers and establishing biomimetics-focused organizations and consortia can contribute to developing common frameworks and best practices [1, 60].

Biological and Technical Constraints: Biomimetic designs often face constraints arising from differences between natural and engineered systems. Biological constraints, such as compatibility with living tissues or regulatory limitations, and technical constraints, such as limitations in material properties or manufacturing processes, need to be considered during the biomimetic design process. Overcoming these constraints requires innovative approaches and interdisciplinary collaborations [17, 61].

Long-Term Performance and Durability: Ensuring biomimetic designs' long-term performance and durability is a challenge. Biomimetic materials and structures must withstand environmental conditions, mechanical stresses, and potential long-term degradation. Understanding the degradation mechanisms, designing for repair and self-healing, and implementing suitable maintenance strategies are important considerations [8, 62].

Funding and Commercialization: Securing funding for biomimetic research and the successful commercialization of biomimetic technologies can be challenging. Biomimetic projects often require interdisciplinary collaborations, long development cycles, and substantial investment. Building strong partnerships with industry, government support, and effective technology transfer strategies are crucial for overcoming funding and commercialization challenges [63, 64].

These references provide insights into the challenges related to biomimetics. They highlight the complexity of biological systems, the need to bridge the gap between biology and engineering, scalability and manufacturing challenges, ethical and sustainability considerations, lack of standardization, biological and technical constraints, long-term performance and durability, and funding and commercialization obstacles. Addressing these challenges is essential for successfully developing and implementing biomimetic technologies.

6.2 Future of Biomimetics

Biomimetics, biomimicry or biologically inspired design, has great potential for addressing complex challenges and driving innovation across various fields. As research in biomimetics advances, several exciting directions and opportunities emerge. Here is a detailed description of the future of biomimetics and research:

Advanced Materials and Manufacturing Techniques: Future research in biomimetics will focus on developing advanced materials and manufacturing techniques inspired by biological systems. This includes bio-inspired composites, self-healing materials, shape-memory materials, and adaptive materials that can mimic the properties and functionalities of natural materials [18, 65].

Robotics and Artificial Intelligence (AI): Integrating biomimetic principles with robotics and AI will lead to the development of more advanced and autonomous systems. Biomimetic robots will exhibit enhanced locomotion, dexterity, and adaptability, inspired by the movement and behaviors of animals. Additionally, AI algorithms will be designed to mimic the information processing and decision-making capabilities observed in biological organisms [66, 67].

Sustainable and Energy-Efficient Design: Biomimetics will play a crucial role in developing sustainable and energy-efficient solutions. By studying nature's energy and resource-efficient strategies, researchers will design buildings, transportation systems, and energy technologies that minimize environmental impact and optimize resource utilization [14, 68].

Healthcare and Biomedical Applications: Future research in biomimetics will continue to revolutionize healthcare and biomedical applications. This includes the development of biomimetic implants, drug delivery systems, tissue engineering scaffolds, and diagnostic tools that can replicate and interact seamlessly with the human body, leading to improved treatments and personalized medicine [36, 65].

Environmental Remediation and Conservation: Biomimetics will contribute to environmental remediation and conservation efforts by developing innovative solutions inspired by natural systems. Researchers will study the mechanisms of pollutant degradation, water filtration, and waste management in nature to design biomimetic systems for cleaning up pollutants, conserving resources, and restoring ecosystems [69, 70].

Multi-modal Sensing and Actuation: Future biomimetic research will focus on developing multi-modal sensing and actuation systems that can replicate the capabilities of biological organisms. These systems will integrate sensing mechanisms inspired by vision, touch, hearing, and other sensory modalities, enabling robots and devices to perceive and interact with their environment more effectively [71, 72].

Computational Modeling and Simulation: Advancements in computational modeling and simulation will play a vital role in the future of biomimetics. Researchers will develop sophisticated models and simulation tools to accurately predict the behavior and performance of biomimetic designs, enabling rapid prototyping, optimization, and refinement of biomimetic solutions [4, 73].

Bioinformatics and Big Data Analysis: As biomimetics research generates vast amounts of data, the future will see increased integration of bioinformatics and big data analysis techniques. These tools will be used to analyze biological data, extract patterns, and identify key principles that can be applied in biomimetic design and innovation [1, 74].

7 Conclusion

The introduction to biomimetics, modeling, and analysis provides a foundation for understanding the principles and applications of this interdisciplinary field. Biomimetics offers innovative solutions inspired by nature's designs and processes, and integrating modeling and analysis techniques enables the translation of biological knowledge into practical applications. With ongoing advancements and interdisciplinary collaborations, biomimetics has the potential to revolutionize various industries and contribute to sustainable and efficient design practices.

Acknowledgements The authors gratefully acknowledge Universiti Putra Malaysia (UPM) for providing opportunities for biomimicry and soft robotics research to flourish and make this insect-inspired small aerial vehicle research a reality. The authors would also like to convey their gratitude to UPM to grant them the funding required to advance in biomimetic research through the Weststar Group (Tan Sri Syed Azman Endowment) industrial research grant; 6338204 – 10801.

References

1. Vincent, J.F.V., Bogatyreva, O.A.: Biomimetics: its practice and theory. J. R. Soc. Interface. **10**(79), 20130304 (2013)
2. Pohl, M., Bülthoff, H.H. (eds.): Biomimetic Research for Architecture and Building Construction: Biological Design and Integrative Structures. Springer (2018)
3. Sarikaya, M., Tamerler, C., Jen, A.K., Schulten, K., Baneyx, F., Schwartz, D.: Molecular biomimetics: nanotechnology through biology. Nat. Mater. **2**(9), 577–585 (2003)
4. Bhushan, B.: Biomimetics: lessons from nature—an overview. Philos. Trans. R. Soc. A Math. Phys. Eng. Sci. **374**(2073), 20150101 (2016)
5. Ceballos, G., Ehrlich, P.R., Dirzo, R.: Biological annihilation via the ongoing sixth mass extinction signaled by vertebrate population losses and declines. Proc. Natl. Acad. Sci. **114**(30), E6089–E6096 (2017)
6. Ge, J., Lei, Y., Liu, Z., Xu, H., Yao, Z.: Recent progress in biomimetic underwater locomotion systems. Biomimetics **6**(4), 58 (2021)
7. Vogel, S.: Cats' Paws and Catapults: Mechanical Worlds of Nature and People. WW Norton & Company (1999)
8. Patek, S.N., Korff, W.L., Caldwell, R.L.: Biomechanics: deadly strike mechanism of a mantis shrimp. Nature **428**(6985), 819–820 (2004)
9. Jung, H., Bhushan, B., Lee, J.: Biomimetic application in mechanical engineering: a review. J. Bionic Eng. **5**(4), 289–304 (2008)
10. Schmitt, O.H.: Biophysics: searching for principles. Annu. Rev. Biophys. Biomol. Struct.. Rev. Biophys. Biomol. Struct. **27**(1), 1–20 (1998)
11. Vincent, J.F.V.: Biomimetics: the use of principles from nature in engineering. Philos. Trans. R. Soc. Lond. Ser. A Math. Phys. Sci. **330**(1624), 287–292 (1990)
12. Benyus, J.M.: Biomimicry: Innovation Inspired by Nature. William Morrow Paperbacks (1997)
13. Sarikaya, M., Tamerler, C., Jen, A.K., Schulten, K., Baneyx, F., Schwartz, D.: Molecular biomimetics: nanotechnology through biology. Nat. Mater. **2**(9), 577–585 (2003)
14. Speck, T., Speck, O.: Bioinspiration and Biomimetics in Architecture: Nature-Analogies-Technology. Springer (2017)
15. Turner, J.S.: Biomimetics in materials science: self-healing, self-lubricating, and self-cleaning materials. Solid State Sci. **18**, 53–59 (2013)
16. Barthelat, F., Yin, Z.: Biomimetic design principles for generating multifunctional materials. MRS Bull. **44**(2), 117–123 (2019)
17. Lauder, G.V.: Bio-inspired design: what can we learn from nature? J. R. Soc. Interface. **13**(121), 20160347 (2016)
18. Barthelat, F., Yin, Z.: Bending and buckling of biological materials. Philos. Trans. R. Soc. A Math. Phys. Eng. Sci. **367**(1893), 3555–3575 (2009)

19. Meyers, M.A., Chen, P.Y., Lin, A.Y.M., Seki, Y.: Biological materials: structure and mechanical properties. Prog. Mater. Sci. **53**(1), 1–206 (2008)
20. Autumn, K., Hsieh, S.T., Dudek, D.M., Chen, J., Chitaphan, C., Full, R.J., Kenny, T.W.: Dynamics of geckos running vertically. J. Exp. Biol. **203**(5), 741–748 (2000)
21. Bhushan, B., Sayer, R.A.: Biomimetics: Lessons from nature for future nanotechnology. J. Vac. Sci. Technol. B: Microelectron. Nanometer Struct. **26**(2), 375–392 (2008)
22. White, S.R., Sottos, N.R., Geubelle, P.H., Moore, J.S., Kessler, M.R., Sriram, S.R., Brown, E.N., et al.: Autonomic healing of polymer composites. Nature **409**(6822), 794–797 (2001)
23. Toohey, K.S., Sottos, N.R., Lewis, J.A., Moore, J.S., White, S.R.: Self-healing materials with microvascular networks. Nat. Mater. **6**(8), 581–585 (2007)
24. Kinoshita, S., Yoshioka, S., Miyazaki, J.: Physics of structural colors. Rep. Prog. Phys. **71**(7), 076401 (2008)
25. Parker, A.R.: The diversity and implications of animal structural colours. J. Exp. Biol. **204**(3), 291–297 (2000)
26. Wood, R.J.: The first takeoff of a biologically inspired at-scale robotic insect. IEEE Trans. Rob. **24**(2), 341–347 (2008)
27. Webb, B.: Are we there yet? Trends in the development of biomimetic robots. Auton. Robot.. Robot. **8**(3), 287–293 (2000)
28. Sitti, M.: Miniature devices: mimicking insect locomotion. Nature **458**(7241), 1121–1122 (2009)
29. Dürr, V., Ebeling, W.: Biomimetic tactile sensor array. Bioinspir. Biomim.. Biomim. **5**(3), 036001 (2010)
30. Rus, D., Tolley, M.T.: Design, fabrication and control of soft robots. Nature **521**(7553), 467–475 (2015)
31. Laschi, C., Mazzolai, B., Cianchetti, M.: Soft robotics: technologies and systems pushing the boundaries of robot abilities. Sci. Robot. **1**(1), eaah3690 (2016)
32. Pylatiuk, C., Döderlein, L.: Biomechanical principles of functional electrical stimulation. J. Artif. OrgansArtif. Organs **10**(4), 185–190 (2007)
33. Kim, S., Laschi, C., Trimmer, B.: Soft robotics: a bioinspired evolution in robotics. Trends Biotechnol.Biotechnol. **31**(5), 287–294 (2013)
34. Torchilin, V.P.: Multifunctional nanocarriers. Adv. Drug Deliv. Rev.Deliv. Rev. **63**(7), 609–613 (2011)
35. Peer, D., Karp, J.M., Hong, S., Farokhzad, O.C., Margalit, R., Langer, R.: Nanocarriers as an emerging platform for cancer therapy. Nat. Nanotechnol.Nanotechnol. **2**(12), 751–760 (2007)
36. Langer, R., Vacanti, J.P.: Tissue engineering. Science **260**(5110), 920–926 (1993)
37. Vrana, N.E., Dupret-Bories, A., Thorpe, A.A., Jiskoot, W., Groll, J., Malda, J.: Integration of biologically inspired and additive manufacturing techniques for medical advancement. Adv. Healthcare Mater. **6**(16), 1700013 (2017)
38. Cheng, H.Y., Grossman, J.C.: Synthetic photosystems for light-driven water splitting: recent advances and future directions. Chem. Soc. Rev. **38**(9), 2522–2535 (2009)
39. Lee, J.H., Kim, D.H.: Biomimetic and bioinspired photonic structures for green building applications. Energies **4**(10), 1613–1632 (2011)
40. Bie, Z., He, Q.: Biomimicry for optimizing the energy performance of building envelopes: a review. Renew. Sustain. Energy Rev. **15**(1), 305–317 (2011)
41. Bonser, R.H.C., Vincent, J.F.V., Jeronimidis, G.: The mechanics of rooting in the staghorn fern platycerium bifurcatum. J. Exp. Bot. **47**(296), 1437–1444 (1996)
42. Zhu, B., Chen, X., Gu, Y., Wei, Q., Guo, Z.: Biomimetics for next generation materials: towards bioinspired synthesis and processing. Mater. Today **20**(8), 460–480 (2017)
43. Huang, X., Jiang, P., Carne-Sánchez, A., Ariga, K.: Advanced biomimetic and bioinspired materials for healthcare, energy, coatings, water-repellency, and food packaging applications. J. Mater. Chem. B **5**(23), 4317–4340 (2017)
44. Bolhuis, P.G., Chandler, M.: Challenges and approaches in computational studies of protein structure and dynamics. J. Phys. Chem. B **114**(28), 9373–9376 (2010)

45. Srinivasan, M.V., Zhang, S.W., Chahl, J.S.: A case study in biomimetic design: the research, development and commercialization history of insect-inspired miniature robots. Bioinspir. Biomim.. Biomim. **4**(1), S1–S11 (2001)
46. Yuen, D.A., Patel, R.V.: From biomimicry to bioinspiration: biologically-inspired engineering. IEEE Robot. Autom. Mag.Autom. Mag. **25**(3), 15–17 (2018)
47. Sitti, M., Kim, S.: Biomimetics for robotic applications. J. Micro-Bio Robot. **12**(1–4), 63–75 (2017)
48. Haraway, D.: Staying with the Trouble: Making Kin in the Chthulucene. Duke University Press (2016)
49. Glavovic, B.C., Smith, T.F.: Engineering nature's mimicry and borrowing: viewpoints from ethics, politics, and science. J. Bioethical Inq. **12**(4), 559–569 (2015)
50. Whittaker, R.H., Likens, G.E.: The biosphere and man. Hum. Ecol. **3**(4), 327–338 (1975)
51. Hosny, A., Parmar, C., Quackenbush, J., Schwartz, R.: Artificial intelligence in radiology. Nat. Rev. Cancer **18**(8), 500–510 (2018)
52. Liang, X., Zhao, J.: Biomimetic intelligence: a digital ecosystem perspective. J. Manag. Anal. **5**(2), 89–108 (2018)
53. Bouligand, Y.: Are cholesteric and smectic structures key architectures for biological materials? Lessons from liquid crystal physics. J. Mater. Chem. B **1**(37), 4810–4820 (2013)
54. Winters, N., Devine-Wright, P., Gregory-Smith, D.: Nature-based solutions for carbon storage: highlighting the social dimensions. J. Environ. Planning Policy Manage. **21**(3), 261–277 (2019)
55. Lackner, S., Bröchler, S.: The future of biomimicry—potentials, barriers, and ways forward. Sustain. Sci. **11**(3), 413–421 (2016)
56. Bullock, J.M., Bonte, D., Pufal, G., da Silva Carvalho, C., Chapman, D.S., García, C., Delgado, M.M.: Human-mediated dispersal and the rewiring of spatial networks. Trends Ecol. Evol.Evol. **35**(12), 1133–1143 (2020)
57. San-Miguel, A., Kowalczyk, P.: The bio-inspired paradigm for engineering optimization: principles, methods and applications. Soft. Comput.Comput. **24**(21), 15935–15956 (2020)
58. Pan, H., Yan, Y., Zhang, H., Gao, W., Yu, X., Yang, X., Xie, T.: Bio-inspired scalable manufacture of highly stretchable carbon nanotube strain sensors. Nature **562**(7728), 570–574 (2018)
59. Keswani, M., Vavylonis, D.: Biomimetic self-organization: fascinating model systems for soft matter physics. Soft Matter **15**(38), 7514–7527 (2019)
60. White, R.J., Tobin, M.J.: Biomimetic micro- and nanoscale technologies: challenges and prospects. Annu. Rev. Biomed. Eng.. Rev. Biomed. Eng. **20**, 59–85 (2018)
61. Peattie, A.M.: Designing for sustainability: a case study in biomimicry. J. Mech. Des. **129**(7), 695–701 (2007)
62. Ghahramani, Z., Urban, M.W.: Self-healing materials: strategies and recent developments. Wiley Interdiscip. Rev.: Comput. Mol. Sci. **5**(3), 245–264 (2015)
63. Clode, D.: The commercialization of biomimetic materials. In: Biomimetic Research for Architecture and Building Construction, pp. 401–412. Springer (2013)
64. Ulijn, R.V., Smith, A.M.: Designing peptide based nanomaterials. Chem. Soc. Rev. **37**(4), 664–675 (2008)
65. Whitesides, G.M., Grzybowski, B.: Self-assembly at all scales. Science **295**(5564), 2418–2421 (2002)
66. Laschi, C., Cianchetti, M.: Soft robotics: new perspectives for robot bodyware and control. Front. Bioeng. Biotechnol. **2**, 3 (2014)
67. Sukhatme, G.S.: Biomimetic robotic systems. Annu. Rev. Control Robot. Auton. Syst. **2**, 391–414 (2019)
68. Rana, K.S., Mandal, S.K.: Recent advances in energy and environmental sustainability through biomimetics. Environ. Sci. Pollut. Res.Pollut. Res. **28**(5), 4904–4918 (2021)
69. Bhushan, B., Nosonovsky, M.: Biomimetic and bio-inspired technologies for sustainable solutions. Philos. Trans. R. Soc. A Math. Phys. Eng. Sci. **368**(1914), 4775–4777 (2010)
70. Pacheco-Torgal, F., Ivanov, V., Karak, N.: Biomimetic-Based Solutions for Energy and Environmental Challenges. Woodhead Publishing (2019)

71. Kim, S., Laschi, C.: Soft robotics: a bioinspired evolution in robotics. Trends Biotechnol.Biotechnol. **37**(9), 983–986 (2019)
72. Kim, S., Spenko, M., Trujillo, S., Heyneman, B., Santos, D., Cutkosky, M.: Smooth vertical surface climbing with directional adhesion. IEEE Trans. Rob. **24**(1), 65–74 (2008)
73. Faria, G.C., Fernandes, P.R., Guedes, J.M.: Computational modeling of biomimetic materials and structures. J. Mech. Behav. Biomed. Mater.Behav. Biomed. Mater. **79**, 146–162 (2018)
74. Saha, M., Ghosh, S., Bhowmik, S., Chaudhuri, S.: Bioinformatics in the era of big data: challenges and opportunities. J. Transl. Med. **18**(1), 1–15 (2020)
75. Teraa, S., Bencherif, M.: From hygrothermal adaptation of endemic plants to meteorosensitive biomimetic architecture: case of Mediterranean biodiversity hotspot in Northeastern Algeria. Environ. Dev. Sustain. **24**, 10876–10901 (2022). https://doi.org/10.1007/s10668-021-01887-y
76. Jakšić, Z., Devi, S., Jakšić, O., Guha, K.: A comprehensive review of bio-inspired optimization algorithms including applications in microelectronics and nanophotonics. Biomimetics 8(3), 278 (2023). https://doi.org/10.3390/biomimetics8030278
77. Kawamura, A., Kohri, M., Morimoto, G., et al.: Full-color biomimetic photonic materials with iridescent and non-iridescent structural colors. Sci. Rep. **6**, 33984 (2016). https://doi.org/10.1038/srep33984
78. Sun, J., Bauman, L., Yu, L., Zhao, B.: Gecko-and-inchworm-inspired untethered soft robot for climbing on walls and ceilings. Cell Rep. Phys. Sci. **4**(2), 101241 (2023). https://doi.org/10.1016/j.xcrp.2022.101241
79. Haidar, Z.S.: Introductory chapter: bioMimetics for HealthCare—innovations inspired by nature. Biomedical Engineering. IntechOpen (2023). https://doi.org/10.5772/intechopen.106328
80. Sheikh, W.T., Asghar, Q.: Adaptive biomimetic facades: enhancing energy efficiency of highly glazed buildings. Front. Arch. Res. **8**(3), 319–331 (2019)https://doi.org/10.1016/j.foar.2019.06.001

High Performance Computing and Its Application in Computational Biomimetics

Mohd. Firdaus bin Abas, Balbir Singh, and Kamarul Arifin Ahmad

Abstract The convergence of High Performance Computing (HPC) and computational biomimetics has ushered in a new era of scientific exploration and technological innovation. This book chapter shows the intricate relationship between HPC and the field of computational biomimetics, demonstrating how the synergistic interplay between these two domains has revolutionized our understanding of nature-inspired design and complex biological processes. Through a comprehensive analysis of cutting-edge research and architecture, this chapter highlights the pivotal role of HPC in simulating, modeling, and deciphering biological phenomena with remarkable accuracy and efficiency. The chapter begins by elucidating the fundamental principles of HPC and computational biomimetics, elucidating how biological systems serve as inspiration for the development of novel technologies and solutions. It subsequently looks into the underlying architecture and capabilities of modern HPC systems, elucidating how their parallel processing prowess enables the simulation of intricate biological processes and the exploration of large-scale biomimetic design spaces. A significant portion of the chapter is devoted to exploring diverse applications of HPC in the field of computational biomimetics. These applications encompass a wide spectrum of disciplines, ranging from fluid dynamics and materials science to robotics and drug discovery. Each application is accompanied by real-world examples that showcase the transformative impact of HPC-driven computational biomimetics on advancing scientific knowledge and engineering innovation.

Mohd. F. Abas · B. Singh (✉) · K. A. Ahmad
Department of Aerospace Engineering, Faculty of Engineering, Universiti Putra Malaysia, Serdang 43400, Selangor, Malaysia
e-mail: balbir.s@manial.edu

Mohd. F. Abas · K. A. Ahmad
Aerospace Malaysia Research Centre, Faculty of Engineering, Universiti Putra Malaysia, 43400 Serdang, Selangor, Malaysia

B. Singh
Department of Aeronautical and Automobile Engineering, Manipal Institute of Technology, Manipal Academy of Higher Education, Manipal, Karnataka 576104, India

K. A. Ahmad et al. (eds.), *High Performance Computing in Biomimetics*, Series in BioEngineering, https://doi.org/10.1007/978-981-97-1017-1_2

Keywords High performance computing, Cray-1 · Scalability · Parallelism · Biomimetics

1 Definition and Overview

High-Performance Computing (HPC) refers to using advanced computing technologies and techniques to solve complex computational problems requiring significant computational power, memory, or storage resources. HPC systems leverage parallel processing, large-scale data management, and advanced algorithms to achieve high-speed and efficient execution of computationally intensive tasks. This detailed description provides an overview of HPC, its key components, applications, and its benefits to various fields of research and industry.

1.1 High-Performance Computing as a System

Hardware Infrastructure (Cluster): HPC systems consist of a cluster of interconnected computing nodes, including multi-core processors, accelerators (e.g., GPUs or FPGAs), high-speed networks, and high-capacity storage systems. These components are designed to deliver high computational throughput and efficient data movement [1, 2].

Software Stack: HPC software includes operating systems, parallel programming models (e.g., MPI, OpenMP), compilers, libraries, and tools designed to exploit parallelism, optimize performance, and manage the complex workflows and resources of HPC applications [3, 4].

Algorithms and Applications: HPC algorithms and applications are developed to leverage the computational capabilities of HPC systems efficiently. These applications span a wide range of domains, including scientific simulations, weather modeling, financial modeling, data analytics, and machine learning, among others [5, 6].

Basically, it works using three main components, i.e., compute, network, and storage, as shown in Fig. 1.

1.2 High Performance Computing: Uses and Benefits

Some of the modern uses of HPC-based platforms are as:

Scientific Simulations: HPC enables scientists to perform complex simulations and modeling, such as climate modeling, astrophysics, molecular dynamics, and

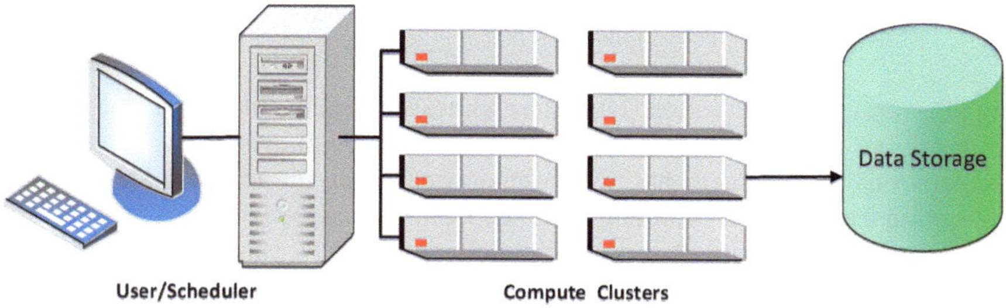

Fig. 1 HPC system architecture and deployment scenario (*Source* https://doi.org/10.3390/sym12061029)

quantum chemistry, allowing for detailed investigations and predictions of complex phenomena [7, 8].

Big Data Analytics: HPC facilitates the processing and analysis of vast amounts of data, enabling data-driven insights and decision-making. Applications include large-scale data mining, bioinformatics, genomics, social network analysis, and financial modeling [9, 10].

Machine Learning and AI: HPC plays a crucial role in training and inference for machine learning and AI algorithms. HPC systems accelerate the processing of large-scale datasets and complex neural networks, enabling advancements in image recognition, natural language processing, and autonomous systems [11, 12].

Benefits of High Performance Computing:

Improved Time-to-Solution: HPC systems allow for the rapid execution of computationally intensive tasks, reducing the time required to obtain results and insights. This is particularly beneficial in time-sensitive applications such as weather prediction, drug discovery, and emergency response [13, 14].

Scalability and Parallelism: HPC systems excel in scaling computations across multiple processing units, enabling the efficient utilization of thousands or even millions of cores. This scalability is crucial for tackling large-scale simulations, data analytics, and AI applications [15, 16].

Innovation and Discovery: HPC fosters innovation by providing researchers and scientists with the computational power to explore new frontiers, validate hypotheses, and make groundbreaking discoveries in various scientific domains [17, 18].

High Performance Computing (HPC) leverages advanced hardware, software, and algorithms to solve complex computational problems in domains such as scientific simulations, big data analytics, and machine learning. HPC offers improved time-to-solution, scalability, and enables innovation and discovery in various fields. The continuous advancements in HPC systems and techniques open new frontiers for computational research and provide solutions to increasingly complex challenges.

2 Evolution of HPC

High Performance Computing (HPC) has witnessed significant advancements since its inception, enabling scientists, researchers, and engineers to tackle complex computational challenges that were once considered intractable. This detailed description explores the evolution of HPC, from its early beginnings to its current state, highlighting key milestones and technologies that have shaped its development.

Early Computing Era: The roots of HPC can be traced back to the early computing era, marked by the emergence of mainframe computers in the 1950s. These large and expensive machines were the primary computing resources for scientific and engineering applications. Notable examples include the IBM 704 and the Control Data Corporation (CDC) 6600, which introduced vector processing capabilities, enabling faster execution of scientific simulations and calculations [15, 19].

Parallel Computing and Supercomputers: In the 1970s and 1980s, parallel computing emerged as a key aspect of HPC. The Cray-1 as shown in Fig. 2, released in 1976, was one of the earliest supercomputers designed specifically for scientific and engineering applications. It employed vector processing and parallelism to achieve high performance. Parallel computing architectures, such as symmetric multiprocessors (SMP) and massively parallel processors (MPP), gained prominence during this era [20, 21].

Distributed Computing and Clusters: The 1990s witnessed a shift towards distributed computing and the rise of cluster computing. Clusters, consisting of commodity off-the-shelf (COTS) computers interconnected through high-speed networks, allowed for cost-effective scalability and improved performance. Beowulf clusters, pioneered by Thomas Sterling and Donald Becker, became popular for scientific computing applications [22, 23].

Massively Parallel Processing and GPUs: The introduction of Graphics Processing Units (GPUs) in the early 2000s revolutionized HPC. Originally designed for rendering graphics, GPUs exhibited massive parallelism and computational power, making them ideal for scientific computations. Researchers and developers began leveraging GPUs for general-purpose computing, leading to the emergence of General-Purpose GPU (GPGPU) programming models such as CUDA and OpenCL [24, 25].

HPC in the Cloud: The advent of cloud computing in the 2010s brought HPC capabilities to a broader audience. Cloud service providers like Amazon AWS, Microsoft Azure etc., started offering HPC resources on-demand, allowing users to access scalable computing power without the need for large upfront investments in hardware and infrastructure. This democratization of HPC has opened up new possibilities for scientific research, data analytics, and simulations [26, 27].

Exascale Computing: The current frontier in HPC is the pursuit of exascale computing, aiming to achieve computing systems capable of performing at least

Fig. 2 Cray-1 on display at the science museum in London, U.K. (*Source* Wikipedia)

one exaflop (10^{18} floating-point operations per second). Exascale computing has the potential to address grand challenges in fields such as climate modeling, astrophysics, and drug discovery. Researchers are exploring novel architectures, energy-efficient designs, and new programming models to realize exascale systems [18, 28]. Figure 3 shows European Processor Initiative architecture and a kind of computing environment based on exascale computing.

Convergence of HPC with AI and Big Data: The convergence of HPC with Artificial Intelligence (AI) and Big Data is another significant trend. HPC technologies, such as accelerators and high-speed interconnects, are being utilized to accelerate AI training and inference workloads. Additionally, HPC techniques are applied to process and analyze vast amounts of Big Data, enabling insights and discoveries in fields like genomics, climate modeling, and high-energy physics [9, 18].

The evolution of HPC from mainframes to supercomputers, parallel computing to distributed systems, and the integration of GPUs, cloud computing, and AI highlights

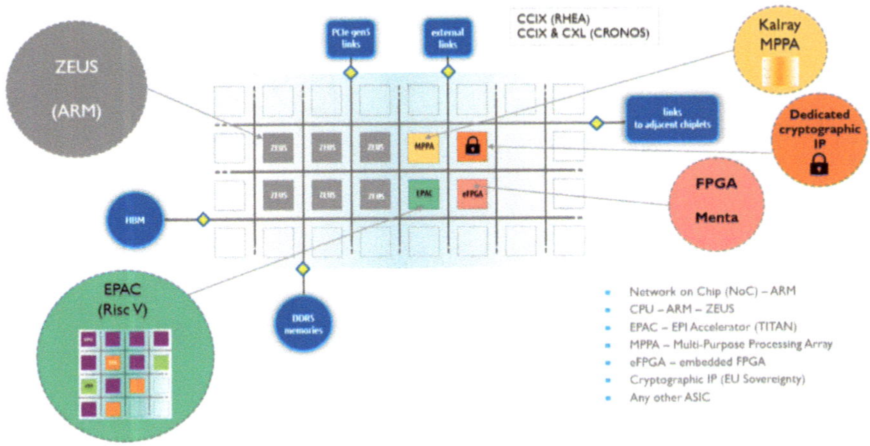

Fig. 3 Exascale based EPI system architechture (*Source* https://doi.org/10.3390/mca25030046)

its continuous growth and adaptability to new challenges. The pursuit of exascale computing and the convergence with AI and Big Data present exciting prospects for the future of HPC, enabling breakthroughs in scientific research, engineering, and data-driven decision-making.

3 Characteristics and Components of HPC Systems

High Performance Computing (HPC) is characterized by its ability to deliver exceptional computational power, enabling the efficient execution of complex and computationally intensive tasks. This detailed description explores the key characteristics and components of HPC, shedding light on the essential features that distinguish HPC systems and contribute to their high-performance capabilities.

3.1 Characteristics of High Performance Computing

Computational Power: HPC systems are designed to deliver high computational power, enabling the rapid execution of large-scale simulations, data analytics, and scientific computations. This computational power is achieved through parallel processing, vector processing, and the use of accelerators such as GPUs or FPGAs [1, 15].

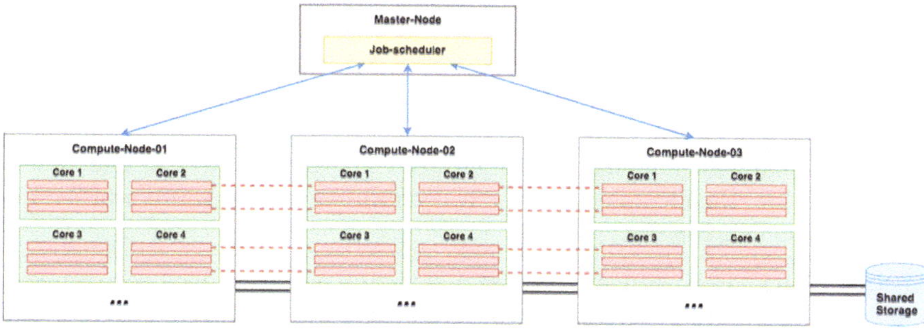

Fig. 4 Parallelism in typical HPC cluster [34]

Scalability: HPC systems exhibit scalability, allowing computational resources to be efficiently scaled up or down based on the requirements of the task at hand. Scalability is crucial for handling large datasets, accommodating increasing computational demands, and achieving optimal performance across a wide range of problem sizes [29, 30].

Parallelism: Parallel processing is a key characteristic of HPC, where computations are divided into smaller tasks that can be executed simultaneously across multiple processing units. This parallelism can be achieved through shared-memory models (e.g., OpenMP) or message-passing models (e.g., MPI), allowing for efficient utilization of multiple cores or nodes in a computing cluster [3, 31] as shown in Fig. 4. Here the job scheduler runs on the master node and controls the highest level of parallelism in running jobs on each node (blue arrows).

High-Speed Interconnects: HPC systems incorporate high-speed interconnects, such as InfiniBand or Ethernet-based technologies, to ensure efficient communication between computing nodes. These interconnects provide low-latency, high-bandwidth connections, enabling fast data transfer and synchronization among parallel processing units [32, 33].

3.2 Components of High Performance Computing

Computing Nodes: HPC systems consist of multiple computing nodes, which can include multi-core processors, accelerators (such as GPUs or FPGAs), and high-capacity memory. These nodes work together to execute computational tasks in parallel, enhancing overall performance and throughput [14, 29].

High-Speed Storage: HPC systems incorporate high-capacitystorage solutions, such as parallel file systems or distributed storage architectures, to handle the large volumes of data generated and processed by HPC applications. These storage systems provide fast access to data and support high-bandwidth data transfers, enabling efficient data input/output operations [35, 36].

Networking Infrastructure: HPC systems rely on high-performance networking infrastructure to facilitate communication and data exchange between computing nodes. This infrastructure includes high-speed networks, switches, and routers that provide low-latency, high-bandwidth connections to enable efficient parallel computing and data transfer [32, 37].

Software Stack: HPC systems rely on a software stack that includes operating systems, parallel programming models (such as MPI or OpenMP), compilers, libraries, and tools. This software stack facilitates the development and execution of parallel applications, optimizing performance, resource management, and scalability [1, 38].

Management and Scheduling Tools: HPC systems require management and scheduling tools to efficiently allocate resources, manage job queues, and ensure optimal utilization of computing nodes. These tools provide job submission interfaces, resource monitoring, and load balancing capabilities to maximize the system's performance [22, 39].

Understanding the characteristics and components of HPC is crucial for designing and deploying high-performance systems that meet the demands of modern computational challenges. The combination of computational power, scalability, parallelism, high-speed interconnects, and a well-designed software stack contributes to the exceptional performance and capabilities of HPC systems.

4 Importance of HPC in Scientific Research and Engineering

High Performance Computing (HPC) plays a crucial role in advancing scientific research and engineering by providing the computational power and resources necessary to tackle complex problems and accelerate the pace of discovery. This detailed description explores the importance of HPC in scientific research and engineering, highlighting its impact across various disciplines and domains.

4.1 Simulation and Modeling

Climate Modeling: HPC enables climate scientists to simulate and model complex climate systems, facilitating predictions and projections of climate patterns, weather events, and long-term climate changes. These simulations aid in understanding climate dynamics, assessing the impact of human activities, and developing strategies for climate mitigation and adaptation [40, 41].

Computational Fluid Dynamics (CFD): HPC allows engineers to perform detailed simulations of fluid flows, enabling the design and optimization of aerodynamic profiles, combustion systems, and fluid flow phenomena. CFD simulations contribute to advancements in aerospace engineering, automotive design, and energy system optimization [42, 43].

Molecular Dynamics and Materials Science: HPC facilitates molecular simulations and materials modeling, providing insights into the behavior and properties of complex materials at the atomic and molecular scale. These simulations contribute to the development of new materials, drug discovery, understanding chemical reactions, and predicting material properties [44, 45] as shown in Fig. 5.

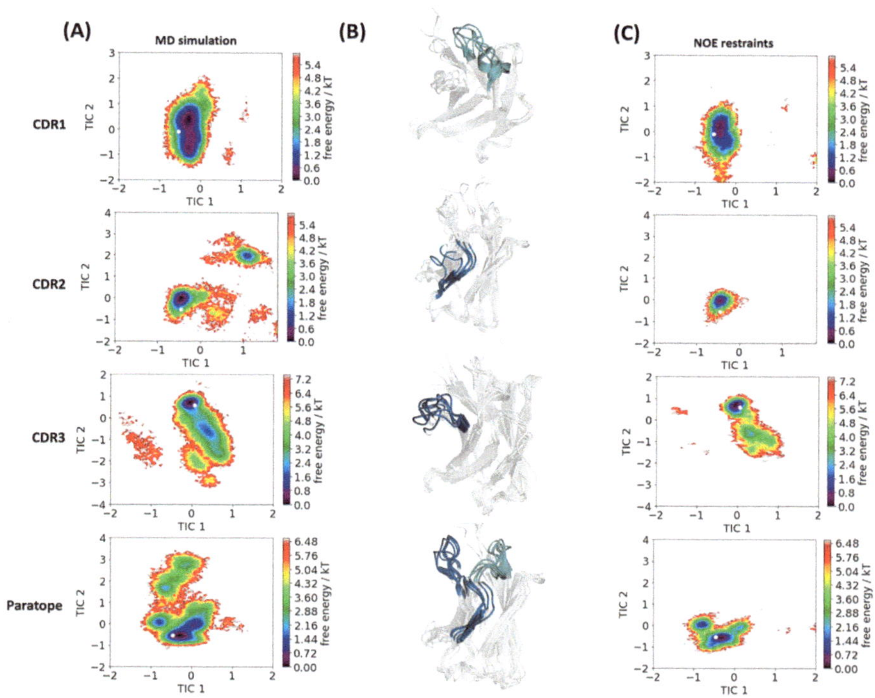

Fig. 5 MD simulations with NOE restraints to compare the respective ensembles in solution [46]

4.2 Big Data Analytics

Genomics and Bioinformatics: HPC enables the analysis of large-scale genomic datasets, facilitating genome sequencing, annotation, and comparative genomics. HPC systems aid in understanding genetic variations, disease mechanisms, and personalized medicine [47, 48].

Astrophysics and Cosmology: HPC enables astrophysicists and cosmologists to simulate and analyze vast amounts of observational data, aiding in the study of the universe's origins, structure, and evolution. HPC systems contribute to modeling galaxy formation, simulating cosmic events, and understanding the fundamental laws of physics [49, 50].

4.3 Optimization and Design

Computational Chemistry and Drug Discovery: HPC enables computational chemistry simulations and virtual screening of large chemical libraries, accelerating the discovery of new drugs and understanding molecular interactions. HPC systems aid in optimizing drug candidates, predicting their efficacy, and reducing the time and cost associated with experimental drug discovery [51, 52].

Engineering Design and Optimization: HPC allows engineers to perform complex simulations, optimization, and virtual prototyping, enabling the design and refinement of innovative products and systems. HPC systems aid in optimizing aerodynamic profiles, structural analysis, and mechanical designs, resulting in improved performance, efficiency, and reliability [53, 54]. Figure 6 shows the simulation for a geometrically adaptable heart valve replacement.

4.4 Data-Intensive Research

Data Analytics and Machine Learning: HPC systems accelerate the processing and analysis of large-scale datasets, enabling data-driven insights and machine learning algorithms. HPC facilitates the training of complex neural networks, deep learning models, and data-intensive analytics for various applications, including image recognition, natural language processing, and pattern recognition [12, 55].

Data-Driven Simulations: HPC enables the integration of experimental data and simulations, enhancing the accuracy and fidelity of scientific modeling and predictions. Data-driven simulations leverage HPC capabilities to assimilate observational data, validate models, and improve understanding of complex phenomena [56, 57].

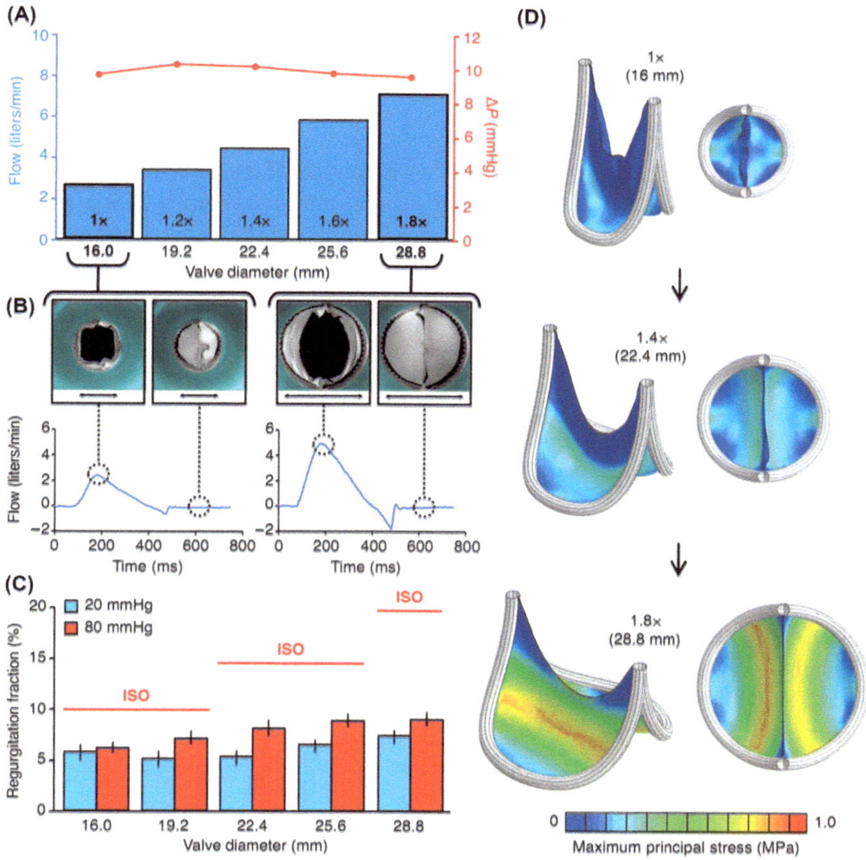

Fig. 6 Flow loop testing and computational modeling of the biomimetic bileaflet valve. From Ref [58]. Copyright (2020). Reprinted with permission from AAAS

High-Performance Computing (HPC) is of utmost importance in scientific research and engineering. HPC facilitates complex simulations, modeling, data analytics, optimization, and design in various fields, including climate modeling, genomics, astrophysics, drug discovery, and engineering. HPC systems accelerate the pace of discovery, provide insights into complex phenomena, and enable data-intensive research, ultimately driving innovation and advancing knowledge across disciplines.

5 HPC Technologies and Architectures

5.1 Parallel Computing

Parallel computing involves the simultaneous execution of multiple tasks or parts of a task, utilizing multiple processing units to achieve improved computational performance. It is a cornerstone in High Performance Computing (HPC), enabling complex simulations and computations to be divided into smaller tasks that can be processed in parallel. This approach greatly enhances computational speed and efficiency, making it essential for applications such as molecular dynamics simulations and biomimetic modeling [59, 60].

5.2 Distributed Computing

Distributed computing involves the use of interconnected computers to solve a single problem or perform a task, distributing the workload among different machines. It facilitates collaboration and sharing of resources across a network, enabling larger-scale computations and data analysis. In the context of computational biomimetics, distributed computing can accelerate simulations of complex biological processes by leveraging the power of multiple nodes [61, 62].

5.3 Grid Computing

Grid computing involves the coordinated use of geographically dispersed computing resources to solve large-scale computational problems. It focuses on resource sharing and collaboration across organizations or institutions. In the context of computational biomimetics, grid computing can enable researchers to access and utilize a distributed network of powerful computers for running simulations and data analysis [63, 64].

5.4 Cluster Computing

Cluster computing involves the interconnected use of multiple computers that work together as a single system to perform high-performance computations. It provides a cost-effective way to achieve high computational power, making it a common choice for many HPC applications, including computational biomimetics. Clusters are especially suited for parallel processing tasks and simulations [65, 66].

5.5 Supercomputing

Supercomputing refers to using extremely powerful computers capable of performing complex calculations at high speeds. These systems are crucial for solving computationally intensive problems in fields like computational biomimetics, where simulations of intricate biological processes require substantial computational resources [67, 68].

5.6 Accelerators and Co-processors (e.g., GPUs, FPGAs)

Accelerators and co-processors such as Graphics Processing Units (GPUs) and Field-Programmable Gate Arrays (FPGAs) are specialized hardware components that can significantly enhance computational performance by offloading specific tasks from the central processing unit (CPU). These components are particularly advantageous for parallel processing applications in computational biomimetics, enabling faster simulations and data analysis [69, 70].

6 Computational Biomimetics

Computational Biomimetics is an interdisciplinary field that draws inspiration from biological systems and processes to develop innovative solutions for various scientific and engineering challenges. It combines principles from biology, computer science, mathematics, and engineering to simulate, model, and replicate natural phenomena. The central concept of computational biomimetics is to uncover the underlying design principles and mechanisms that nature has evolved over millions of years, and then apply these insights to create novel technologies and designs that are efficient, adaptable, and sustainable [71, 72]. Biomimetic design and engineering involve translating biological principles into practical solutions. Computational biomimetics plays a pivotal role in this process by providing tools and methodologies to analyze and mimic natural structures, behaviors, and functions. By harnessing computational methods such as finite element analysis, molecular dynamics simulations, and computational fluid dynamics, researchers and engineers can replicate complex biological systems at various scales, from molecular interactions to macroscopic behaviors.

This process includes studying the form-function relationships in nature, understanding how organisms have evolved to solve specific problems efficiently, and then applying these insights to create innovative products, materials, and technologies. Examples of biomimetic design range from developing lightweight and strong structural materials inspired by bone structures to designing agile and efficient robotic systems based on animal locomotion [73, 74].

7 Importance of Computational Biomimetics

Computational biomimetics holds immense importance in various scientific and technological domains due to its potential to drive innovation and address complex challenges. Its significance can be understood from the following perspectives.

7.1 Advancing Scientific Understanding

Computational biomimetics enables researchers to understand the intricate details of biological systems and processes that are often challenging to study experimentally. By simulating and modeling these systems, scientists can gain deeper insights into the underlying mechanisms, which can lead to new discoveries in fields such as biology, physiology, and ecology.

7.2 Innovation in Engineering and Technology

The application of biomimetic principles in engineering has led to the development of groundbreaking technologies and products that are more energy-efficient, sustainable, and adaptable. This includes designing aerodynamic structures inspired by bird flight, self-healing materials inspired by the human skin, and autonomous robots mimicking animal behaviors.

7.3 Sustainability and Environmental Conservation

Biomimetic designs often lead to solutions that are optimized for resource utilization and environmental compatibility. By emulating nature's strategies for efficiency, waste reduction, and adaptation, computational biomimetics contributes to the development of sustainable technologies that minimize negative impacts on the environment.

7.4 Interdisciplinary Collaboration

Computational biomimetics bridges the gap between various disciplines, fostering collaboration between biologists, engineers, physicists, and computer scientists. This cross-disciplinary approach encourages the exchange of ideas and expertise, leading to more holistic problem-solving and innovative outcomes [75, 76].

In summary, computational biomimetics leverages the power of simulation, modeling, and interdisciplinary collaboration to unravel the mysteries of nature and apply its ingenious solutions to address contemporary scientific and engineering challenges. This approach holds promise for shaping a more sustainable and technologically advanced future.

8 Role of HPC in Computational Biomimetics

8.1 Simulation and Modeling

High Performance Computing (HPC) plays a crucial role in advancing the field of computational biomimetics by enabling complex simulations and accurate modeling of biological systems and processes. Many biological phenomena involve intricate interactions at multiple scales, ranging from molecular dynamics to macroscopic behaviors. HPC systems provide the computational power needed to simulate these interactions with high fidelity, allowing researchers to explore the behavior of biomolecules, study biomechanics, and simulate ecological systems. Through molecular dynamics simulations, HPC facilitates the understanding of protein folding, molecular interactions, and drug binding, which are essential for drug discovery and development. Similarly, biomechanical simulations using HPC help engineers design biomimetic structures and materials that replicate the efficiency and resilience of natural systems [77, 78].

8.2 Data Analysis and Processing

Biological research generates vast amounts of data from sources such as genomics, proteomics, and imaging techniques. HPC systems are essential for managing, analyzing, and extracting meaningful insights from these data sets. Machine learning algorithms and data-driven approaches can be applied to identify patterns, classify biological entities, and predict outcomes.

In computational biomimetics, HPC aids in processing large datasets derived from biological observations and simulations. For instance, analyzing movement patterns of animals or studying gene expression profiles can be computationally intensive tasks that benefit from the parallel processing capabilities of HPC [12, 79].

8.3 Optimization and Design

HPC-driven optimization techniques are crucial for designing biomimetic structures, materials, and systems. Evolutionary algorithms, genetic algorithms, and other optimization methods can explore vast design spaces to identify solutions that mimic nature's efficiency and functionality. HPC allows researchers to run numerous simulations in parallel, systematically testing various design parameters and configurations to arrive at optimal solutions.

Whether it's designing energy-efficient buildings inspired by termite mounds or creating streamlined underwater vehicles inspired by fish, HPC's parallel processing capabilities empower researchers to fine-tune designs and iterate rapidly [80, 81].

8.4 Visualization and Virtual Reality

HPC enhances the visualization of complex biological and biomimetic systems, allowing researchers to gain intuitive insights into their behavior. Three-dimensional renderings, animations, and virtual reality simulations enable scientists and engineers to explore and interact with intricate models. This is especially useful for conveying complex concepts and fostering deeper understanding among researchers and the public [82, 83]. In computational biomimetics, visualization techniques powered by HPC aid in demonstrating the results.

So, High Performance Computing serves as a cornerstone of computational biomimetics by providing the computational resources needed for accurate simulations, efficient data analysis, optimization, and immersive visualization. This synergy empowers researchers and engineers to unravel nature's mysteries, replicate its ingenious solutions, and drive innovation across diverse domains.

9 HPC Applications in Computational Biomimetics

Biomimetics has revolutionized material and structural design, and High Performance Computing (HPC) is pivotal in simulating and optimizing these innovations. Some of the applications of HPC in Computational Biomimetics include modeling the mechanical properties of natural materials. HPC facilitates simulations of the mechanical properties of biological materials like bones, shells, and spider silk. By mimicking the hierarchical structure of these materials, researchers can understand their exceptional strength, flexibility, and toughness. This knowledge informs the design of innovative synthetic materials [84, 85]. Using HPC, researchers simulate the behavior of biological composites like nacre and wood, enabling the creation of lightweight yet durable synthetic materials. These materials find applications in aerospace, automotive, and construction industries [86, 87]. Biomimetic robotics draws inspiration from nature's mechanisms for locomotion and behavior, with HPC enabling sophisticated simulations. HPC-driven simulations model animal locomotion, helping researchers understand complex movements (aerodynamics) like bird flight and insect navigation. These insights inform the design of agile and efficient robotic systems [88, 89]. HPC assists in optimizing the design and behavior of biomimetic robots by simulating interactions with their environment. This approach enhances their adaptability and efficiency [90, 91]. HPC also drives progress in understanding genetic and molecular processes through simulations and data analysis. It accelerates the analysis of vast genomic data, aiding in identifying genes, regulatory elements, and potential disease markers [92, 93]. HPC-driven simulations predict protein structures and folding mechanisms, aiding drug design and disease understanding [94, 95]. It basically accelerates drug discovery by simulating molecular interactions and predicting drug behavior. HPC enables virtual screening of large compound libraries against target proteins, identifying potential drug candidates efficiently [96, 97]. HPC models drug interactions with target proteins, predicting binding affinity and guiding drug design [98, 99]. HPC further helps model complex environmental systems and optimize sustainable solutions. HPC simulates ecosystems to understand interactions among species, aiding in conservation efforts [100, 101]. HPC optimizes building designs for energy efficiency, drawing inspiration from natural ventilation and thermal regulation mechanisms [102, 103]. In conclusion, High Performance Computing plays a vital role in various applications within computational biomimetics, from materials and robotics to genomics and conservation, to name a few. By providing the computational power needed for simulations, modeling, and data analysis, HPC accelerates innovation and deepens our understanding of biological and natural systems. Figure 7 depicts one such application from insect aerodynamic study for mimicking small-scale insect robotics using IBM method on HPC platform.

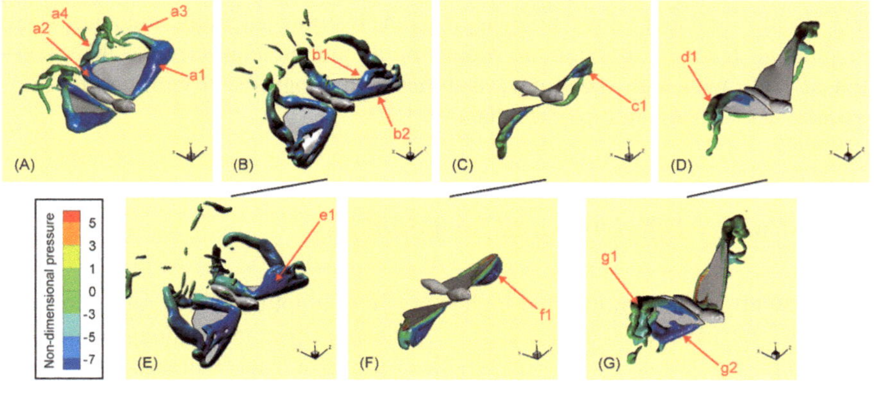

Fig. 7 Vortex structures and associated pressure for various cases at selected time-instances [104]

10 Challenges and Future Directions

10.1 Scalability and Performance Optimization

One of the primary challenges in the application of High Performance Computing (HPC) to computational biomimetics is ensuring scalability and optimizing performance. As computational models and simulations become more complex, they require distributed computing resources to handle the increased computational load. Ensuring that algorithms and simulations can effectively utilize the parallel processing capabilities of HPC clusters is essential. This involves optimizing code, load balancing, and minimizing communication overhead to achieve efficient scalability.

Future Directions: Researchers are exploring techniques like adaptive mesh refinement, parallel algorithm development, and hybrid computing models that combine CPU and accelerator resources for improved performance. Additionally, advancements in parallel programming frameworks and tools will continue to play a significant role in addressing scalability challenges [105, 106].

10.2 Big Data and Data Management

The influx of data from diverse sources in computational biomimetics poses challenges related to storage, processing, and analysis. Managing and extracting meaningful insights from large biological datasets require efficient data storage, retrieval, and analysis pipelines. HPC systems need to handle the data deluge while maintaining data integrity and ensuring data security.

Future Directions: Developing scalable data management frameworks, integrating data visualization tools, and employing machine learning techniques for data preprocessing and feature extraction are promising directions. Furthermore, the integration of HPC with high-speed data analytics technologies will facilitate real-time processing of streaming biological data [9, 107].

10.3 Energy Efficiency and Sustainability

HPC systems are known for their high energy consumption, which raises concerns about their environmental impact and cost. Balancing the computational demands of biomimetic simulations with energy efficiency is a significant challenge. Researchers must develop strategies to reduce power consumption while maintaining the necessary computational performance.

Future Directions: The development of energy-efficient algorithms, hardware designs, and cooling solutions will be crucial. Exploring renewable energy sources, optimizing resource allocation, and employing dynamic voltage and frequency scaling are also areas of future focus to enhance the sustainability of HPC in computational biomimetics [108, 109].

10.4 Integration of HPC with Machine Learning and Artificial Intelligence

Integrating HPC with machine learning (ML) and artificial intelligence (AI) techniques holds great potential for advancing computational biomimetics. However, combining these technologies presents challenges related to data preprocessing, algorithm scalability, and model interpretability. Future Directions: Researchers are exploring hybrid models that leverage HPC's computational power for training and inference of ML/AI models. This integration can enhance biomimetic design, pattern recognition, and data-driven discovery by combining simulation-based insights with data-driven predictions [110, 111].

10.5 Cloud Computing and HPC

Integrating HPC with cloud computing offers flexibility in resource allocation but also introduces challenges related to data security, latency, and cost management. Cloud-based HPC solutions need to address these issues while providing seamless access to computational resources.

Future Directions: Advances in hybrid cloud solutions, improved data encryption methods, and the development of cost-effective, specialized cloud instances for HPC workloads will drive the integration of cloud computing and HPC. Researchers are exploring strategies to optimize resource provisioning and data movement in cloud-based HPC environments [112, 113]. In summary, the challenges in High Performance Computing applications for computational biomimetics span scalability, data management, energy efficiency, integration with ML/AI, and cloud computing. Addressing these challenges will shape the future of HPC-driven biomimetic research and innovation, contributing to the advancement of both scientific understanding and technological solutions.

11 Case Studies: HPC in Computational Biomimetics

11.1 *Case Study 1: Modeling the Flight of Birds/Insects for Aircraft Design*

Birds have evolved efficient flight mechanisms over millions of years, making them a rich source of inspiration for aircraft design. HPC plays a pivotal role in simulating bird flight dynamics to uncover the aerodynamic principles that enable birds to achieve remarkable maneuvers and energy-efficient flight.

Case Study: Researchers used HPC clusters to simulate the flight of birds, creating computational models that mimic their wing motions and airflow interactions. By analyzing the aerodynamics and fluid dynamics involved in bird flight, they gained insights into lift generation, drag reduction, and wing morphing. These insights informed the design of more agile and fuel-efficient flying robots [34, 114, 115]. This case study utilizes high performance computing to simulate the flapping flight of Hummingbird and to understand unsteady aerodynamics around the wings in flight. The analysis was performed using high-fidelity IBM based solver on the HPC platform and processed using specific software. The wake topology both in downstroke and upstroke in one full stroke sycle (t/T) is visualized and shown in Fig. 8.

11.2 *Case Study 2: Simulating Biomolecular Interactions for Drug Discovery*

Understanding the interactions between biomolecules, such as proteins and ligands, is essential for drug discovery. HPC enables the simulation of these complex interactions, allowing researchers to predict binding affinities and design effective drugs.

Case Study: Researchers employed HPC clusters to perform molecular docking simulations, predicting the binding of drug candidates to target proteins. These simulations simulated the dynamics of protein–ligand interactions and calculated binding energies. The results guided drug design and aided in selecting compounds with high binding affinities for further experimental testing [96, 98]. Nature's design principles have inspired the development of novel biomaterials with applications in medicine. HPC accelerates the design and optimization of these materials, which can mimic natural properties for improved biocompatibility and functionality.

Researchers employed HPC to simulate the behavior of bio-inspired materials such as bone scaffolds or tissue implants. These simulations allowed them to optimize the materials' mechanical properties, porosity, and degradation rates. By mimicking the hierarchical structures found in natural tissues, HPC-enabled design strategies yielded biomaterials with enhanced strength, adaptability, and biocompatibility [86,

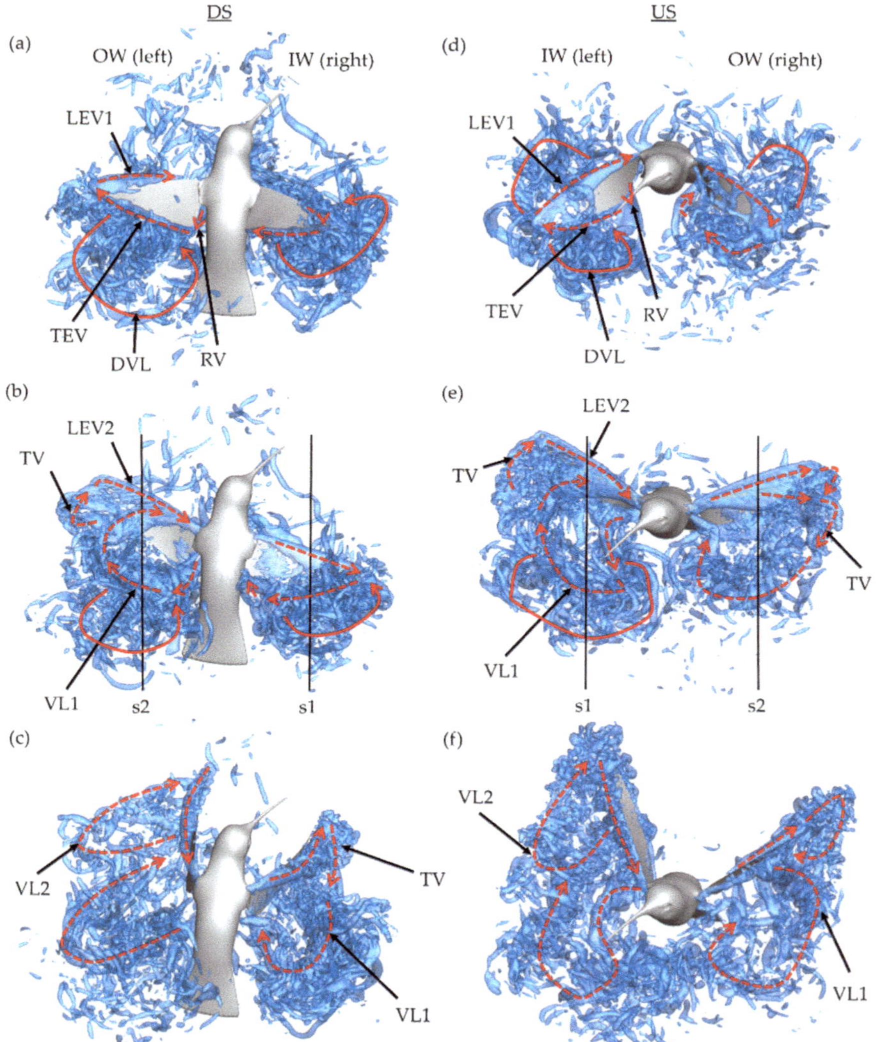

Fig. 8 Wake topology generated by the hummingbird's third DS at **a** t/T* = 0.16, **b** t/T* = 0.32, and **c** t/T* = 0.48 and third stroke US at **d** t/T* = 0.60, **e** t/T* = 0.72, and **f** t/T* = 0.98 [115]

87]. In conclusion, case studies demonstrate how High Performance Computing (HPC) transforms computational biomimetics by enabling in-depth simulations, accurate predictions, and innovative designs. From aircraft aerodynamics to drug discovery and biomaterials, HPC-driven investigations are pushing the boundaries of scientific understanding and engineering innovation. By harnessing nature's solutions and simulating their intricacies, researchers and engineers are revolutionizing industries and paving the way for a more efficient and sustainable future.

12 Conclusions

High Performance Computing (HPC) plays a vital role in advancing scientific research and engineering, and its application in computational biomimetics is no exception. This chapter provides an in-depth exploration of HPC technologies and architectures, along with an overview of computational biomimetics. The chapter highlights the crucial role that HPC plays in simulation, modeling, data analysis, optimization, and design in biomimetics. It looks into various applications of HPC in computational biomimetics, including bio-inspired materials and structures, biomimetic robotics and prosthetics, bioinformatics and genomics, drug discovery and development, and environmental modeling and conservation. Additionally, the chapter discusses the challenges and future directions of HPC in computational biomimetics, presents case studies that showcase HPC's application in this field and concludes with a summary of key points and future prospects.

Acknowledgements The authors gratefully acknowledge the contributions of Universiti Putra Malaysia (UPM) in providing opportunities for this Book Chapter to be a success through the university's Geran Putra—Inisiatif Putra Muda (GP-IPM) research grant; GP-IPM/2022/9730400.

References

1. Dongarra, J., et al.: The international exascale software project roadmap. Int. J. High Perform. Comput. Appl.Comput. Appl. **25**(1), 3–60 (2011)
2. Hwu, W.W., Kirk, D.B.: The landscape of parallel computing research: a view from Berkeley. Comput. Sci. Eng. **19**(2), 80–90 (2017)
3. Chapman, B., et al.: Using OpenMP: Portable Shared Memory Parallel Programming. MIT Press (2007)
4. Quinn, M.J.: Parallel Programming in C with MPI and OpenMP. McGraw-Hill (2003)
5. Gao, W., Ovchinnikov, S.: High-performance computing in finance. ACM Comput. Surv.Comput. Surv. **50**(3), 43 (2017)
6. Dongarra, J., et al.: The international exascale software project roadmap. Int. J. High Perform. Comput. Appl.Comput. Appl. **28**(3), 201–290 (2014)
7. Kuhlman, C.J., Reed, D.A.: Computing the universe: simulating the cosmos from grand challenge to desktop. Sci. Am. **286**(2), 42–49 (2002)
8. Coveney, P.V., Highfield, R.R.: The computation-powered revolution: building the virtual universe. J. Comput. Sci.Comput. Sci. **45**, 101151 (2020)
9. Dean, J., Ghemawat, S.: MapReduce: simplified data processing on large clusters. Commun. ACM. ACM **51**(1), 107–113 (2008)
10. Jagadish, H.V., et al.: Big data and its technical challenges. Commun. ACM. ACM **57**(7), 86–90 (2014)

11. Jouppi, N.P., et al.: In-datacenter performance analysis of a tensor processing unit. In: Proceedings of the 44th Annual International Symposium on Computer Architecture, pp. 1–12 (2017)
12. LeCun, Y., Bengio, Y., Hinton, G.: Deep learning. Nature **521**(7553), 436–444 (2015)
13. Bangerth, W., Heister, T., Heltai, L., Kronbichler, M., Maier, M.: Algorithms and data structures for massive parallelism on large-scale machines. Philos. Trans. R. Soc. A Math. Phys. Eng. Sci. **374**(2068), 20150189 (2016)
14. Foster, I., Kesselman, C.: The Grid: Blueprint for a New Computing Infrastructure. Morgan Kaufmann (1999)
15. Gustafson, J.L.: Reevaluating Amdahl's law. Commun. ACM. ACM **31**(5), 532–533 (1988)
16. Foster, I., Lusk, E.: The globus project: a status report. In: Proceedings of the 7th IEEE Symposium on High Performance Distributed Computing, pp. 4–10 (1995)
17. Margo, D.W.: Supercomputers: charting the future of cyberinfrastructure. Issues Sci. Technol. **31**(4), 43–51 (2015)
18. Dongarra, J., et al.: The international exascale software project roadmap. Int. J. High Perform. Comput. Appl.Comput. Appl. **34**(1), 3–73 (2020)
19. Campbell-Kelly, M., Aspray, W.: Computer: A History of the Information Machine. Westview Press (1996)
20. Saraf, P.R., et al.: Scalable Parallel Computing: Technology, Architecture. McGraw-Hill, Programming (1990)
21. Sterling, T., et al.: High Performance Computing: Modern Systems and Practices. Morgan Kaufmann (1994)
22. Buyya, R., et al.: A case for economy grid architecture for service-oriented grid computing. J. Concurr. Comput.: Pract. Exp. **17**(2–4), 337–355 (1999)
23. Becker, D., Sterling, T.: How to Build a Beowulf: A Guide to the Implementation and Application of PC Clusters. The MIT Press (1995)
24. Owens, J.D., Luebke, D., Govindaraju, N., Harris, M., Krüger, J., Lefohn, A.E., Purcell, T.J.: A survey of general-purpose computation on graphics hardware. Comput. Graph. Forum **26**(1), 80–113 (2007)
25. Kirk, D.B., Hwu, W.W.: Programming Massively Parallel Processors: A Hands-on Approach. Morgan Kaufmann (2012)
26. Fox, G., Williams, R.: Cloud Computing and Distributed Systems (No. UCB/EECS-2010-10). University of California, Berkeley, EECS Department (2010)
27. Armbrust, M., et al.: A view of cloud computing. Commun. ACM. ACM **53**(4), 50–58 (2010)
28. Bergman, K., et al.: Exascale Computing Study: Technology Challenges in Achieving Exascale Systems. Technical Report, U.S. Department of Energy (2008)
29. Sterling, T., et al.: High Performance Computing: Modern Systems and Practices. Morgan Kaufmann (2012)
30. Bangerth, W., et al.: Algorithms and data structures for massive parallelism on large-scale machines. Philos. Trans. R. Soc. A Math. Phys. Eng. Sci. **374**(2068), 20150189 (2016)
31. Snir, M., et al.: MPI: The Complete Reference. The MIT Press (1996)
32. Kim, H.S., Gupta, A.: High-speed interconnects in high-performance computing: a review. IEEE Trans. Parallel Distrib. Syst.Distrib. Syst. **25**(1), 3–14 (2014)
33. Hoefler, T., et al.: Scientific computing's productivity grid: parallelization strategies for a multicore world. IEEE Comput.Comput. **43**(4), 51–59 (2010)
34. Alnasir, J.J.: Fifteen quick tips for success with HPC, i.e., responsibly BASHing that Linux cluster. PLoS Comput. Biol. **17**(8), e1009207 (2021). https://doi.org/10.1371/journal.pcbi.1009207
35. Carns, P.H., et al.: PVFS: a parallel file system for Linux clusters. In: Proceedings of the 4th Annual Linux Showcase and Conference (2000)
36. Gibson, G.A., et al.: A cost-effective, high-bandwidth storage architecture. In: Proceedings of the 5th International Conference on Architectural Support for Programming Languages and Operating Systems, pp. 92–103 (1997)

37. Al-Fares, M., Loukissas, A., Vahdat, A.: A scalable, commodity data center network architecture. ACM SIGCOMM Comput. Commun. Rev. **38**(4), 63–74 (2008)
38. Gropp, W., et al.: Using MPI: Portable Parallel Programming with the Message-Passing Interface. MIT Press (1996)
39. Liu, C., Layton, R.A.: Task scheduling in high-performance computing systems. IEEE Trans. Parallel Distrib. Syst.Distrib. Syst. **24**(7), 1340–1351 (2013)
40. Collins, W.D., et al.: The community climate system model version 3 (CCSM3). J. Clim.Clim. **19**(11), 2122–2143 (2006)
41. Houghton, J.T., et al.: Climate Change 2001: The Scientific Basis. Cambridge University Press (2001)
42. Ferziger, J.H., Peric, M.: Computational Methods for Fluid Dynamics. Springer Science & Business Media (2012)
43. Drikakis, D., Fureby, C.: High-Order Methods for Computational Physics. Cambridge University Press (2012)
44. Frenkel, D., Smit, B.: Understanding Molecular Simulation: From Algorithms to Applications. Academic Press (2001)
45. Rappe, A.K., et al.: UFF, a full periodic table force field for molecular mechanics and molecular dynamics simulations. J. Am. Chem. Soc. **118**(22), 11225–11236 (1996)
46. Fernández-Quintero, M.L., DeRose, E.F., Gabel, S.A., Mueller, G.A., Liedl, K.R.: Nanobody paratope ensembles in solution characterized by MD simulations and NMR. Int. J. Mol. Sci. **23**(10), 5419 (2022). https://doi.org/10.3390/ijms23105419
47. Schadt, E.E., Friend, S.H.: Computational approaches to genomics. Science **323**(5918), 591–594 (2009)
48. Aluru, S., Tang, J.: Big data analytics in genomics. In: Big Data Analytics in Bioinformatics and Healthcare, pp. 41–63. CRC Press (2018)
49. Hockney, R.W., Eastwood, J.W.: Computer Simulation Using Particles. CRC Press (1988)
50. Springel, V., et al.: Simulations of the formation, evolution and clustering of galaxies and quasars. Nature **435**(7042), 629–636 (2005)
51. Lavecchia, A.: Machine-learning approaches in drug discovery: methods and applications. Drug Discov. TodayDiscov. Today **20**(3), 318–331 (2015)
52. Kitchen, D.B., Glen, R.C.: A review of in silico tools for the design of bioactive compounds: towards a paradigm shift in drug discovery. J. Chem. Inf. Model. **57**(8), 1347–1354 (2017)
53. Martins, J.R.R.A., Lambe, A.B.: Multidisciplinary design optimization: a survey of architectures. AIAA J. **51**(9), 2049–2075 (2013)
54. Balaji, P., et al.: Advances in high-performance computing for CFD simulations. In: High-Performance Computing for Computational Science—VECPAR 2014, pp. 209–233. Springer (2016)
55. Hastie, T., Tibshirani, R., Friedman, J.: The Elements of Statistical Learning: Data Mining, Inference, and Prediction. Springer Science & Business Media (2009)
56. Reichstein, M., et al.: Deep learning and process understanding for data-driven earth system science. Nature **566**(7743), 195–204 (2019)
57. Ghanem, R., & Higdon, D. (2007). Handbook of Uncertainty Quantification. Springer Science & Business Media.
58. Sophie, C.H., et al.: A geometrically adaptable heart valve replacement. Sci. Transl. Med. **12**, eaay4006(2020). https://doi.org/10.1126/scitranslmed.aay4006
59. Quinn, M.J.: Parallel Programming in C with MPI and OpenMP. McGraw-Hill Education (2004)
60. Pacheco, P.: An Introduction to Parallel Programming. Morgan Kaufmann (2011)
61. Coulouris, G., Dollimore, J., Kindberg, T., Blair, G.: Distributed Systems: Concepts and Design. Pearson Education (2011)
62. Tanenbaum, A.S., van Steen, M.: Distributed Systems: Principles and Paradigms. Pearson (2016)
63. Foster, I., Kesselman, C.: The Grid 2: Blueprint for a New Computing Infrastructure. Morgan Kaufmann (2004)

64. Berman, F., Fox, G., Hey, A. (eds.): Grid Computing: Making the Global Infrastructure a Reality. Wiley (2003)
65. Buyya, R., Goscinski, A.: Cluster, cloud and grid computing: a comprehensive survey. Futur. Gener. Comput. Syst.. Gener. Comput. Syst. **46**, 3–4 (2015)
66. Hwang, K., Dongarra, J.: Distributed and Cloud Computing: From Parallel Processing to the Internet of Things. Morgan Kaufmann (2016)
67. Dongarra, J., Meuer, H., Strohmaier, E. (eds.): TOP500 Supercomputer Sites: Performance, Statistics, and Analysis. Springer (2010)
68. Hockney, R.W., Jesshope, C.R.: Parallel Computers 2: Architecture, Programming, and Algorithms. CRC Press (1988)
69. Kirk, D.B., Hwu, W.M.: Programming Massively Parallel Processors: A Hands-on Approach. Morgan Kaufmann (2016)
70. Smith, A.D., Nettles, S.M.: FPGAs for Software Programmers. Addison-Wesley Professional (2019)
71. Vincent, J.F.V., Mann, D.L.: Systematic technology transfer from biology to engineering. Philos. Trans. R. Soc. Lond. Ser. A: Math. Phys. Eng. Sci. **360**(1791), 159–173 (2002)
72. Laflamme, S., Blouin, J.: Modeling flexible multibody systems with contact and friction, application to biological systems. Multibody Sys.Dyn.Sys.Dyn. **9**(3), 283–309 (2003)
73. Benyus, J.M.: Biomimicry: Innovation Inspired by Nature. William Morrow Paperbacks (1997)
74. Pahl, G., Beitz, W.: Engineering Design: A Systematic Approach. Springer (1996)
75. Speck, T., Speck, O.: Biomimetics: learning from nature. Biologist **51**(3), 109–114 (2004)
76. Vincent, J.F.V.: Smart Structures and Materials. In Biomimetics: Nature-Based Innovation, pp. 175–191. Springer (2012)
77. Karplus, M., McCammon, J.A.: Molecular dynamics simulations of biomolecules. Nat. Struct. Mol. Biol. **9**(9), 646–652 (2002)
78. Miller, K.: Challenges and strategies in modeling complex biological systems. Multiscale Model. Simul. **16**(2), 631–646 (2018)
79. Baldi, P., Brunak, S.: Bioinformatics: The Machine Learning Approach. MIT Press (2001)
80. Holland, J.H.: Adaptation in Natural and Artificial Systems. University of Michigan Press (1975)
81. Dorigo, M., Stützle, T.: Ant Colony Optimization. MIT Press (2004)
82. Ware, C., Franck, G.: Evaluating stereo and motion cues for visualizing information nets in three dimensions. ACM Trans. Graph. **15**(2), 121–140 (1996)
83. Suh, Y.K., Radcliffe, D.F.: CAD visualization for biomimetic design. In: Biomimetic Design Method for Innovation and Sustainability, pp. 71–94. Springer (2011)
84. Meyers, M.A., Chen, P.Y., Lin, A.Y.M., Seki, Y.: Biological materials: structure and mechanical properties. Prog. Mater. Sci. **53**(1), 1–206 (2008)
85. Espinosa, H.D., Fischer, F.D.: Biomimetic design in nanotechnology: theoretical approaches and examples. Nanotechnol. Rev.. Rev. **1**(1), 101–131 (2011)
86. Wegst, U.G.K., Bai, H., Saiz, E., Tomsia, A.P., Ritchie, R.O.: Bioinspired structural materials. Nat. Mater. **14**(1), 23–36 (2015)
87. Liu, Y., Chen, X., Ding, Y.: Biomimetic design and fabrication of lightweight and strong materials. Bioinspir. Biomim.. Biomim. **13**(1), 011001 (2018)
88. Lentink, D., Dickinson, M.H.: Bioinspired flight control. Philos. Trans. R. Soc. B: Biol. Sci. **364**(1521), 3521–3538 (2009)
89. Goldman, D.I., Revzen, S., Full, R.J.: Active tails enhance arboreal acrobatics in geckos. Proc. Natl. Acad. Sci. **110**(46), 18716–18721 (2013)
90. Pfeifer, R., Bongard, J.: How the Body Shapes the Way we Think: A New View of Intelligence. MIT Press (2006)
91. Cutkosky, M.R.: Robotic grasping and contact: a review. Robot. Auton. Syst.Auton. Syst. **54**(4), 345–353 (2005)
92. Schatz, M.C., Langmead, B.: The DNA data deluge. Nat. Biotechnol.Biotechnol. **30**(5), 423–425 (2013)

93. O'Leary, N.A., Wright, M.W., Brister, J.R., Ciufo, S., Haddad, D., McVeigh, R., Pruitt, K.D., et al.: Reference sequence (RefSeq) database at NCBI: current status, taxonomic expansion, and functional annotation. Nucl.C Acids Res. **44**(D1), D733–D745 (2016)
94. Dill, K.A., MacCallum, J.L.: The protein–folding problem, 50 years on. Science **338**(6110), 1042–1046 (2012)
95. Moult, J.: A decade of CASP: progress, bottlenecks and prognosis in protein structure prediction. Curr. Opin. Struct. Biol.. Opin. Struct. Biol. **15**(3), 285–289 (2005)
96. Kitchen, D.B., Decornez, H., Furr, J.R., Bajorath, J.: Docking and scoring in virtual screening for drug discovery: methods and applications. Nat. Rev. Drug Discov.Discov. **3**(11), 935–949 (2004)
97. Friesner, R.A., Banks, J.L., Murphy, R.B., Halgren, T.A., Klicic, J.J., Mainz, D.T., Perry, J.K., et al.: Glide: A new approach for rapid, accurate docking and scoring. 1. Method and assessment of docking accuracy. J. Med. Chem. **47**(7), 1739–1749 (2004)
98. Durrant, J.D., McCammon, J.A.: Molecular dynamics simulations and drug discovery. BMC Biol. **9**(1), 71 (2011)
99. Gilson, M.K., Zhou, H.X.: Calculation of protein–ligand binding affinities. Annu. Rev. Biophys. Biomol. Struct.. Rev. Biophys. Biomol. Struct. **36**, 21–42 (2007)
100. DeAngelis, D.L., Mooij, W.M.: Individual-based modeling of ecological and evolutionary processes. Annu. Rev. Ecol. Evol. Syst.. Rev. Ecol. Evol. Syst. **36**, 147–168 (2005)
101. Grimm, V., Railsback, S.F.: Pattern-oriented modelling: a 'multi-scope' for predictive systems ecology. Philos. Trans. R. Soc. B: Biol. Sci. **367**(1586), 298–310 (2012)
102. Bahaj, A.S., James, P.A.: Urban energy generation: the added value of photovoltaics in social housing. Appl. Energy **84**(3), 256–268 (2007)
103. Siddiqui, O., Rehman, S.: Biomimicry: inspiration for energy-efficient building design. Sustain. Cities Soc. **37**, 1–12 (2018)
104. Zheng, L., Hedrick, T.L., Mittal, R.: Time-varying wing-twist improves aerodynamic efficiency of forward flight in butterflies. PLoS ONE **8**(1), e53060 (2013). https://doi.org/10.1371/journal.pone.0053060
105. Gropp, W., Lusk, E., Skjellum, A.: Using MPI: Portable Parallel Programming with the Message-Passing Interface. MIT Press (1996)
106. Dongarra, J., Foster, I., Fox, G.: Sourcebook of Parallel Computing. Morgan Kaufmann (1997)
107. Berman, F., Fox, G., Hey, A.: Grid Computing: Making The Global Infrastructure a Reality. Wiley (2003)
108. Feng, W.C., Feng, J.J.: Energy-aware scheduling for HPC data centers: a survey. ACM Comput. Surv. (CSUR) **48**(1), 9 (2015)
109. Ortega, D.G., Sipper, M.: Computational biomimetics: taking computer science from nature to practice. IEEE Trans. Evol. Comput.Evol. Comput. **11**(3), 279–295 (2007)
110. Zhang, Y., Chen, J., Zhou, X.S.: Deep learning on high-performance computing architectures. IEEE Trans. Neural Netw. Learn. Syst. **31**(10), 3622–3634 (2020)
111. Golkar, A., Yoon, H.J., Niar, S.: A survey on parallel machine learning algorithms on GPU, CPU, and cluster systems. ACM Comput. Surv. (CSUR) **50**(6), 1–35 (2017)
112. Armbrust, M., Fox, A., Griffith, R., Joseph, A.D., Katz, R., Konwinski, A., Zaharia, M., et al.: A view of cloud computing. Commun. ACM **53**(4), 50–58 (2010)
113. Barker, A., Srinivasan, A., Mueller, F.: Parallel, distributed, and cloud computing technologies for bioinformatics. Brief. Bioinform.Bioinform. **13**(6), 639–647 (2012)
114. Rayner, J.M.V.: Aviation's golden fleece: using nature to inspire design. Endeavour **22**(2), 74–78 (1998)
115. Menzer, A., Ren, Y., Guo, J., Tobalske, B.W., Dong, H.: Wing kinematics and unsteady aerodynamics of a hummingbird pure yawing maneuver. Biomimetics **7**(3), 115 (2022). https://doi.org/10.3390/biomimetics7030115

Bio-inspired Computing and Associated Algorithms

Balbir Singh and Manikandan Murugaiah

Abstract This chapter explores the fascinating intersection of biology and computer science, where nature's design principles are harnessed to solve complex computational problems. This chapter provides an overview of bio-inspired computing techniques, including genetic algorithms, neural networks, swarm intelligence, and cellular automata. It goes into the core concepts of each approach, highlighting their biological counterparts and demonstrating their applications across various domains. Furthermore, this chapter discusses the evolution of bio-inspired algorithms, emphasizing their adaptation to contemporary computing paradigms such as machine learning and artificial intelligence. It examines how these algorithms have been employed to address real-world challenges, ranging from optimization problems and pattern recognition to robotics and autonomous systems. In addition to theoretical insights, the chapter offers practical guidance on implementing bio-inspired algorithms, including algorithmic design considerations and the integration of bio-inspired approaches with traditional computing methods. It also discusses the ethical and societal implications of bio-inspired computing, touching upon topics like algorithm bias and data privacy.

Keywords Bio-inspired compute · Algorithms · Cuckoo search · Gray Wolf · Particle swarm

---

B. Singh (✉) · M. Murugaiah
Department of Aeronautical and Automobile Engineering, Manipal Institute of Technology, Manipal Academy of Higher Education, Manipal, Karnataka 576104, India
e-mail: balbir.s@manipal.edu

B. Singh
Department of Aerospace Engineering, Faculty of Engineering, Universiti Putra Malaysia, 43400 Serdang, Selangor, Malaysia

K. A. Ahmad et al. (eds.), *High Performance Computing in Biomimetics*, Series in BioEngineering, https://doi.org/10.1007/978-981-97-1017-1_3

1 Introduction to Bio-inspired Computing

1.1 Definition and Overview

Bio-inspired computing, also known as nature-inspired computing or biologically inspired computing is a burgeoning interdisciplinary field that draws inspiration from biological systems to develop innovative computational techniques and algorithms. At its core, bio-inspired computing seeks to harness the efficiency, adaptability, and problem-solving capabilities observed in biological organisms to solve complex real-world problems in various domains. These problems may include optimization, pattern recognition, robotics, and more. Bio-inspired computing encompasses a wide array of computational paradigms, including genetic algorithms, neural networks, swarm intelligence, and cellular automata. These paradigms simulate and replicate processes found in nature to perform tasks that can be challenging for traditional computing methods. For instance, genetic algorithms mimic the process of natural selection and evolution to solve optimization problems, while neural networks model the structure and function of biological brains to perform tasks like image recognition. Figure 1 shows the traditional and computational thinking that has evolved as such leading to development of bio-inspired structures and machines [1].

1.2 Inspiration from Biological Systems

The inspiration for bio-inspired computing comes from the observation that biological systems, honed by millions of years of evolution, exhibit remarkable capabilities in adaptation, learning, and optimization. For instance:

Genetic Algorithms: These algorithms are inspired by the process of natural selection and genetics. They use a population of potential solutions, evolve and breed

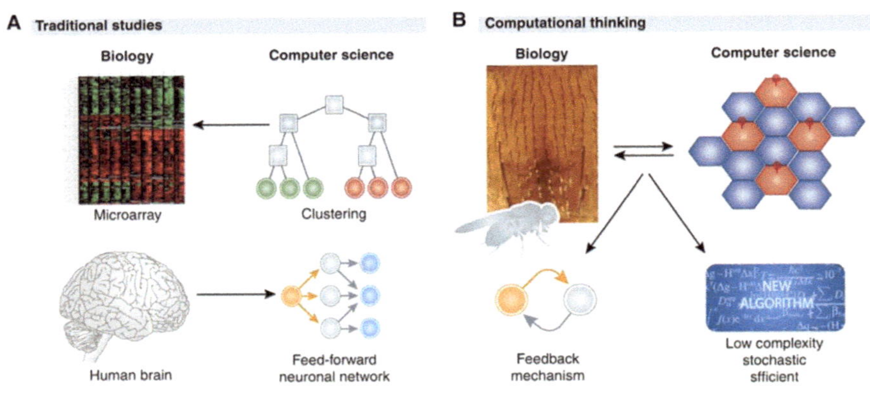

Fig. 1 **a** Traditional studies versus **b** computational thinking [1]

them through selection mechanisms, and iteratively improve these solutions over generations to find optimal or near-optimal solutions to complex problems.

Neural Networks: Neural networks are inspired by the structure and functioning of biological brains. They consist of interconnected artificial neurons that can process information, recognize patterns, and learn from data. Deep learning, a subset of neural networks, has achieved remarkable success in tasks like image and speech recognition.

Swarm Intelligence: This paradigm takes inspiration from the collective behavior of social organisms such as ants, bees, and birds. Algorithms like ant colony optimization and particle swarm optimization simulate the collaborative behavior of these organisms to solve optimization problems.

Cellular Automata: Cellular automata are inspired by the behavior of cells in living organisms. These are discrete mathematical models where cells change their state based on a set of rules and their neighbors' states. Cellular automata have applications in modeling complex systems, simulating physical processes, and generating random numbers. Figure 2 shows these algorithms evolved in years with complexity of problem with time [2]

1.3 Relationship Between Bio-inspired Computing and Artificial Intelligence

Bio-inspired computing is closely related to the field of artificial intelligence (AI), and the two fields often intersect. While AI focuses on creating intelligent agents that can perceive their environment, reason, and make decisions, bio-inspired computing

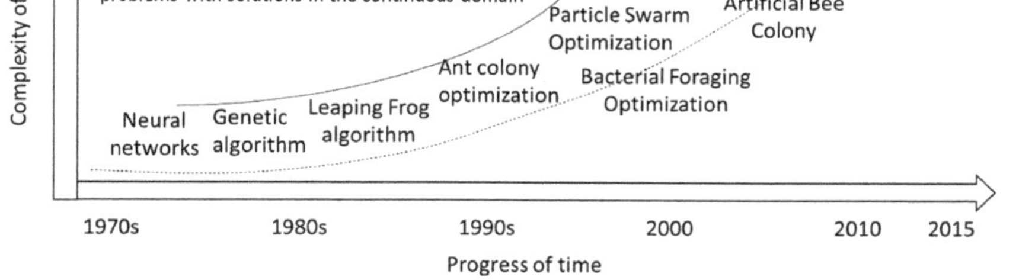

Fig. 2 Bio-inspired algorithms evolved in years with complexity of problem with time. Adapted from ref [2], copyright (2016), with permission from Elsevier

contributes by offering innovative algorithms and models to achieve these goals. Many machine learning techniques, such as neural networks and evolutionary algorithms, have their roots in bio-inspired computing. These algorithms play a pivotal role in training models to perform tasks like image and speech recognition, natural language processing, and autonomous decision-making. Bio-inspired algorithms are extensively used in robotics to create robots that can navigate uncertain environments, mimic animal behaviors, and adapt to changing conditions. Swarm robotics, for example, leverages swarm intelligence principles to control groups of robots for various applications. Bio-inspired computing also contributes to cognitive computing, which aims to develop systems that can mimic human-like cognitive functions such as perception, reasoning, and problem-solving. Neural networks and deep learning models are instrumental in this area [3–9]. Collectivly, bio-inspired computing is a dynamic and evolving field that capitalizes on the wisdom of nature to develop novel computational solutions. Its integration with AI and its application across diverse domains make it a promising avenue for addressing complex real-world challenges.

2 Biological Inspiration for Computing

2.1 Neural Networks and Artificial Neurons

Neural networks, inspired by the structure and function of biological brains, have become a cornerstone of modern artificial intelligence (AI). These networks consist of interconnected artificial neurons that can process information, recognize patterns, and learn from data. Each artificial neuron receives input signals, applies a weighted sum and an activation function, and passes the output to subsequent neurons. Neural networks have been instrumental in solving complex problems in areas such as image and speech recognition, natural language processing, and autonomous decision-making. Recent advancements in deep learning, a subset of neural networks, have demonstrated remarkable performance in various tasks. For example, deep convolutional neural networks (CNNs) have achieved state-of-the-art results in image classification and object detection, while recurrent neural networks (RNNs) and transformer models have revolutionized natural language processing [10, 11].

2.2 Evolutionary Algorithms and Genetic Algorithms

Genetic algorithms (GAs) are a class of evolutionary algorithms inspired by the process of natural selection and genetics. They start with a population of potential solutions to a problem and iteratively evolve and breed these solutions through selection mechanisms. GAs use concepts such as mutation, crossover, and selection to mimic the survival of the fittest, gradually improving the solutions over generations. Genetic algorithms have found applications in optimization problems, parameter tuning, and machine learning, among others [3, 12]. Recent research in genetic algorithms has focused on improving their efficiency and scalability. Hybrid approaches that combine GAs with other optimization techniques have been proposed to address complex and large-scale optimization problems.

2.3 Swarm Intelligence and Ant Colony Optimization

Swarm intelligence is a field inspired by the collective behavior of social organisms such as ants, bees, and birds. Ant Colony Optimization (ACO) is a notable bio-inspired algorithm that mimics the foraging behavior of ants to solve optimization problems. In ACO, artificial ants explore a solution space, deposit pheromones, and communicate with each other to find optimal or near-optimal solutions. These algorithms have been applied to a wide range of problems, including the traveling salesman problem, network routing, and scheduling.

Recent advancements in ACO and swarm intelligence research have focused on adapting these algorithms for dynamic and distributed environments, as well as incorporating machine learning techniques to enhance their performance [5, 13].

2.4 Cellular Automata and Self-organization

Cellular automata (CA) are discrete mathematical models inspired by the behavior of cells in living organisms. These models consist of a grid of cells, each of which can be in a finite number of states. Cells change their state based on a set of rules and the states of their neighboring cells. Cellular automata have been used to model complex systems, simulate physical processes, and generate random numbers [6]. Recent research in cellular automata has explored their applications in simulating and understanding various natural phenomena, including the spread of diseases, traffic flow, and self-organizing systems.

2.5 DNA Computing and Molecular Computing

DNA computing and molecular computing are cutting-edge fields that draw inspiration from the molecular processes in living organisms to perform computation. DNA molecules are used to store and process information, and operations are performed through biochemical reactions. These approaches have the potential to solve complex problems, particularly in areas like cryptography and optimization [14]. Recent advances in DNA and molecular computing have focused on improving the scalability and reliability of these systems. Researchers are exploring novel methods for error correction and expanding the range of problems that can be tackled using molecular computing [14, 15].

3 Neural Networks and Deep Learning

3.1 Introduction to Neural Networks

Neural networks, a core component of deep learning, are computational models inspired by the human brain's neural structure. These networks consist of interconnected nodes, or neurons, organized into layers: an input layer, one or more hidden layers, and an output layer. Each connection between neurons is associated with a weight, and neurons use activation functions to process inputs and produce outputs. Neural networks excel at capturing complex relationships in data, making them a vital tool in various fields [10, 16]. Recent developments in neural networks have seen the emergence of deep learning, which involves neural networks with many hidden layers (deep neural networks). This depth enables neural networks to learn hierarchical representations of data, making them exceptionally effective in tasks such as image recognition, natural language processing, and reinforcement learning.

3.2 Perceptrons and Multilayer Neural Networks

Perceptrons are the simplest form of neural networks, consisting of a single layer of binary threshold units. They can model linearly separable functions but are limited in their capacity to handle complex problems. However, when these perceptrons are stacked into layers to form multilayer neural networks, they become capable of approximating more complex, nonlinear functions. This multilayer architecture, with the introduction of activation functions like the sigmoid or ReLU (Rectified Linear Unit), is the foundation for modern neural networks. Recent advancements include the development of novel activation functions, initialization techniques, and regularization methods to improve the training and performance of deep neural networks [17, 18].

3.3 Training Algorithms (E.G., Backpropagation)

Training neural networks involves optimizing the network's weights and biases to minimize a loss function, which quantifies the difference between predicted and actual outputs. *Backpropagation* is a foundational algorithm for training neural networks. It computes gradients of the loss with respect to the network's parameters and uses gradient descent variants to update these parameters iteratively. Research in training algorithms has led to improvements in convergence speed and stability. Techniques like batch normalization, adaptive learning rates, and advanced optimizers (e.g., Adam and RMSprop) have made training deep networks more efficient [19, 20].

3.4 *Convolutional Neural Networks (CNNs)*

Convolutional Neural Networks (CNNs) are specialized neural networks designed for processing grid-like data, such as images and videos. They employ convolutional layers that automatically learn features hierarchically, from simple edges and textures to complex patterns and objects. CNNs have revolutionized computer vision tasks, achieving superhuman performance in image classification, object detection, and segmentation. Advancements in CNNs include architectures like ResNet, Inception, and efficient networks that balance accuracy and computational efficiency. Transfer learning, where pre-trained CNNs are fine-tuned for specific tasks, has also become a standard practice [18, 21]. Figure 3 shows the typical outline of a CNN.

3.5 *Recurrent Neural Networks (RNNs) and Long Short-Term Memory (LSTM)*

Recurrent Neural Networks (RNNs) are designed for sequential data, making them suitable for tasks like natural language processing, speech recognition, and time-series prediction. RNNs maintain hidden states that capture temporal dependencies in the data. *Long Short-Term Memory (LSTM)* networks are a specialized type of RNN that can capture long-range dependencies and mitigate the vanishing

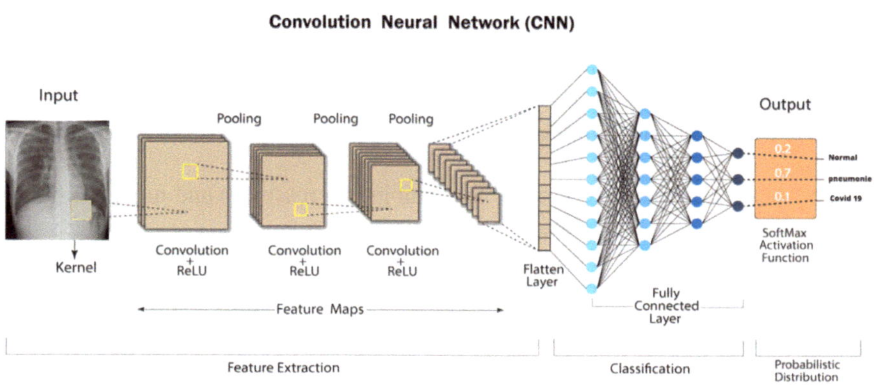

Fig. 3 Convolutional neural network (CNN) [23]

gradient problem. Recent developments in RNNs include attention mechanisms, gated recurrent units (GRUs), and transformer models. These enhancements have greatly improved the ability of RNNs and LSTMs to handle sequential data with varying lengths and complexities [11, 22].

3.6 Deep Learning Applications and Success Stories

Deep learning has achieved groundbreaking success across various domains. In computer vision, deep neural networks have powered autonomous vehicles, facial recognition systems, and medical image analysis. In natural language processing, transformer-based models have enabled state-of-the-art language translation, chatbots, and sentiment analysis. Deep reinforcement learning has excelled in training agents for games, robotics, and recommendation systems. Recent applications include the use of deep learning in drug discovery, finance, climate modeling, and healthcare, where it has demonstrated the potential to transform industries and address complex, real-world challenges [24, 25].

4 Evolutionary Algorithms and Genetic Algorithms

4.1 Principles of Evolutionary Computation

Evolutionary computation is a family of optimization and search algorithms inspired by the principles of biological evolution. These algorithms operate on populations of potential solutions to a problem, mimicking the process of natural selection, genetic variation, and survival of the fittest. The fundamental principles of evolutionary computation include:

- Population: Evolutionary algorithms maintain a population of candidate solutions rather than a single solution.
- Fitness Evaluation: Each candidate solution is evaluated using a fitness function that quantifies how well it solves the problem.
- Selection: Solutions with higher fitness have a greater chance of being selected for reproduction.
- Reproduction: Selected solutions are combined (crossover) and modified (mutation) to create new candidate solutions.
- Termination Criteria: Evolutionary algorithms run for a specified number of generations or until a stopping condition is met.

Recent developments in evolutionary computation involve the exploration of hybrid algorithms that combine evolutionary techniques with machine learning or other optimization methods to improve performance and efficiency [26, 27].

4.2 *Genetic Algorithms: Chromosomes, Genes, and Fitness Evaluation*

Genetic algorithms (GAs) are a subset of evolutionary algorithms that use the concepts of chromosomes, genes, and fitness evaluation to search for optimal or near-optimal solutions. In GAs, a candidate solution is represented as a chromosome, which is typically a binary or real-valued string. Each position on the chromosome corresponds to a gene, representing a potential solution component. The fitness evaluation function assesses the quality of a solution based on its chromosome representation. The goal is to maximize or minimize the fitness value, depending on the optimization problem [3, 12]. Recent advancements in genetic algorithms include adaptive operator selection, where the choice of selection, crossover, and mutation operators evolves over time, as well as parallel and distributed GAs that exploit multiple processors or compute nodes to enhance scalability [3, 12].

4.3 *Selection, Crossover, and Mutation Operators*

Selection, crossover, and mutation are the key genetic operators in genetic algorithms: The selection operator determines which individuals from the current population will be chosen as parents to produce the next generation. Common selection methods include roulette wheel selection, tournament selection, and rank-based selection. Crossover combines genetic material from two or more parents to create one or more offspring. Different crossover techniques, such as one-point crossover, two-point crossover, and uniform crossover, can be applied depending on the problem. Mutation introduces small random changes to an individual's chromosome. It helps maintain genetic diversity in the population. Mutation rates and strategies can vary based on the problem domain. Recent research in genetic algorithms has focused on designing adaptive and dynamic operator selection strategies that adjust the probabilities of selection, crossover, and mutation during the optimization process to improve convergence and solution quality [3, 26].

4.4 *Genetic Programming and Evolutionary Strategies*

Genetic programming (GP) is an extension of genetic algorithms that evolves computer programs or mathematical expressions to solve problems. In GP, individuals are represented as syntax trees, where nodes represent operations and terminals represent variables or constants. Evolution in GP involves applying genetic operators (crossover and mutation) to these tree structures. Evolutionary strategies (ES) are another variant of evolutionary algorithms that focus on optimizing real-valued parameters. ES is particularly suited for continuous optimization problems and is

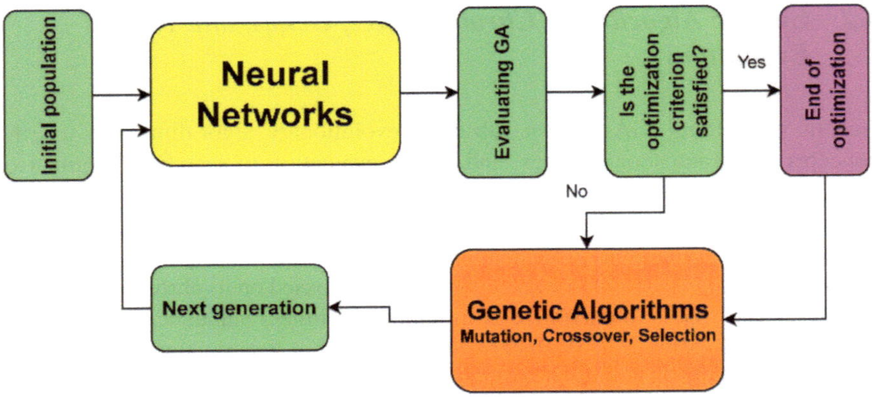

Fig. 4 The schematic flow diagram of NN linkage with GA [30]

used in applications such as optimization of neural network hyperparameters and robotics control. Recent advances in genetic programming include the development of more efficient tree-based representations and automatic program simplification techniques. In evolutionary strategies, researchers have explored covariance matrix adaptation and various adaptation mechanisms to improve convergence rates [28, 29]. It is also possible to create a visual demonstration of NN learning called SnakeAI by author Greer Viau using a program in Processing shows the topology of an NN and its configuration using a simple diagram of neurons and their connections as shown in Fig. 4 [30].

4.5 Applications of Evolutionary Algorithms

Evolutionary algorithms find applications in a wide range of domains, including: GAs are employed to solve complex optimization problems in engineering, finance, logistics, and more. Evolutionary strategies and genetic programming are used for hyperparameter tuning, feature selection, and evolving machine learning models. Evolutionary algorithms are used to optimize robot control strategies, design robotic structures, and perform autonomous robot navigation. GAs help in solving biological problems such as protein folding, gene regulatory network inference, and sequence alignment. Evolutionary algorithms have been used in game AI to evolve strategies for board games, video games, and other competitive environments. They are applied in portfolio optimization, risk management, and trading strategy development. Latest applications include the use of evolutionary algorithms in real-time optimization for autonomous vehicles, energy-efficient resource allocation in cloud computing, and evolving neural network architectures for deep learning [31, 32]. In conclusion, evolutionary algorithms, especially genetic algorithms, have made significant contributions to solving complex optimization and search problems across

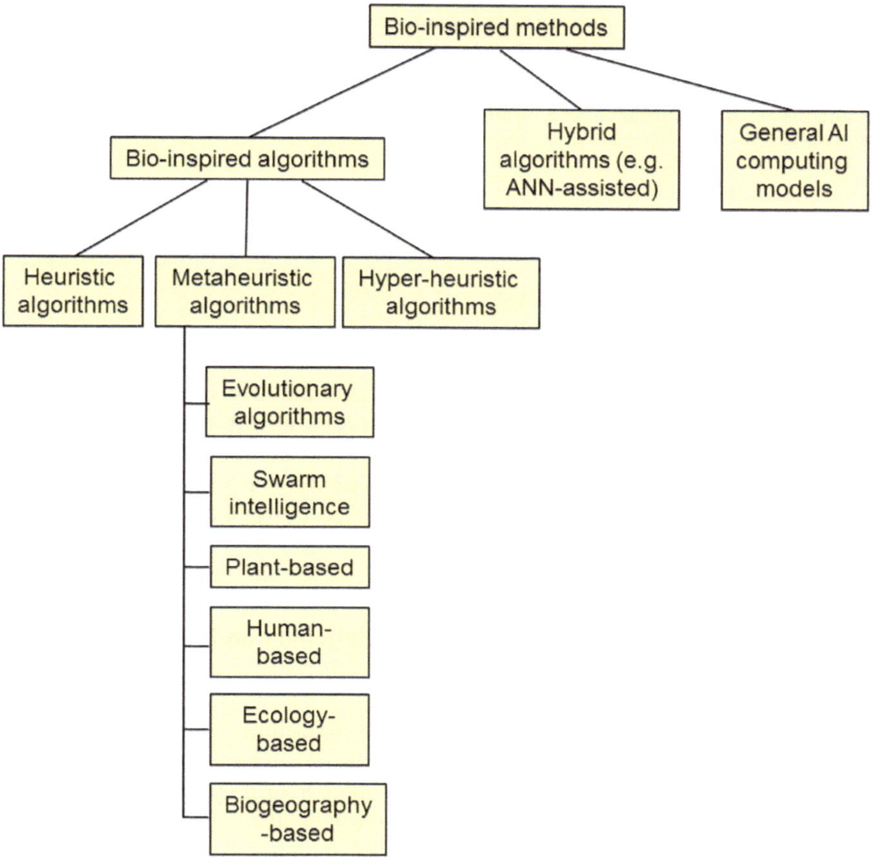

Fig. 5 General classification of bio-inspired computing methods [33]

various domains. Recent research focuses on enhancing their efficiency, scalability, and adaptability, making them powerful tools for addressing contemporary challenges in optimization and machine learning. Let us discuss about these bio-inspired optimization algorithms as per Fig. 5, in detail below.

5 Metaheuristic Algorithms

Metaheuristic algorithms are a class of optimization algorithms that are used to find approximate solutions to complex optimization problems. These algorithms are inspired by natural processes, social behavior, and other phenomena, and they are designed to efficiently explore large solution spaces. Metaheuristic algorithms are particularly useful when traditional optimization techniques are ineffective or computationally expensive. In this detailed description, we will explore various categories of metaheuristic algorithms, including Evolutionary algorithms, Swarm Intelligence, Plant-based, and Human-based metaheuristic algorithms [34–38].

5.1 Evolutionary Algorithms

5.1.1 Genetic Algorithms (GAs)

Genetic algorithms are inspired by the process of natural selection. They operate by maintaining a population of potential solutions and iteratively evolving them over generations. Solutions in the population are represented as chromosomes, and genetic operators like mutation and crossover are used to create new candidate solutions. Selection mechanisms favor better solutions for reproduction, mimicking the survival of the fittest in nature. Figure 3 shows a simplified flowchart of a typical genetic algorithm.

5.1.2 Memetic Algorithms: Combining Evolutionary and Local Search Methods

Memetic algorithms combine elements of genetic algorithms with local search techniques. They seek to improve the overall search process by applying local optimization to individuals within the population. This hybrid approach enhances the algorithm's ability to explore the solution space efficiently.

Memetic algorithms are hybrid optimization techniques that combine the global exploration capabilities of evolutionary algorithms with the local refinement of traditional local search methods. These algorithms aim to strike a balance between exploration and exploitation. Typically, a population of candidate solutions is evolved using evolutionary operators, and then local search methods are applied to each individual in the population to improve their quality.

Recent research in memetic algorithms focuses on adaptive and efficient local search strategies, handling large-scale problems, and leveraging parallel computing. Keywords for finding the latest research include "memetic algorithms," "hybrid optimization algorithms," and "metaheuristics with local search."

5.1.3 Differential Evolution

Differential evolution is a population-based optimization algorithm that uses the difference between candidate solutions to generate new ones. It maintains a population of potential solutions and evolves them by perturbing their differences. Differential evolution is known for its simplicity and effectiveness in solving a wide range of optimization problems [39–43].

5.2 *Swarm Intelligence*

Swarm intelligence algorithms are inspired by the collective behavior of social organisms. They leverage the interactions and cooperation among individuals in a swarm to find optimal solutions. Table 1 provides the selected list of swarm intelligence algorithms from 1995 to 2023 [33].

5.2.1 Particle Swarm Optimization (PSO)

PSO is inspired by the social behavior of birds and fish. It consists of particles (representing potential solutions) that move through the solution space to find the best solution. Particles adjust their positions based on their own best-known solution and the best-known solution within the swarm. Particle Swarm Optimization (PSO) is a population-based optimization algorithm inspired by the social behavior of birds flocking or fish schooling. It was developed by James Kennedy and Russell Eberhart in 1995. PSO is used to find optimal solutions to optimization and search problems, especially in multidimensional spaces.

The algorithm is particularly useful for problems where the objective function is continuous, differentiable, and has a well-defined fitness landscape. Each solution candidate in the search space is represented as a particle. A particle's position represents a possible solution to the optimization problem. Each particle has a velocity that determines how it moves through the search space. The fitness function evaluates how good a particle's current position is in terms of solving the optimization problem. It assigns a fitness value to each particle, indicating its quality. Each particle remembers its best position in the search space so far, along with the corresponding fitness value. This is called the personal best position (pBest). The best position among all particles in the swarm is known as the global best position. It represents the best solution found by any particle in the swarm (gBest). The algortithm steps include:

- Initialization: Initialize a swarm of particles randomly within the search space. Assign initial velocities to the particles.
- Evaluation: Calculate the fitness value for each particle in the swarm based on the fitness function.
- Update Personal Best: For each particle, update its personal best position and fitness value if the current position is better than the previous personal best.
- Update Global Best: Identify the particle with the best fitness value among all the particles in the swarm and set it as the global best position.
- Velocity Update: Update the velocity of each particle to adjust its movement. The velocity is updated using the following formula:

$$\text{v\_i}(t+1) = \text{w} * \text{v\_i}(t) + \text{c}1 * \text{rand}() * (\text{pBest\_i} - \text{x\_i}(t)) \\ + \text{c}2 * \text{rand}() * (\text{gBest} - \text{x\_i}(t))$$

Table 1 Selected swarm intelligence algorithms [33]

Algorithm name	Abbr	Proposed by, Year	References
Particle Swarm Optimization	PSO	Eberhart, Kennedy, 1995	[44]
Whale Optimization Algorithm	WOA	Mirjalili, Lewis, 2016	[45]
Gray Wolf Optimizer	GWO	Mirjalili, Mirjalili, and Lewis, 2014	[46]
Artificial Bee Colony Algorithm	ABCA	Karaboga, 2005	[47]
Ant Colony Optimization	ACO	Dorigo, 1992	[48]
Artificial Fish Swarm Algorithm	AFSA	Li, Qian, 2003	[49]
Firefly Algorithm	FA	Yang, 2009	[50]
Fruit Fly Optimization Algorithm	FFOA	Pan, 2012	[51]
Cuckoo Search Algorithm	CS	Yang and Deb, 2009	[52]
Bat Algorithm	BA	Yang, 2010	[53]
Bacterial Foraging	BFA	Passino, 2002	[54]
Social Spider Optimization	SSO	Kaveh et al., 2013	[55]
Locust Search Algorithm	LS	Cuevas et al., 2015	[56]
Symbiotic Organisms Search	SOS	Cheng and Prayogo, 2014	[57]
Moth-Flame Optimization	MFOA	Mirjalili, 2015	[58]
Honey Badger Algorithm	HBA	Hashim et al., 2022	[59]
Elephant Herding Optimization	EHO	Wang, Deb, Coleho, 2015	[60]
Grasshopper Algorithm	GOA	Saremi, Mirjalili, Lewis, 2017	[61]
Harris Hawks Optimization	HHO	Heidari et al., 2019	[62]
Orca Predation Algorithm	OPA	Jiang, Wu, Zhu, Zhang, 2022	[63]
Starling Murmuration Optimizer	SMO	Zamani, Nadimi-Shahraki, Gandomi, 2022	[64]
Serval Optimization Algorithm	SOA	Dehghani, Trojovský, 2022	[65]
Coral Reefs Optimization Algorithm	CROA	Salcedo-Sanz et al., 2014	[66]
Krill Herd Algorithm	KH	Gandomi, Alavi, 2012	[67]
Gazelle Optimization Algorithm	GOA	Agushaka, Ezugwu, Abualigah, 2023	[68]

where

'v_i(t+1)' is the updated velocity of particle 'I' at time 't+1'.

'w' is the inertia weight, which controls the trade-off between exploration and exploitation.

'c1' and 'c2' are acceleration coefficients, controlling the particle's attraction to its personal best and the global best, respectively.

'rand()' generates a random number between 0 and 1.

'pBest_i' is the personal best position of particle 'I'.

'gBest' is the global best position among all particles.

'x_i(t)' is the current position of particle 'I' at time 't'.

- Position Update: Update the position of each particle using its new velocity:

$$x\_i(t + 1) = x\_i(t) + v\_i(t + 1)$$

- Termination Condition: Repeat steps 2 to 6 for a certain number of iterations or until a termination condition is met (e.g., a satisfactory solution is found).

PSO is relatively easy to implement and understand compared to some other optimization algorithms. It balances exploration (searching for new areas of the solution space) and exploitation (refining known good solutions). PSO does not require the computation of derivatives of the objective function, making it suitable for problems with non-differentiable or complex fitness landscapes. It has certain disadvantages like it does not guarantee global convergence to the optimal solution. The algorithm may get stuck in local optima. The performance of PSO can be sensitive to the choice of parameters, such as swarm size, inertia weight, and acceleration coefficients. Handling constraint optimization problems in PSO can be challenging. In some cases, PSO may converge prematurely before finding the global optimum [44, 69].

5.2.2 Ant Colony Optimization: Behavior of Ant Colonies

Ant Colony Optimization (ACO) is inspired by the foraging behavior of ants. It uses artificial ants to explore the solution space. Ants deposit pheromone trails to communicate and influence the search. Over time, ACO converges towards optimal solutions by emphasizing paths with higher pheromone levels. Ant Colony Optimization (ACO) is a nature-inspired optimization algorithm that is based on the foraging behavior of ants. It was developed by Marco Dorigo in the early 1990s as a way to solve complex optimization problems. ACO has been successfully applied to a wide range of combinatorial optimization problems, including the traveling salesman problem, job scheduling, and network routing, among others.

The fundamental idea behind ACO is to mimic the behavior of real ants as they search for food in their environment. Ants are social insects that communicate with each other through the deposition of pheromones on the ground. They leave pheromone trails as they move, and other ants can detect and follow these trails to find food sources. Over time, the pheromone trails evaporate, so the shorter and more efficient paths accumulate more pheromone and are more likely to be followed by other ants. Here is a detailed overview of the key components and steps involved in the Ant Colony Optimization Algorithm:

1. Initialization:
   - Create a population of artificial ants, each representing a potential solution to the optimization problem.
   - Initialize pheromone levels on all possible paths or edges in the solution space.
2. Ant Movement:
   - Each ant starts from a randomly chosen initial solution.
   - At each step, an ant selects the next move based on a combination of pheromone levels and a heuristic function. The heuristic function guides the ant towards potentially better solutions.
   - The ant continues to move through the solution space until a termination criterion is met, such as a fixed number of iterations or a time limit.
3. Pheromone Update:
   - After all ants have completed their tours, the pheromone levels on each path are updated.
   - The amount of pheromone deposited on a path is typically proportional to the quality of the solution found by the ant. Better solutions result in more pheromone being deposited.
   - Pheromone evaporates over time to simulate the natural evaporation process. This helps avoid stagnation and encourages exploration of new paths.
4. Exploration vs. Exploitation:
   - ACO balances exploration (searching for new solutions) and exploitation (reinforcing known good solutions) through the combination of pheromone levels and the heuristic information.
   - The pheromone level biases ants towards following paths with higher pheromone concentrations, while the heuristic information guides them towards promising solutions.
5. Termination:

   The algorithm continues to iterate through the ant movement and pheromone update steps until a termination criterion is met. This criterion could be a maximum number of iterations, a convergence threshold, or a time limit.
6. Solution Extraction:

   The best solution found by the ants is extracted as the final solution to the optimization problem.

ACO can be further enhanced with variations and extensions, such as the introduction of elite ants, local search heuristics, and dynamic pheromone updates. These modifications are often problem-specific and aim to improve the algorithm's performance [4, 5, 48].

5.2.3 Whale Optimization Algorithm

The Whale Optimization Algorithm is inspired by the social behavior of humpback whales. It includes exploration and exploitation phases and employs multiple search agents to navigate the solution space. The algorithm's main strength lies in its ability to handle multimodal and complex optimization problems. The Whale Optimization Algorithm (WOA) is a nature-inspired optimization algorithm that is based on the social behavior and hunting strategy of humpback whales. It was proposed by Seyedali Mirjalili and Andrew Lewis in 2016 and is designed to solve complex optimization problems. WOA is part of a family of metaheuristic algorithms that draw inspiration from the behaviors of animals and natural phenomena to find solutions to optimization problems as shown in Fig. 6. Here's a detailed explanation of the key components and steps involved in the Whale Optimization Algorithm [45, 69].

1. Initialization:

   Initialize a population of potential solutions, often referred to as "whales," randomly or using a specific initialization strategy.
2. Encoding:

   Represent each potential solution in a suitable format, depending on the nature of the optimization problem. This could be binary encoding, real-valued encoding, or another appropriate representation.
3. Fitness Evaluation:

   Evaluate the fitness of each whale in the population based on the objective function of the optimization problem. The objective function quantifies how well a solution performs.
4. Exploration and Exploitation:

   - WOA balances exploration (searching for new solutions) and exploitation (refining known solutions) through two main mechanisms: "exploration" and "exploitation" phases.
   - During the exploration phase, whales mimic the behavior of humpback whales as they search for food in a large area. They perform random movements to explore new regions of the solution space.
   - In the exploitation phase, whales mimic the behavior of humpback whales when they encircle a school of fish to capture prey efficiently. This phase involves three key operators:
     a. Encircling Prey: Whales move closer to the current best solution in the population.

b. Spiral Update: Whales perform a spiraling motion to adjust their positions within a bounded search space.
c. Breaching: Some whales leap out of the water, which introduces randomness and diversity into the search process.

5. Updating the Best Solution:

After both exploration and exploitation phases, update the best solution found so far.

6. Termination:

The algorithm continues to iterate through the exploration and exploitation phases until a termination criterion is met. Common termination criteria include a maximum number of iterations, a convergence threshold, or a time limit.

7. Solution Extraction:

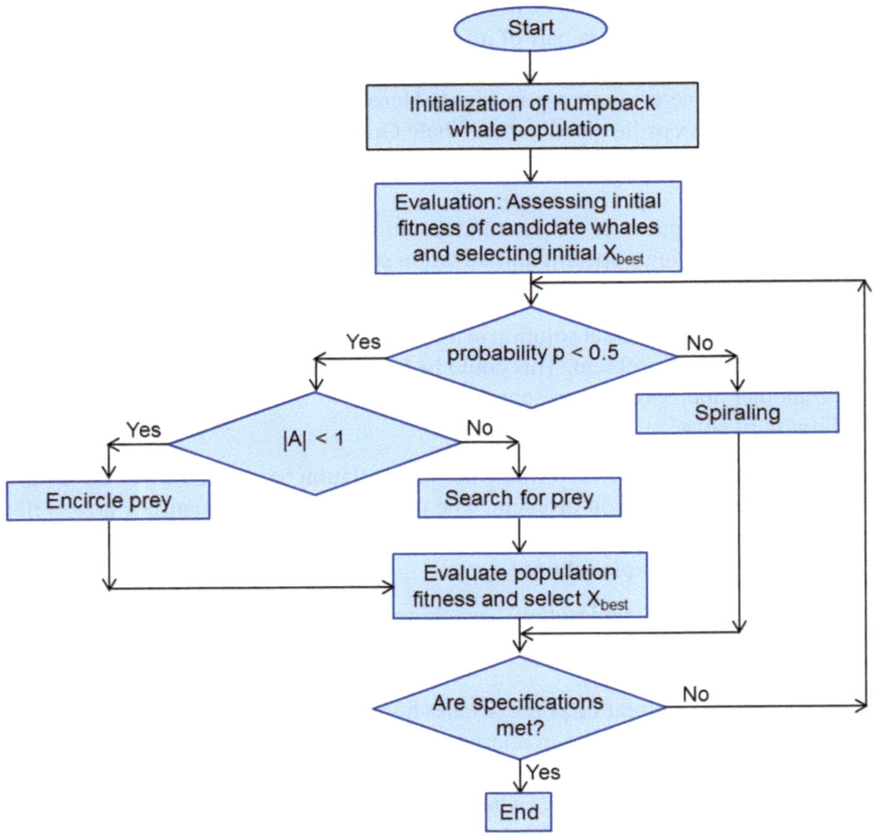

Fig. 6 WOA Flowchart [33]

The best solution found by the algorithm is extracted as the final solution to the optimization problem.

WOA's exploration and exploitation phases, inspired by the hunting behaviors of humpback whales, make it capable of efficiently exploring the search space to discover promising solutions while also fine-tuning those solutions for improved quality. The algorithm's performance can be influenced by parameters such as the number of whales, the rate of exploration and exploitation, and the specific encoding and search space constraints tailored to the problem at hand [45].

5.2.4 Grey Wolf Optimizer

The Grey Wolf Optimizer is based on the hierarchical social structure of grey wolf packs. It simulates the hunting behavior of wolves and uses alpha, beta, and delta individuals to represent solutions. The algorithm mimics the hunting strategy, including exploration, exploitation, and encircling prey. The Grey Wolf Optimizer (GWO) is a population-based metaheuristic optimization algorithm that was inspired by the hunting behavior of grey wolves in nature. It was first proposed by Mirjalili et al. in 2014. GWO is designed to solve complex optimization problems by simulating the social hierarchy and hunting strategies of grey wolves to find the optimal solutions. It has gained popularity in various fields for its simplicity and effectiveness in finding near-optimal solutions. Here is a detailed description of the Grey Wolf Optimizer:

1. Initialization:
   - GWO starts by initializing a population of candidate solutions, where each solution represents a potential solution to the optimization problem. These solutions are often referred to as "wolves."
   - The population size is typically determined by the user and is a crucial parameter that affects the algorithm's performance.
2. Pack Structure:

   In GWO, the population of wolves is organized into a hierarchical structure, similar to real wolf packs. There are three main categories of wolves:
   - Alpha wolf: The leader of the pack and represents the best solution found so far.
   - Beta wolf: The second-best solution.
   - Delta wolf: The third-best solution.
   - Other wolves: The rest of the population.
3. Grey Wolf Behavior:
   - GWO simulates the hunting behavior of grey wolves, which is characterized by cooperation and competition within the pack.
   - Each wolf's position in the solution space represents a potential solution to the optimization problem.

- Wolves update their positions iteratively in an attempt to find the optimal solution.

4. Exploration and Exploitation:
   - GWO balances exploration and exploitation by using the positions of alpha, beta, and delta wolves to guide the search.
   - Alpha wolf leads the pack and guides the search towards promising areas in the solution space.
   - Beta and delta wolves assist in exploration and exploitation by following the alpha wolf and exploring nearby regions.
5. Update Rules:
   - In each iteration, the positions of the wolves are updated based on their current positions and the positions of the alpha, beta, and delta wolves.
   - The position update for each wolf is determined using mathematical formulas that mimic the social interactions and hunting behavior of grey wolves.
6. Fitness Evaluation:
   - After updating the positions of the wolves, the fitness of each wolf's solution is evaluated using the objective function of the optimization problem.
   - Wolves strive to minimize or maximize this fitness value, depending on whether it is a minimization or maximization problem.
7. Termination Criterion:

   GWO continues iterating until a termination criterion is met. Common termination criteria include a maximum number of iterations or reaching a satisfactory solution.
8. Solution Extraction:

   The algorithm returns the best solution found, which is the position of the alpha wolf when the termination criterion is met.
9. Parameter Tuning:

   Like many optimization algorithms, GWO may require parameter tuning to achieve optimal performance for a specific problem. Parameters such as the population size, the number of iterations, and exploration–exploitation parameters can be adjusted based on the problem's characteristics [46].

5.2.5 Firefly Optimization Algorithm

Inspired by the flashing behavior of fireflies, this algorithm models the attractiveness of solutions as the brightness of fireflies. Fireflies move toward brighter neighbors in the solution space, aiming to converge toward optimal solutions. Firefly Optimization Algorithm (FOA) is a nature-inspired metaheuristic optimization algorithm that draws inspiration from the behavior of fireflies to solve optimization problems. FOA was proposed by Xin-She Yang in 2010 and is part of the broader field of swarm intelligence, which includes algorithms like particle swarm optimization and ant colony optimization. This algorithm is particularly useful for solving continuous optimization problems. The basic steps of the Firefly Optimization Algorithm are as follows [50, 70].

1. Initialization: Generate an initial population of fireflies, each representing a potential solution to the optimization problem. The brightness of each firefly is determined by evaluating the fitness of the corresponding solution.
2. Light intensity update: Calculate the attractiveness of each firefly to others based on their fitness (brightness) and the distance between them. This is typically done using a mathematical formula that simulates the attraction and attenuation of light. Fireflies with higher fitness and closer proximity are more attractive.
3. Movement: Each firefly moves towards a more attractive firefly, adjusting its position in the solution space. The movement is guided by the attractiveness and a randomization factor, which introduces randomness into the search.
4. Termination: Repeat the light intensity update and movement steps for a specified number of iterations or until a convergence criterion is met.
5. Solution selection: After the algorithm terminates, the best solution found during the search is selected as the final result.

FOA has been used to solve a wide range of optimization problems, including function optimization, parameter tuning in machine learning, and optimization in engineering and economics. It is known for its ability to efficiently explore solution spaces, escape local optima, and converge to high-quality solutions. However, like many optimization algorithms, the performance of FOA can be influenced by parameter settings and the choice of optimization problem, so fine-tuning may be required for specific applications. Figure 7 shows an enhanced FOA to address the parameter selection and adaptation strategy in the standard firefly algorithm. For example, for the solution of the Levy's Problem.

5.2.6 Bat Optimization Algorithm

The Bat Algorithm is inspired by the echolocation behavior of bats. Bats emit sounds to locate prey and adjust their positions accordingly. In the optimization context, the algorithm models the bat's flight behavior and adjusts the solution space exploration using echolocation-inspired mechanisms. The Bat Algorithm is a nature-inspired metaheuristic optimization algorithm that was introduced by Xin-She Yang in 2010. It is based on the echolocation behavior of bats. The algorithm is designed to solve optimization problems by simulating the hunting behavior of bats as they search for prey. Below is a simplified pseudocode for the Bat Optimization Algorithm [71].

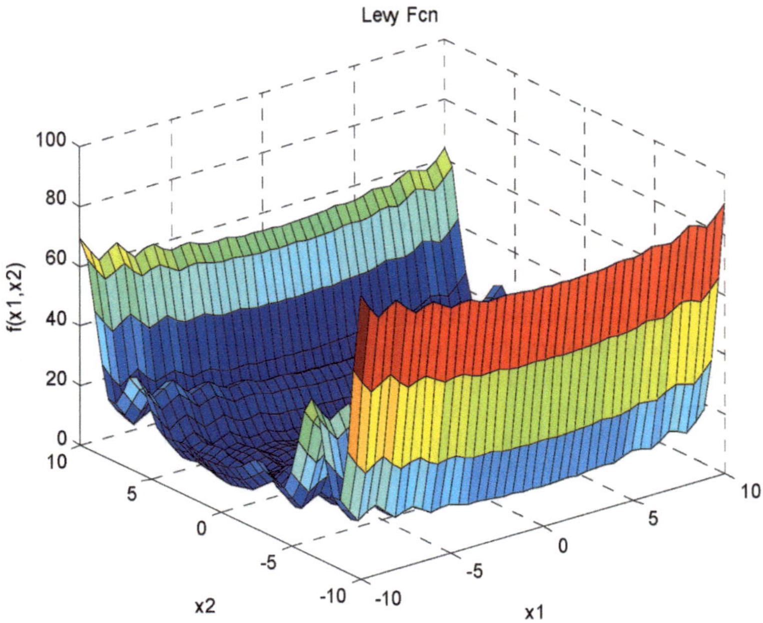

Fig. 7 Levy's Problem in three dimensions, multimodal functions [70]

Step 1: Initialize parameter settings, including the population size *n*, initial impulse loudness A_0, initial impulse emission rate r_0, maximum frequency $Q_{\max}$, minimum frequency $Q_{\min}$, max number of iterations *N*, and fitness evaluation function *Fitness* (*x*)
Step 2: While ($t <$ the max number of iterations) Calculate the frequency Q_i, location S_i, speed V_i, and fitness value $Fitness_i$ of each bat
Step 3: If ($rand > r_i$) 1. Obtain an optimal solution *BestS* in this iteration 2. Calculate the local solutions around the optimal solution
Step 4: End if Produce new solutions by change randomly
Step 5: if ($rand < A_i$ and $Fitness\ (x_i) < Fitness\ (x_*)$) 1. Accept the new solutions 2. Decrease the impulse loudness A_i and increase the impulse emission rate r_i
Step 6: End if Sort all bats and obtain the optimal solution *BestS* in this iteration
Step 7: End while

In the pseudocode above, each "bat" represents a potential solution to the optimization problem. The bats move in the solution space by adjusting their positions based on their current solutions and the solutions of other bats. The algorithm also includes randomness and exploration in the search process through the use of loudness, frequency, and random movements. Please note that the specific parameters (e.g., pulse rate, loudness, frequency, and decay rate) and the stopping criterion may

Fig. 8 Echolocation hunting of the bat [71]

need to be adjusted based on the problem being solved. The Bat Algorithm is versatile and can be adapted to different optimization tasks. Figure 8 shows how the bat relies on echoes for accurate positioning, which allows it to fly freely and hunt accurately in a completely dark [53, 72–74].

5.2.7 Orca Predation Algorithm

The Orca Predation Algorithm is inspired by the hunting behavior of killer whales (orcas). It simulates the cooperative hunting strategy of orcas to explore and exploit the solution space. The algorithm is particularly effective for dynamic and constrained optimization problems.

The OPA (Orca Predation Algorithm) emulates the hunting tactics of killer whales (Orcinus orca), which exhibit intelligence on par with humans. During their hunts, these creatures collaborate in groups, utilizing echolocation to locate their prey. They communicate through sonar signals, sharing information and orchestrating their assaults with remarkable precision. This coordinated approach distinguishes them as top-tier ocean predators, even preying on sharks. The exceptional efficiency of their hunting methods inspired the development of the OPA.

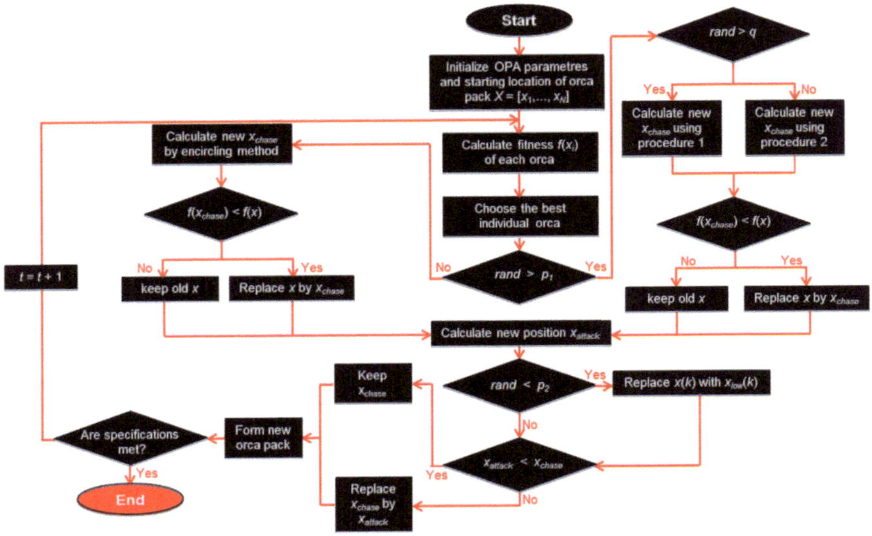

Fig. 9 OPA algorithm flowchart [33], [63]

The algorithm dissects the process into three mathematical sub-models: (1) the quest for prey, (2) steering and encircling the target, and (3) launching an attack. To strike a balance between exploration and exploitation phases, distinct weight coefficients are assigned to various stages of prey pursuit and encirclement. Algorithm parameters are fine-tuned to achieve this equilibrium. Specifically, the positions of superior, average, and randomly selected killer whales are identified. This approach allows the algorithm to approach optimal solutions during the attack phase while maintaining a rich diversity of individual killer whales [63, 75, 76].

Figure 9 displays the flowchart of the Orca Predation Algorithm, which has been adapted from the diagram presented in reference [63] with a few adjustments [33]. The parameter "xlow" signifies the lower limit of the problem. Despite its apparent complexity upon initial inspection, the algorithm is relatively straightforward from a mathematical perspective, resulting in a notably high computational speed.

5.2.8 Starling Murmuration Optimizer

This algorithm is inspired by the flocking behavior of starlings. It models the coordination and synchronization observed in starling murmurations to optimize solutions as shown in Fig. 10. Individual agents in the swarm interact with their neighbors to find better solutions collectively.

The Starling Murmuration Optimizer (SMO) was presented in 2022 by Zamani et al. [64] as a metaheuristic algorithm. SMO is a population-based approach that incorporates a dynamic multi-flock structure. Notably, it introduces three novel

Fig. 10 Starlings Murmuration [*Source* Walter Baxter (https://commons.wikimedia.org/wiki/File:Starling_murmuration.jpg) "Starling murmuration", https://creativecommons.org/licenses/by-sa/2.0/legalcode]

search strategies: separating, diving, and whirling [33]. Figure 11 shows the flowchart of the SMO algorithm [64, 77].

The starlings that remained after the separation dynamically construct the multi-flock with members f1, f2, ..., fk. The quality Qq(t) of the qth flock is calculated by

$$Qq(t) = \sum k\,i = 1 \sum n\,j = 1sfij(t)1n \sum n\,i = 1sfqi(t)$$

to select either the diving (exploration) or the whirling (exploitation) search strategy. The diving strategy explores the search space using a new quantum random dive operator, and the whirling strategy exploits the neighborhood of promising regions using a new cohesion force operator [78]. Here, sfij(t) is the fitness value of the ith starling from the jth flock fj; k is the number of flocks in a murmuration M. The average quality of all flocks is denoted as μq.

5.2.9 Artificial Bee Colony Algorithm

Inspired by the foraging behavior of honey bees, this algorithm uses employed bees, onlooker bees, and scout bees to explore and exploit the solution space. Employed bees search around known food sources, while onlooker bees select food sources

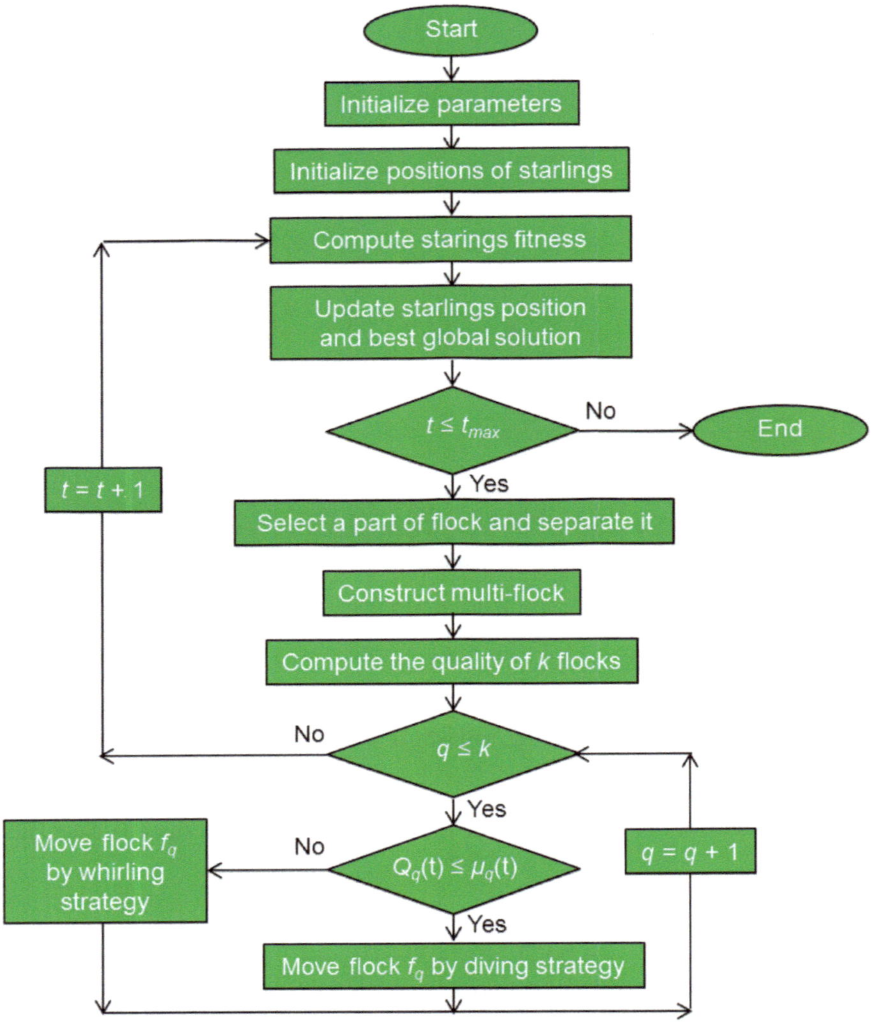

Fig. 11 Starlings Murmuration algorithm flowchart [33]

based on their quality. The Artificial Bee Colony (ABC) algorithm is a population-based optimization algorithm inspired by the foraging behavior of honeybees. It was first introduced by Derviş Karaboğa in 2005. The ABC algorithm is used to find optimal solutions to various optimization problems. Here's a brief description of the Artificial Bee Colony Algorithm and pseudocode:

The ABC algorithm is based on the foraging behavior of bees, where there are three main types of bees: employed bees, onlooker bees, and scout bees.

1. Employed Bees: Employed bees explore the search space and share information about the quality of food sources with other employed bees. Each employed bee is associated with a specific solution.
2. Onlooker Bees: Onlooker bees choose food sources based on their quality information and may become employed bees. The probability of an onlooker bee selecting a food source is determined by the quality of that source.
3. Scout Bees: Scout bees identify food sources that are no longer productive and need replacement. When a food source is abandoned, a scout bee looks for a new one in the search space.

```
Initialize the population of employed bees and onlooker bees with random solutions
Evaluate the quality of each food source using the objective function

Repeat until a stopping criterion is met:
  Employed Bee Phase:
  For each employed bee:
    Select a neighbor food source and generate a new solution
    Evaluate the quality of the new solution

  Calculate the probabilities of food sources for onlooker bees based on quality

  Onlooker Bee Phase:
  For each onlooker bee:
    Select a food source based on the probabilities
    Generate a new solution for the selected food source
    Evaluate the quality of the new solution

  Scout Bee Phase:
  If a food source is exhausted (no improvement), replace it with a new random solution

  Update the best solution found so far

End loop

Return the best solution found
```

The ABC algorithm iteratively refines its population of solutions by exploring the search space and sharing information about the quality of food sources. It balances exploration and exploitation, making it suitable for a wide range of optimization problems. The algorithm continues until a stopping criterion is met, such as a maximum number of iterations or reaching a desired solution quality [47].

5.2.10 Cuckoo Search Algorithm

The Cuckoo Search Algorithm is inspired by the brood parasitism behavior of some cuckoo species. It uses a population of cuckoos to lay eggs (representing potential solutions) in host nests. The algorithm focuses on preserving and improving the quality of eggs in the nests. Cuckoo Search is a nature-inspired optimization algorithm introduced by Xin-She Yang and Suash Deb in 2009. This algorithm is inspired by the brood parasitism of some species of cuckoo birds, which lay their eggs in the nests of other bird species. The Cuckoo Search Algorithm is used to find optimal solutions to optimization problems. Here's a description of the Cuckoo Search Algorithm and pseudocode:

The Cuckoo Search Algorithm is based on the idea of the cuckoo bird laying its eggs in the nests of other birds. In this context, each egg represents a potential solution to an optimization problem, and each nest represents a potential solution space. The algorithm's main components are as follows:

1. Eggs (Solutions): A population of solutions (or eggs) is generated randomly or using other methods. Each egg represents a potential solution to the optimization problem.
2. Nests (Solution Spaces): Each nest corresponds to a specific location in the solution space, where a cuckoo can lay an egg. Nests can be thought of as a set of potential solutions.
3. Lévy Flight: Cuckoos lay eggs in nests with a probability determined by the quality of the nest. The Lévy flight mechanism, which is a random walk following a Lévy distribution, is often used to control the movement of cuckoos in search of better nests.
4. Nest Replacement: In the process of laying eggs in nests, cuckoos replace the eggs in poor-quality nests with better eggs to improve the overall quality of the population.

```
Initialize the population of employed bees and onlooker bees with random solutions
Evaluate the quality of each food source using the objective function

Repeat until a stopping criterion is met:
  Employed Bee Phase:
  For each employed bee:
    Select a neighbor food source and generate a new solution
    Evaluate the quality of the new solution

  Calculate the probabilities of food sources for onlooker bees based on quality

  Onlooker Bee Phase:
  For each onlooker bee:
    Select a food source based on the probabilities
    Generate a new solution for the selected food source
    Evaluate the quality of the new solution

  Scout Bee Phase:
  If a food source is exhausted (no improvement), replace it with a new random solution

  Update the best solution found so far

End loop

Return the best solution found
```

The Cuckoo Search Algorithm uses a combination of exploration through Lévy flights and exploitation by replacing nests with better solutions to find the optimal solution to an optimization problem. The algorithm continues until a stopping criterion is met, such as a maximum number of iterations or reaching a desired solution quality. This book contains one full dedicated chapter to the Cuckoo Search Algorithm and its deployment using high-performance computing [52].

5.2.11 Pheromone-Based Communication and Stigmergy

Several swarm intelligence algorithms, such as ACO and some variants of PSO, rely on pheromone-based communication and stigmergy. Pheromone trails left by individuals or agents in the solution space are used to guide the search and convey information about the quality of solutions. Pheromone-based communication and stigmergy are concepts often associated with swarm intelligence and collective behavior, where individuals within a group communicate indirectly through the environment, using chemical signals (pheromones in the case of ants, for example) to coordinate their actions. Stigmergy refers to the process of interaction or coordination through the modification of the environment. Here's a description of pheromone-based communication and stigmergy with simplified pseudocode:

1. Pheromone-Based Communication: Pheromone-based communication is a form of indirect communication where individuals in a group leave and respond to chemical markers (pheromones) in the environment. Pheromones can indicate information such as the location of resources, the presence of threats, or the path to a target. In the context of swarm intelligence, this is often used in algorithms like Ant Colony Optimization (ACO) to find optimal solutions in complex problem spaces.
2. Stigmergy: Stigmergy is a broader concept that encompasses pheromone-based communication. It refers to the way interactions among individuals are mediated through modifications to the environment. This modification can be through the deposition of pheromones, but it could also involve physical changes, like the construction of trails or structures.

```
Initialize a population of ants and a problem space
Initialize pheromone levels on paths in the problem space

Repeat until a stopping criterion is met:
  For each ant:
    Choose a path to explore based on pheromone levels and heuristics
    Follow the chosen path, updating pheromone levels

  Evaporate pheromone levels on all paths to simulate decay
  Deposit pheromone on paths that were taken by ants based on the quality of the solutions
found

End loop

Return the best solution found
```

```
Initialize a population of agents and the environment
Initialize environmental markers, e.g., pheromones or structures

Repeat until a stopping criterion is met:
  For each agent:
    Observe the environment and decide on an action based on environmental markers
    Perform the action, potentially modifying the environment

  Update environmental markers based on agent interactions with the environment

End loop
```

In this pseudocode, agents interact with the environment, leaving markers or making changes, and then the environment, in turn, affects the agents' decisions in a self-organized manner. Stigmergy can be applied to various scenarios beyond pheromone-based communication, such as in robotics, where robots modify their surroundings during construction tasks, or in social insects like termites building nests.

5.3 *Plant-Based Algorithms*

In this subsection, you'll find metaheuristic optimization algorithms that draw inspiration from the characteristics of plant life. There are many plant-based optimization algorithms like Flower Pollination Algorithm (FPA), Invasive Weed Optimization (IWO), Plant Propagation Algorithm (PPA), Plant Growth Optimization (PGO), Tree Seed Algorithm (TSA) and Paddy Field Algorithm (PFA). Let us discuss about widely published and cited Flower Pollination Algorithm (FPA).

5.3.1 Flower Pollination Algorithm (FPA)

The Flower Pollination Algorithm is inspired by the pollination process of flowering plants. It models the transfer of information (solutions) between flowers and uses pollination operators to guide the search. Flowers with higher fitness attract pollinators more effectively. The Flower Pollination Algorithm (FPA) is a metaheuristic optimization algorithm inspired by the pollination behavior of flowers and the transfer of pollen between flowers by pollinators like bees. It was introduced by Xin-She Yang in 2012. The FPA is used for solving optimization problems.

In the basic version of the algorithm, it assumes that each plant produces a single flower, and each of these flowers yields just one pollen grain. Consequently, the flower or its pollen grain serves as the candidate solution as shown in Fig. 12. Movement within the search space is achieved through biotic cross-pollination, and the motion of each pollen grain is characterized by Lévy flight, a random walk method that has also been adopted by other metaheuristics [100]. The Flower Pollination Algorithm (FPA) facilitates the sharing of information and the selection of improved solutions,

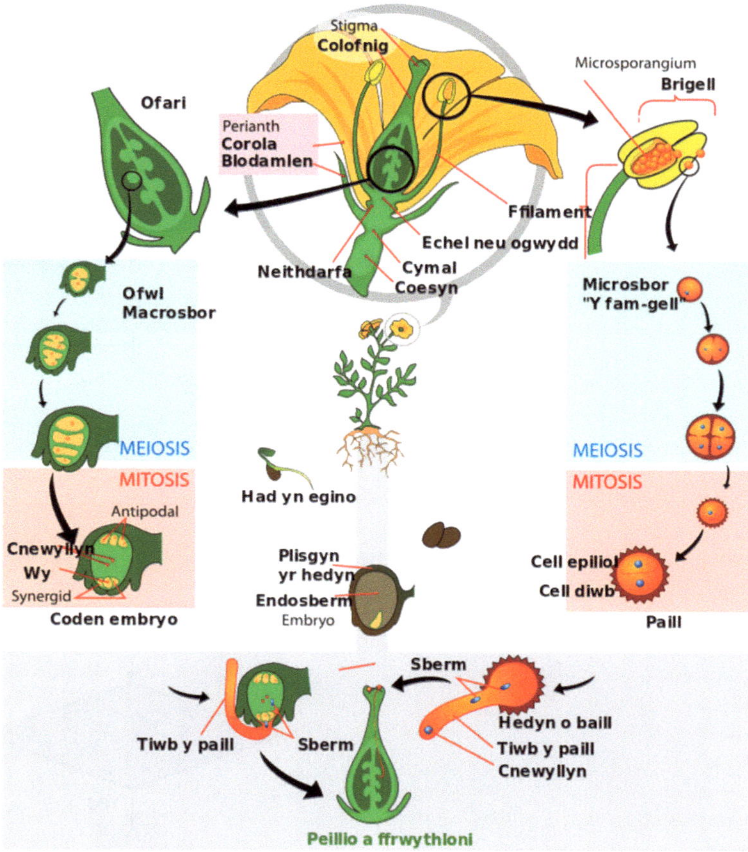

Fig. 12 Flower angiosperm life cycle [*Source* Llywelyn2000 (https://commons.wikimedia.org/wiki/File:Angiosperm_life_cycle_diagram-cy.svg), https://creativecommons.org/licenses/by-sa/4.0/legalcode]

thereby fostering the exchange of "knowledge" among various flowers or candidate solutions and enhancing the exploration phase.

Typically, an FPA procedure begins by initializing an initial population of flowers or pollen grains, where the positions of these pollen grains represent potential candidate solutions. In keeping with the natural pollination process, FPA permits flowers to interact and exchange information in multiple ways to seek better solutions. A flower is chosen for reproduction based on its fitness or objective function value and becomes the pollinating flower (the source). It then perturbs its position in the search space using a random mechanism, which is regulated by a randomization factor, ultimately generating a new solution. This can involve randomly selecting either local or global pollination. The fitness of the offspring solution is compared to that of the pollinator flower, and depending on their fitness values, the original flower

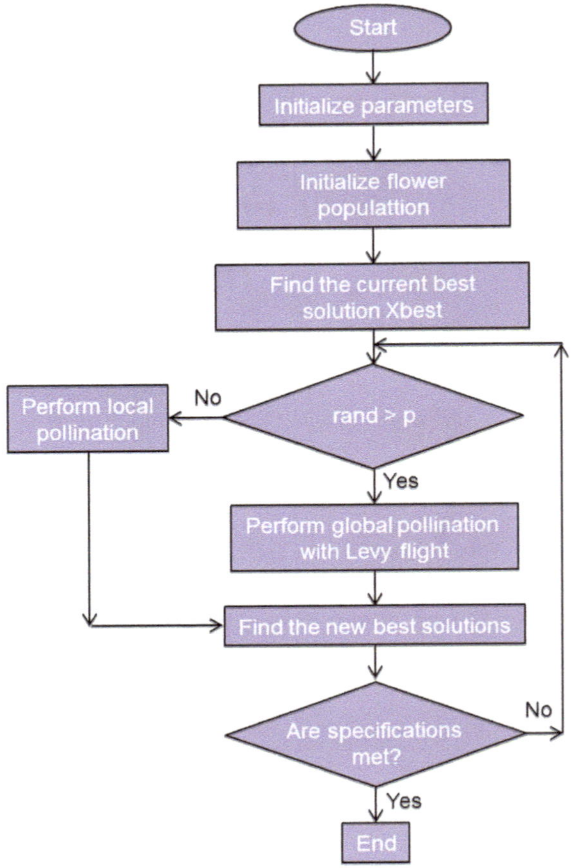

Fig. 13 Flowchart of FPA algorithm [33]

is either replaced or retained. These steps are iterated for the entire population of flowers. This process accomplishes exploration through random perturbations, while exploitation is driven by the selection process based on fitness criteria. The algorithm concludes when a satisfactory solution is found or when a maximum number of iterations is reached. Additionally, Fig. 13 provides a visual representation in the form of a flowchart illustrating the FPA [79–82].

5.4 *Human-Based*

5.4.1 Artificial Immune Systems (AISs)

Artificial Immune Systems are inspired by the human immune system's ability to recognize and eliminate pathogens. They use immune-inspired algorithms to detect anomalies or optimize solutions. AISs include mechanisms like clonal selection and antibody affinity maturation. Artificial immune systems (AISs) are a category of algorithms designed to mimic the functionality of the human and vertebrate immune systems. They are widely recognized as popular optimization algorithms, with approximately 22,000 papers published on AISs as of May 2023. Depending on their usage, they can be categorized as both metaphor-based metaheuristic algorithms and machine learning techniques. They fall under the umbrella of metaheuristic algorithms because they are employed for optimization tasks, explore the search space, and iteratively enhance their solutions to approximate optimal outcomes. Simultaneously, they are considered machine learning techniques because they incorporate rule-based learning from data and employ adaptive mechanisms, utilizing feedback information. The distinguishing feature of AISs is their bio-inspired and metaphor-based nature, setting them apart from conventional machine learning methods. In essence, AISs inherently combine metaheuristic algorithms with machine learning, where these two domains intersect [83–85].

The operation of an AIS algorithm is founded on various functions of the immune system. When encountering antigens (foreign agents targeted by the immune system), immune cells trigger an immune response, leading to the production of antibodies or the activation of immune cells. In AIS algorithms, the immune response mirrors the evolution-based adaptation of antibodies to enhance their fitness or affinity to antigens. This can involve modifying existing antibodies or generating new ones through processes like mutation or recombination. Given that AISs encompass a range of procedures, they encompass various algorithms. Among these, the most well-known and widely used ones include the clonal selection algorithm, the artificial immune network algorithms, the negative selection algorithm, the dendritic cell algorithm, the danger theory, the humoral immune response, the pattern recognition receptor model, and the artificial immune recognition system. Some of these categories within the AIS algorithm group consist of sub-algorithms or related approaches. Some of the most renowned AIS algorithms are [33].

1. Clonal selection algorithms (CSAs)
2. Artificial immune network algorithms
3. The negative selection algorithm
4. The dendritic cell algorithm
5. Artificial immune recognition system

5.4.2 Tabu Search Algorithm (TSA)

Tabu Search is a local search algorithm that maintains a short-term memory (tabu list) to avoid revisiting previously explored solutions. It iteratively explores the neighborhood of the current solution while avoiding tabu moves, aiming to escape local optima. As previously noted, the classification of the Tabu (or taboo) search algorithm varies in the literature, leading to different categorizations among researchers, even within the realm of mathematics-based nature-inspired algorithms [86]. However, due to its primary foundation in anthropological customs, it is most frequently grouped under algorithms based on human social behavior [99]. The term "tabu" (sometimes spelled as "taboo") originates from Tongan culture and signifies something sacred or forbidden to touch, a concept found in various forms in almost all human societies.

Tabu search stands apart as a single-solution-based metaheuristic optimization algorithm [87], distinguishing it from the population-based nature of other metaheuristics discussed here. It is specifically designed for solving combinatorial optimization problems through the application of local search techniques. When seeking improved potential solutions, it explores neighboring solutions within the search space. Notably, this algorithm employs memory to remember previously visited solutions and prohibits revisiting them. In essence, these solutions become "tabu." Additionally, solutions that violate user-defined rules or criteria (referred to as aspiration criteria) also fall under the tabu category. The algorithm maintains a list of forbidden solutions as a form of memory. Another distinctive feature of this algorithm is its ability to, in situations where no superior solution is available (e.g., when trapped in a local minimum), accept a suboptimal solution. In other words, it relaxes its fundamental requirement of always seeking an improved solution, enabling it to act as a local search method that can escape local minima and continue the pursuit of a global optimum. Figure 14 illustrates the flowchart of the same algorithm. In summary, metaheuristic algorithms encompass a wide range of optimization techniques inspired by nature and human behaviors. These algorithms have been applied successfully to various optimization problems in fields such as engineering, computer science, economics, and biology, among others. Their versatility and adaptability make them valuable tools for tackling complex and challenging optimization tasks [88].

6 Neuroevolution: Combining Neural Networks and Evolutionary Algorithms

Neuroevolution is a fascinating approach that combines neural networks and evolutionary algorithms to optimize neural network architectures and parameters. Instead of manually designing neural networks or using traditional training methods like backpropagation, neuroevolution evolves neural network structures and weights to achieve specific tasks. The process involves creating a population of neural networks, evaluating their performance on a given task, and then using evolutionary operators such as mutation and crossover to produce the next generation of networks.

Recent advancements in neuroevolution focus on scalability, efficiency, and handling large neural networks. Researchers have explored various techniques to improve the convergence rate and explore the design space more effectively. Keywords to search for recent articles include "neuroevolution," "neuroevolution of augmenting topologies (NEAT)," and "evolutionary deep learning" [85].

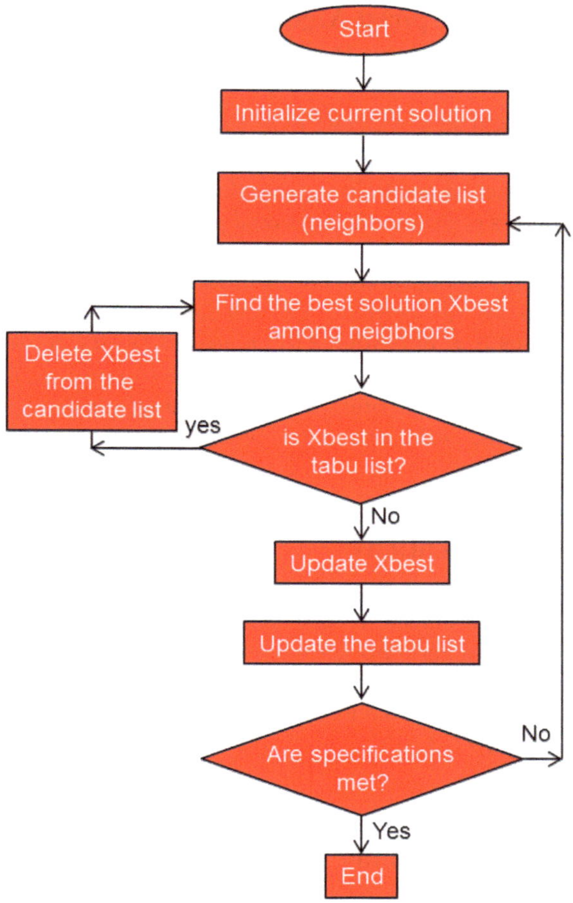

Fig. 14 Flowchart of TSA algorithm [33]

7 Bionic Optimization: Integrating Biological Principles into Optimization Algorithms

Bionic optimization is a category of bio-inspired optimization that specifically draws from biological principles to solve complex problems. It encompasses a wide range of approaches inspired by natural systems, such as the behavior of ants, bees, birds, and genetic processes. These principles are translated into algorithms to tackle real-world optimization challenges [51, 54, 55, 58–61].

Recent research in bionic optimization explores novel applications and adaptations of biological principles. For instance, researchers are developing algorithms inspired by the immune system, social insect behaviors, and ecological dynamics. Keywords to search for include "bionic optimization," "biologically inspired algorithms," and "ecological algorithms" [62, 65, 66, 68, 72, 73].

8 Bio-inspired Algorithms for Optimization, Scheduling, and Pattern Recognition

Bio-inspired algorithms have found applications in various domains beyond optimization, including scheduling and pattern recognition. These algorithms leverage the inherent parallelism, adaptability, and robustness observed in natural systems to address complex computational problems [86, 89–92].

Recent research in this area looks into customized bio-inspired algorithms for specific applications. For example, in scheduling, algorithms inspired by swarm behavior can optimize job schedules in manufacturing environments. In pattern recognition, bio-inspired techniques like neural networks and genetic algorithms are adapted to improve image and speech recognition. Keywords to search for include "bio-inspired algorithms for scheduling," "pattern recognition using bio-inspired methods," and "applications of bio-inspired algorithms" [87, 93–95].

9 Applications

There are numerous success stories and practical applications of hybrid and bio-inspired algorithms across various industries. Researchers and practitioners have employed these techniques to solve challenging optimization problems in fields such as finance, healthcare, transportation, microelectronics, nanofluids, and robotics [69, 96, 97, 33].

10 Conclusions

Bio-inspired computing draws inspiration from biological systems and applies their principles to develop innovative algorithms and computational models. This chapter provides an extensive exploration of various bio-inspired computing approaches, including neural networks, evolutionary algorithms, swarm intelligence, cellular automata, and DNA computing. Each section explains the underlying concepts, algorithms, and applications associated with these approaches. The chapter also discusses hybrid approaches that combine multiple bio-inspired techniques and highlights their significance in optimization, scheduling, and pattern recognition tasks. Furthermore, the chapter addresses the challenges and future directions of bio-inspired computing, emphasizing scalability, explainability, integration with other AI techniques, and ethical considerations. The chapter concludes with a summary of key points, the impact of bio-inspired computing, and future possibilities in this rapidly evolving field.

References

1. Nemade, N., Rane, R.D.: A review on bio-inspired computing algorithms and application. IOSR J. Comput. Eng. (IOSR-JCE), 12–19 (2016)
2. Kar, A.K.: Bio inspired computing—a review of algorithms and scope of applications. Expert Syst. Appl. **59**, 20–32 (2016).https://doi.org/10.1016/j.eswa.2016.04.018
3. Goldberg, D.E.: Genetic Algorithms in Search, Optimization, and Machine Learning. Addison-Wesley (1989)
4. Haykin, S.: Neural Networks: A Comprehensive Foundation. Prentice Hall (1998)
5. Dorigo, M., Stützle, T.: Ant Colony Optimization. MIT Press (2004)
6. Wolfram, S.: Cellular Automata as Simple Self-Organizing Systems. California Institute of Technology (1983)
7. Russel, S.J., Norvig, P.: Artificial Intelligence: A Modern Approach. Pearson (2018)
8. Arkin, R.C.: Behavior-Based Robotics. MIT Press (1998)
9. Pinker, S.: The Stuff of Thought: Language as a Window into Human Nature. Penguin Books (2007)
10. LeCun, Y., Bengio, Y., Hinton, G.: Deep learning. Nature **521**(7553), 436–444 (2015)
11. Vaswani, A., Shazeer, N., Parmar, N., Uszkoreit, J., Jones, L., Gomez, A. N., ... & Polosukhin, I.: Attention is all you need. Advances in neural information processing systems (2017)
12. Deb, K.: Multi-objective Optimization Using Evolutionary Algorithms. Wiley (2001)
13. Blum, C., Merkle, D.: Swarm intelligence: introduction and applications. Nat. Comput. **7**(3), 267–278 (2008)
14. Adleman, L.M.: Molecular computation of solutions to combinatorial problems. Science **266**(5187), 1021–1024 (1994)
15. Seelig, G., Soloveichik, D., Zhang, D.Y., Winfree, E.: Enzyme-free nucleic acid logic circuits. Science **314**(5805), 1585–1588 (2006)
16. Goodfellow, I., Bengio, Y., Courville, A., & Bengio, Y.: Deep Learning (Vol. 1). MIT Press Cambridge (2016)
17. Bishop, C.M.: Neural Networks for Pattern Recognition. Oxford University Press (1995)
18. He, K., Zhang, X., Ren, S., Sun, J.: Deep residual learning for image recognition. In: Proceedings of the IEEE Conference on Computer Vision and Pattern Recognition (2016)
19. Rumelhart, D.E., Hinton, G.E., Williams, R.J.: Learning representations by back-propagating errors. Nature **323**(6088), 533–536 (1985)
20. Kingma, D.P., Ba, J.: Adam: A method for stochastic optimization (2014). arXiv preprint arXiv:1412.6980
21. Krizhevsky, A., Sutskever, I., Hinton, G.E.: ImageNet classification with deep convolutional neural networks. In: Advances in Neural Information Processing Systems (2012)
22. Hochreiter, S., Schmidhuber, J.: Long short-term memory. Neural Comput. **9**(8), 1735–1780 (1997)
23. Rguibi, Z., Hajami, A., Zitouni, D., Elqaraoui, A., Bedraoui, A.: CXAI: explaining convolutional neural networks for medical imaging diagnostic. Electronics **11**(11), 1775 (2022). https://doi.org/10.3390/electronics11111775
24. Silver, D., Huang, A., Maddison, C. J., Guez, A., Sifre, L., van den Driessche, G., ... & Hassabis, D.: Mastering chess and shogi by self-play with a general reinforcement learning algorithm. Nature **529**(7587), 484–489 (2016)
25. Rajpurkar, P., Irvin, J., Zhu, K., Yang, B., Mehta, H., Duan, T., ... & Langlotz, C.P.: CheXNet: Radiologist-level pneumonia detection on chest X-rays with deep learning (2017). arXiv preprint arXiv:1711.05225
26. Eiben, A.E., Smith, J.E.: Introduction to Evolutionary Computing. Springer (2015)
27. Yang, X.S.: Nature-Inspired Optimization Algorithms. Elsevier (2014)
28. Koza, J.R.: Genetic Programming: On the Programming of Computers by Means of Natural Selection. MIT Press (1992)
29. Beyer, H.G., Schwefel, H.P.: Evolution strategies–A comprehensive introduction. Nat. Comput. **1**(1), 3–52 (2002)

30. Kotyrba, M., Volna, E., Habiballa, H., Czyz, J.: The influence of genetic algorithms on learning possibilities of artificial neural networks. Computers **11**(5), 70 (2022). https://doi.org/10.3390/computers11050070
31. Back, T.: Evolutionary Algorithms in Theory and Practice: Evolution Strategies, Evolutionary Programming, Genetic Algorithms. Oxford University Press (1996)
32. Deb, K., et al.: A review on multi-objective optimization in manufacturing industries. Eng. Optim. **48**(6), 841–871 (2016)
33. Jakšić, Z., Devi, S., Jakšić, O., Guha, K.: A comprehensive review of bio-inspired optimization algorithms including applications in microelectronics and Nanophotonics. Biomimetics **8**(3), 278 (2023). https://doi.org/10.3390/biomimetics8030278
34. Vasant, P., Weber, G.-W., Dieu, V.N. (Eds.): Handbook of Research on Modern Optimization Algorithms and Applications in Engineering and Economics; IGI Global: Hershey, PA, USA (2016)
35. Fávero, L.P., Belfiore, P.: Data Science for Business and Decision Making. Academic Press, Cambridge, MA, USA (2018)
36. Montoya, O.D., Molina-Cabrera, A., Gil-González, W.: A Possible Classification for Metaheuristic Optimization Algorithms in Engineering and Science. Ingeniería 27, 1 (2022)
37. Ma, Z., Wu, G., Suganthan, P.N., Song, A., Luo, Q.: Performance assessment and exhaustive listing of 500+ nature-inspired metaheuristic algorithms. Swarm Evol. Comput. **77**, 101248 (2023)
38. Del Ser, J., Osaba, E., Molina, D., Yang, X.-S., Salcedo-Sanz, S., Camacho, D., Das, S., Suganthan, P.N., Coello Coello, C.A., Herrera, F.: Bio-inspired computation: where we stand and what's next. Swarm Evol. Comput. **48**, 220–250 (2019)
39. Holland, J.H.: Adaptation in Natural and Artificial Systems. University of Michigan Press, Ann Arbor, MI, USA (1975)
40. Wilson, A.J., Pallavi, D.R., Ramachandran, M., Chinnasamy, S., Sowmiya, S.: A review on memetic algorithms and its developments. Electr. Autom. Eng. **1**, 7–12 (2022)
41. Bilal; Pant, M., Zaheer, H., Garcia-Hernandez, L., Abraham, A.: Differential evolution: a review of more than two decades of research. Eng. Appl. Artif. Intell. **90**, 103479 (2020)
42. Sivanandam, S.N., Deepa, S.N., Sivanandam, S.N., Deepa, S.N.: Genetic Algorithms; Springer, Berlin/Heidelberg, Germany (2008)
43. Dawkins, R.: The Selfish Gene. Oxford University Press, Oxford, UK (1976)
44. Sengupta, S., Basak, S., Peters, R.A.: Particle swarm optimization: a survey of historical and recent developments with hybridization perspectives. Mach. Learn. Knowl. Extr. **1**, 157–191 (2019)
45. Mirjalili, S., Lewis, A.: The whale optimization algorithm. Adv. Eng. Softw. **95**, 51–67 (2016)
46. Mirjalili, S., Mirjalili, S.M., Lewis, A.: Grey wolf optimizer. Adv. Eng. Softw. **69**, 46–61 (2014)
47. Karaboga, D., Gorkemli, B., Ozturk, C., Karaboga, N.: A comprehensive survey: Artificial bee colony (ABC) algorithm and applications. Artif. Intell. Rev. **42**, 21–57 (2014)
48. Dorigo, M., Stützle, T.: Ant colony optimization: overview and recent advances. In: Gendreau, M., Potvin, J.-Y., (Eds.), Handbook of Metaheuristics, pp. 311–351. Springer International Publishing, Cham, Switzerland (2019)
49. Neshat, M., Sepidnam, G., Sargolzaei, M., Toosi, A.N.: Artificial fish swarm algorithm: a survey of the state-of-the-art, hybridization, combinatorial and indicative applications. Artif. Intell. Rev. **42**, 965–997 (2014)
50. Fister, I., Fister, I., Yang, X.-S., Brest, J.: A comprehensive review of firefly algorithms. Swarm Evol. Comput. **13**, 34–46 (2013)
51. Ranjan, R.K., Kumar, V.: A systematic review on fruit fly optimization algorithm and its applications. Artif. Intell. Rev. (2023)
52. Yang, X.-S., Deb, S.: Cuckoo search: recent advances and applications. Neural Comput. Appl. **24**, 169–174 (2014)
53. Agarwal, T., Kumar, V.: A systematic review on bat algorithm: theoretical foundation, variants, and applications. Arch. Comput. Methods Eng. **29**, 2707–2736 (2022)

54. Selva Rani, B., Aswani Kumar, C.: A comprehensive review on bacteria foraging optimization technique. In: Dehuri, S., Jagadev, A.K., Panda, M. (Eds.) Multi-objective Swarm Intelligence: Theoretical Advances and Applications, pp. 1–25. Springer, Berlin/Heidelberg, Germany (2015)
55. Luque-Chang, A., Cuevas, E., Fausto, F., Zaldívar, D., Pérez, M.: Social spider optimization algorithm: modifications, applications, and perspectives. Math. Probl. Eng. 2018, 6843923 (2018)
56. Cuevas, E., Fausto, F., González, A.: Locust search algorithm applied to multi-threshold segmentation. In: Cuevas, E., Fausto, F., González, A., (Eds.) New Advancements in Swarm Algorithms: Operators and Applications, pp. 211–240. Springer International Publishing: Cham, Switzerland (2020)
57. Ezugwu, A.E., Prayogo, D.: Symbiotic organisms search algorithm: theory, recent advances and applications. Expert Syst. Appl. **119**, 184–209 (2019)
58. Shehab, M., Abualigah, L., Al Hamad, H., Alabool, H., Alshinwan, M., Khasawneh, A.M.: Moth–flame optimization algorithm: Variants and applications. Neural Comput. Appl. **32**, 9859–9884 (2020)
59. Hashim, F.A., Houssein, E.H., Hussain, K., Mabrouk, M.S., Al-Atabany, W.: Honey Badger Algorithm: New metaheuristic algorithm for solving optimization problems. Math. Comput. Simul. **192**, 84–110 (2022)
60. Li, J., Lei, H., Alavi, A.H., Wang, G.-G.: Elephant herding optimization: variants, hybrids, and applications. Mathematics **8**, 1415 (2020)
61. Abualigah, L., Diabat, A.: A comprehensive survey of the Grasshopper optimization algorithm: results, variants, and applications. Neural Comput. Appl. **32**, 15533–15556 (2020)
62. Alabool, H.M., Alarabiat, D., Abualigah, L., Heidari, A.A.: Harris hawks optimization: a comprehensive review of recent variants and applications. Neural Comput. Appl. **33**, 8939–8980 (2021)
63. Jiang, Y., Wu, Q., Zhu, S., Zhang, L.: Orca predation algorithm: a novel bio-inspired algorithm for global optimization problems. Expert Syst. Appl. **188**, 116026 (2022)
64. Zamani, H., Nadimi-Shahraki, M.H., Gandomi, A.H.: Starling murmuration optimizer: A novel bio-inspired algorithm for global and engineering optimization. Comput. Methods Appl. Mech. Eng. **392**, 114616 (2022)
65. Dehghani, M., Trojovský, P.: Serval optimization algorithm: a new bio-inspired approach for solving optimization problems. Biomimetics **7**, 204 (2022)[PubMed]
66. Salcedo-Sanz, S.: A review on the coral reefs optimization algorithm: new development lines and current applications. Prog. Artif. Intell. **6**, 1–15 (2017)
67. Wang, G.-G., Gandomi, A.H., Alavi, A.H., Gong, D.: A comprehensive review of krill herd algorithm: Variants, hybrids and applications. Artif. Intell. Rev. **51**, 119–148 (2019)
68. Agushaka, J.O., Ezugwu, A.E., Abualigah, L.: Gazelle optimization algorithm: A novel nature-inspired metaheuristic optimizer. Neural Comput. Appl. **35**, 4099–4131 (2023)
69. Laskar, N.M., Guha, K., Chatterjee, I., Chanda, S., Baishnab, K.L., Paul, P.K.: HWPSO: a new hybrid whale-particle swarm optimization algorithm and its application in electronic design optimization problems. Appl. Intell. **49**, 265–291 (2019)
70. Sababha, M., Zohdy, M., Kafafy, M.: The enhanced firefly algorithm based on modified exploitation and exploration mechanism. Electronics **7**(8), 132 (2018). https://doi.org/10.3390/electronics7080132
71. Ge, D., Zhang, Z., Kong, X., Wan, Z.: Extreme learning machine using bat optimization algorithm for estimating state of health of lithium-ion batteries. Appl. Sci. **12**(3), 1398 (2022). https://doi.org/10.3390/app12031398
72. Yang, X.-S.: A new metaheuristic bat-inspired algorithm. In: González, J.R., Pelta, D.A., Cruz, C., Terrazas, G., Krasnogor, N. (Eds.) Nature Inspired Cooperative Strategies for Optimization (NICSO 2010), pp. 65–74. Springer, Berlin/Heidelberg, Germany (2010)
73. Ahmmad, S.N.Z., Muchtar, F.: A review on applications of optimization using bat algorithm. Int. J. Adv. Trends Comput. Sci. Eng. **9**, 212–219 (2020)

74. Yang, X.-S., He, X.: Bat algorithm: literature review and applications. Int. J. Bio-Inspired Comput. **5**, 141–149 (2013)
75. Golilarz, N.A., Gao, H., Addeh, A., Pirasteh, S.: ORCA optimization algorithm: a new meta-heuristic tool for complex optimization problems. In: Proceedings of the 17th International Computer Conference on Wavelet Active Media Technology and Information Processing (ICCWAMTIP), Chengdu, China, 18–20 December 2020; pp. 198–204
76. Drias, H., Bendimerad, L.S., Drias, Y.: A three-phase artificial orcas algorithm for continuous and discrete problems. Int. J. Appl. Metaheuristic Comput. **13**, 1–20 (2022)
77. Cavagna, A., Cimarelli, A., Giardina, I., Parisi, G., Santagati, R., Stefanini, F., Viale, M.: Scale-free correlations in starling flocks. Proc. Natl. Acad. Sci. USA **107**, 11865–11870 (2010)
78. Talbi, E.-G.: Combining metaheuristics with mathematical programming, constraint programming and machine learning. Ann. Oper. Res. **240**, 171–215 (2016)
79. Abdel-Basset, M., Shawky, L.A.: Flower pollination algorithm: a comprehensive review. Artif. Intell. Rev. **52**, 2533–2557 (2019)
80. Yang, X.-S.: Flower pollination algorithm for global optimization. In: Durand-Lose, J., Jonoska, N. (Eds.), Unconventional Computation and Natural Computation. pp. 240–249. Springer, Berlin/Heidelberg, Germany (2012)
81. Yang, X.-S., Karamanoglu, M., He, X.: Flower pollination algorithm: a novel approach for multiobjective optimization. Eng. Optim. **46**, 1222–1237 (2014)
82. Valenzuela, L., Valdez, F., Melin, P.: Flower pollination algorithm with fuzzy approach for solving optimization problems. In: Melin, P., Castillo, O., Kacprzyk, J. (Eds.), Nature-Inspired Design of Hybrid Intelligent Systems, pp. 357–369. Springer International Publishing, Cham, Switzerland (2017)
83. Dasgupta, D., Yu, S., Nino, F.: Recent advances in artificial immune systems: models and applications. Appl. Soft Comput. **11**, 1574–1587 (2011)
84. Bernardino, H.S., Barbosa, H.J.C.: Artificial immune systems for optimization. In: Chiong, R. (Ed.) Nature-Inspired Algorithms for Optimisation, pp. 389–411. Springer, Berlin/Heidelberg, Germany (2009)
85. Tang, C., Todo, Y., Ji, J., Lin, Q., Tang, Z.: Artificial immune system training algorithm for a dendritic neuron model. Knowl. Based Syst. **233**, 107509 (2021)
86. Ray, T., Liew, K.M.: Society and civilization: an optimization algorithm based on the simulation of social behavior. IEEE Trans. Evol. Comput. **7**, 386–396 (2003)
87. Husseinzadeh Kashan, A.: League championship algorithm (LCA): an algorithm for global optimization inspired by sport championships. Appl. Soft Comput. **16**, 171–200 (2014)
88. Laguna, M.: Tabu search. In: Martí, R., Pardalos, P.M., Resende, M.G.C. (Eds.), Handbook of Heuristics, pp. 741–758. Springer International Publishing, Cham, Switzerland (2018)
89. Sadollah, A., Sayyaadi, H., Yadav, A.: A dynamic metaheuristic optimization model inspired by biological nervous systems: neural network algorithm. Appl. Soft Comput. **71**, 747–782 (2018)
90. Mousavirad, S.J., Ebrahimpour-Komleh, H.: Human mental search: a new population-based metaheuristic optimization algorithm. Appl. Intell. **47**, 850–887 (2017)
91. Bozorgi, A., Bozorg-Haddad, O., Chu, X.: Anarchic society optimization (ASO) algorithm. In: Bozorg-Haddad, O. (Ed.), Advanced Optimization by Nature-Inspired Algorithms, pp. 31–38. Springer, Singapore (2018)
92. Abdel-Basset, M., Mohamed, R., Chakrabortty, R.K., Sallam, K., Ryan, M.J.: An efficient teaching-learning-based optimization algorithm for parameters identification of photovoltaic models: Analysis and validations. Energy Convers. Manag. **227**, 113614 (2021)
93. Askari, Q., Younas, I., Saeed, M.: Political optimizer: a novel socio-inspired meta-heuristic for global optimization. Knowl. Based Syst. **195**, 105709 (2020)
94. Ibrahim, A., Anayi, F., Packianather, M., Alomari, O.A.: New hybrid invasive weed optimization and machine learning approach for fault detection. Energies **15**, 1488 (2022)
95. Gupta, D., Sharma, P., Choudhary, K., Gupta, K., Chawla, R., Khanna, A., Albuquerque, V.H.C.D.: Artificial plant optimization algorithm to detect infected leaves using machine learning. Expert Syst. **38**, e12501 (2021)

96. Mohamed, H., Korany, R.M., Mohamed Farhat, O.H., Salah, S.A.O.: Optimal design of vertical silicon nanowires solar cell using hybrid optimization algorithm. J. Photonics Energy **8**, 022502 (2017)
97. Cruz-Duarte, J.M., Amaya, I., Ortiz-Bayliss, J.C., Conant-Pablos, S.E., Terashima-Marín, H., Shi, Y.: Hyper-Heuristics to customise metaheuristics for continuous optimisation. Swarm Evol. Comput **66**, 100935 (2021)
98. Hizarci, H., Demirel, O., Turkay, B.E.: Distribution network reconfiguration using time-varying acceleration coefficient assisted binary particle swarm optimization. Eng. Sci. Technol. Int. J. **35**, 101230 (2022)
99. Muazu, A.A., Hashim, A.S., Sarlan, A.: Review of nature inspired metaheuristic algorithm selection for combinatorial t-way testing, in IEEE Access, **10**, 27404–27431 (2022). https://doi.org/10.1109/ACCESS.2022.3157400
100. Abdel-Basset, M., Shawky, L.A.: Flower pollination algorithm: a comprehensive review. Artif. Intell. Rev. **52**, 2533–2557 (2019). https://doi.org/10.1007/s10462-018-9624-4

Cloud Computing Infrastructure, Platforms, and Software for Scientific Research

Prateek Mathur

Abstract Cloud computing has emerged as a transformative technology for scientific research, offering unprecedented access to scalable infrastructure, platforms, and software resources. This chapter explores the role of cloud computing in supporting scientific research endeavors across various domains. It looks into the infrastructure, platforms, and software offerings available in the cloud, highlighting their impact on data analysis, collaboration, and innovation in the scientific community. Through a comprehensive review of case studies and real-world applications, this chapter demonstrates how cloud computing is revolutionizing the way researchers conduct experiments, analyze data, and share findings. The advantages and challenges of utilizing cloud resources for scientific research are discussed, providing insights into optimizing cost-effectiveness, security, and scalability. Ultimately, this chapter underscores the crucial role of cloud computing in advancing scientific knowledge and accelerating the pace of discovery.

Keywords Google cloud · Amazon Web Services · Virtual machine · IaaS · SaaS · PaaS

1 Introduction to Cloud Computing

1.1 Definition and Overview

Cloud computing is the on-demand delivery of IT resources over the Internet. These resources include tools and applications like data storage, servers, databases, networking, and software, which are hosted at a remote data center managed by a cloud services provider on pay-as-you-go pricing. Cloud Service Provider (CSP) provide cloud services enable users to store files and applications on remote servers

P. Mathur (✉)
AON Consulting Private Ltd, Lotus Business Park, Sector 127, Noida, Uttar Pradesh 201313, India
e-mail: prateek.mathur.2@aon.com

K. A. Ahmad et al. (eds.), *High Performance Computing in Biomimetics*, Series in BioEngineering, https://doi.org/10.1007/978-981-97-1017-1_4

and then access all the data via the Internet. This means the user is not required to be in a specific place to gain access to it, allowing the user to work remotely. One of the many advantages of cloud computing is that you only pay for what you use. This allows organizations to scale faster and more efficiently without the burden of having to buy and maintain their own physical data centers and servers.

Cloud Computing Services

1. **Infrastructure as a Service (IaaS)**

Infrastructure as a service (IaaS) offers on-demand access to IT infrastructure services, including compute, storage, networking, and virtualization. Clients can avoid the need to purchase software or servers, and instead procure these resources in an outsourced, on-demand service.

2. **Software-As-A-Service (SaaS)**

SaaS provides you with a complete product that is run and managed by the service provider. Application software that are hosted in the cloud are access by the users via a web browser, a dedicated desktop client, or an API that integrates with a desktop or mobile operating system. A SaaS solution is often an end-user application, where both the service and the infrastructure is managed and maintained by the cloud service provider.

3. **Platform as a Service (PaaS)**

PaaS provides software developers with on-demand platform—hardware, complete software stack, infrastructure, and even development tools—for running, developing, and managing applications without the cost, complexity, and inflexibility of maintaining that platform on-premises. PaaS is designed to make it easier for developers to quickly create web or mobile apps, without worrying about setting up or managing the underlying infrastructure of servers, storage, network, and databases needed for development. Cloud computing is a paradigm in information technology that has revolutionized the way organizations and individuals access and utilize computing resources. It involves delivering computing services such as servers, storage, databases, networking, software, analytics, and intelligence over the internet, often referred to as "the cloud." Cloud computing offers an innovative model for on-demand access to a shared pool of configurable computing resources that can be rapidly provisioned with minimal management effort [1, 2].

Types of Cloud Computing

Public Cloud: Public cloud is a multi-tenant environment—the cloud provider's data center infrastructure is shared by all public cloud customers. Public clouds provide their services on servers and storage on the Internet. These are operated by third-party companies, who handle and control all the hardware, software, and the general infrastructure. Clients access services through accounts that can be accessed by just about anyone.

Private cloud: A private cloud is typically hosted on-premises in the customer's data center. They are built, managed, and owned by a single organization and privately hosted in their own data centers, commonly known as "on-premises" or "on-prem." Many companies choose private cloud over public cloud because private cloud is an easier way to meet their regulatory compliance requirements.

Hybrid cloud: Hybrid cloud is just what it sounds like—a combination of public and private cloud environments. By allowing data and applications to move between private and public clouds, a hybrid cloud gives your business greater flexibility and more deployment options and helps optimize your existing infrastructure, security, and compliance.

1.2 Evolution of Cloud Computing

Cloud computing evolved primarily from various computing technologies such as distributed systems and peripherals, virtualization, web 2.0, service orientation, and utility computing. The concept of Cloud Computing came into existence in the year 1950 with implementation of mainframe computers, accessible via thin/static clients. The cloud computing has evolved from the concepts of grid, utility and SaaS. The development towards cloud computing started in the late 1980s with the concept of grid computing. Grid computing also named as On Demand Computing centers around moving a workload to the area of the required computing assets, which are for the most part remote and are promptly accessible for utilize. A grid is a group of servers where huge task could be separated into smaller tasks which will be keep running in parallel frameworks. Starting here of view, a grid could really be seen as only one virtual server and oblige applications to fit in with the grid programming interfaces [3].

In the 1990s, the idea of virtualization was extended beyond virtual servers to higher levels of abstraction. An approach also known as payper-use or metered services increasingly common in enterprise computing and is sometimes used for the consumer market for Internet service, file sharing, web site access and other applications. More recently software as an service (SaaS) has raised the level of virtualization to the application, with a plan of action of charging not by the resources devoured but rather by the estimation of the application to supporters [4].

At present, cloud computing has become mainstream, with a vast ecosystem of providers and services catering to a wide range of industries and applications (Fig. 1).

1.3 Characteristics and Benefits of Cloud Computing

There are many characteristics of Cloud Computing here are few of them:

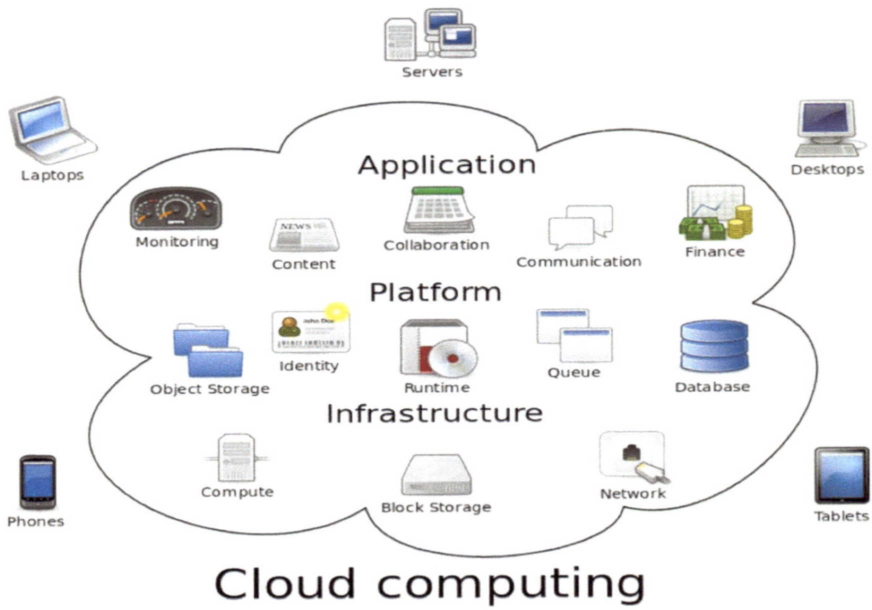

Fig. 1 The environment of cloud computing [1]

Flexibility: Cloud Computing lets users access data or services using internet-enabled devices such as smartphones and laptops. You can instantly access anything you want in the cloud with just a click, making working with data and sharing it simple. Many organizations these days prefer to store their work on cloud systems, as it makes collaboration easy and saves them a lot of costs and resources.

Scalability: A key characteristic and benefit of cloud computing is its rapid scalability. Continuous business expansion demands a rapid expansion of cloud services. One of the most versatile features of Cloud Computing is that it is scalable. Not only does it can expand the number of servers, or infrastructure, according to the demand, but it also offers a great number of features that cater to the needs of its users. Scalability further improves cloud computing's cost-effectiveness and suitability for business use. The usage can be scaled down when the demand is low and can be exponentially increased when the demand is at its peak.

Multi-tenancy: Cloud computing providers can support multiple tenants (users or organizations) on a single set of shared resources. This can be done by implementing a multiple-tenant model, where a cloud service provider can share resources among clients, providing each client with services as per their requirements. In Public clouds, you share the same resources with other organizations and users as well, while in private clouds, the computing resources are used exclusively by one user or organization.

Broad network access: The Computing services are generally provided over standard networks and heterogeneous devices. The client can access the cloud data or transfer the data to the cloud from any place just with a device and internet connection. Cloud providers save that large network access by monitoring and guaranteeing different measurements that reflect how clients access cloud resources and data: latency, access time, data throughput, etc.

On-Demand Self-Service: It is one of the significant and essential features of Cloud Computing. It enables the client to constantly monitor the server uptime, abilities, and allotted network storage. The users can monitor their consumption and can select and use the tools and resources they require right away from the cloud portal itself. This helps users make better decisions and makes them responsible for their consumption.

Flexible pricing models: Cloud providers offer a variety of pricing models, including pay-per-use, subscription-based, and spot pricing, allowing users to choose the option that best suits their needs. This cloud characteristic helps in reducing the IT expenditure of the organizations. There is no covered-up or additional charge which needs to be paid. The administration is economical, and often, some space is allotted for free.

Security: Data security in cloud computing is a major concern among users. Cloud service providers store encrypted data of users and provide additional security features such as user authentication and security against breaches and other threats. CSP invest heavily in security measures to protect their user's data and ensure the privacy of sensitive information.

Measured services: Cloud resources and services such as storage, bandwidth, processing power, networking capabilities, intelligence, software and services, development tools, analytics, etc. used by the consumer are monitored and analyzed by the service providers. It enables both the provider and the client to monitor and report what services have been used and for what purpose. This helps in monitoring billing and ensuring the optimum usage of resources.

Resilience: Cloud computing services are typically designed with redundancy and fault tolerance in mind, which ensures high availability and reliability. A Cloud service provider must be prepared against any disasters or unexpected circumstances since a lot is at stake. Disaster management earlier used to pose problems for service providers but now due to a lot of investments and advancements in this field, clouds have become a lot more resilient [5].

1.4 Role of Cloud Computing in Research

Researchers who want to quickly share data from many locations or who want access to open datasets may find the cloud to be a good fit. Data that is already discoverable, accessible, interoperable, and reusable can benefit from extra collaboration options thanks to cloud processing and sharing. Researchers that need scalability for changing data volumes can benefit from the flexibility of cloud computing resources. The ability to obtain services for fields like artificial intelligence and machine learning on demand makes the cloud alluring for individuals who require intricate data visualization or speedy analytics results. Cloud computing has become an indispensable tool for scientific research [6, 7]:

Data Storage and Analysis: Researchers can store and analyze massive datasets using cloud-based tools and services, such as AWS S3 and Amazon EC2.

Collaboration: Cloud platforms enable global collaboration by providing a centralized location for data and applications accessible to research teams worldwide.

High-Performance Computing (HPC): Cloud providers offer HPC resources for simulations and complex calculations, eliminating the need for expensive on-premises clusters.

Machine Learning and AI: Cloud services, like AWS SageMaker and Google AI Platform, provide access to powerful machine learning and AI tools.

Big Data Processing: Researchers can leverage cloud platforms like Google BigQuery and Azure HDInsight for big data processing.

Scientific Modeling: Cloud resources support scientific modeling and simulations in various fields, from climate modeling to drug discovery.

Data Security and Compliance: Cloud providers offer robust security measures and compliance certifications to protect sensitive research data.

2 Amazon Web Services (AWS)

2.1 Overview of AWS Services and Offerings

Amazon Web Services (AWS) is one of the world's leading cloud service providers, offering a vast array of services across categories like computing, storage, databases, machine learning, analytics, IoT, and more. Some notable AWS services include Amazon EC2 (Elastic Compute Cloud), Amazon S3 (Simple Storage Service), Amazon RDS (Relational Database Service), AWS Lambda, and Amazon SageMaker. Launched in 2006, Amazon Web functions (AWS) started offering crucial infrastructure functions to companies as web services, which is now commonly referred to as cloud computing. The capacity to leverage a new business model and convert capital infrastructure costs into variable costs is the main advantage of cloud computing and AWS. Servers and other IT resources no longer need to be planned and purchased by businesses weeks or months in advance. Businesses can use AWS to access resources when they're needed, providing outcomes more quickly and at a lesser cost by utilizing Amazon's knowledge and economies of scale [8].

Today, Amazon Web Services offers a highly dependable, scalable, affordable cloud infrastructure platform that supports millions of enterprises in 190 different countries.

1. Amazon EC2 (Elastic Compute Cloud)

A safe, scalable cloud platform from Amazon is called EC2. Its goal is to provide developers with simple access and usability for web-scale cloud computing while granting complete control over your compute resources. Quickly deploy apps without having to make an upfront hardware investment, and launch virtual servers at scale as needed.

2. Amazon RDS (Relational Database Services)

With the help of Amazon Relational Database Service (Amazon RDS), setting up, managing, and scaling databases in the cloud is simple. Automate time-consuming operations like hardware provisioning, database configuration, patching, and backups in a way that is cost-effective and appropriate for your requirements. RDS supports six well-known database engines, including Amazon Aurora, PostgreSQL, MySQL, MariaDB, Oracle database, and SQL server, on a variety of database instances that are designed for performance and memory. You may quickly replicate or migrate your current databases to Amazon RDS by using the AWS Database Migration Service. Visit the RDS page on Amazon.

3. Amazon S3 (Simple Storage Service)

At its heart, Amazon S3 enables object storage and offers industry-leading scalability, data accessibility, security, and performance. Large amounts of data may be stored and protected using S3 by businesses of all sizes for a variety of use cases, including websites, applications, backup, and more. The simple administration capabilities of Amazon S3 allow for flexible access limits and effortless data organization.

4. Amazon Lambda

Lambda allows you to run code without the hassle of server ownership or management. You only pay for the time your code is running. With Lambda, you can run code for any application or backend utility without the need for administration. Simply upload your code and let Lambda take care of the rest. This provides efficient software scaling and high availability.

5. Amazon Cognito

AWS Cognito regulates a control get to dashboard for on-boarding clients through sign-up, and sign-in highlights to their web and portable apps. AWS Cognito scales to millions of clients and offers sign-in back with social character suppliers counting Facebook, Google, and Amazon, together with venture character suppliers by means of SAML 2.0.

6. Amazon VPC (Virtual Private Cloud)

Amazon VPC empowers you to set up a sensibly confined area of the AWS Cloud where you'll be able send AWS assets at scale in a virtual environment. VPC gives you add up to control over your environment, which incorporates the choice to select your possess IP address run, creation of subsets, and course of action of course tables and arrange get to focus. Effectively customize the arrange arrangement of your VPC with adaptable dashboard administration controls planned for greatest convenience. For illustration, clients can dispatch public-facing subnet for web servers with web get to.

7. Amazon Kinesis

Pick up convenient bits of knowledge by leveraging Amazon Kinesis to gather, prepare, and analyze information in real-time, making a difference you respond rapidly. Key highlights interior AWS Kinesis are cost-efficient preparing of spilling information at scale, and the choice to select instruments best fit for your application. Ingest real-time information, counting video, sound, application records, site movement, and IoT telemetry information for machine learning and other apps. With Kinesis, clients can track, analyze, and handle information in real-time, empowering moment reaction capabilities.

8. Amazon Auto-scaling

The AWS Auto-scaling arrangement screens your apps and naturally tunes capacity to maintain unfaltering, unsurprising execution at the most reduced conceivable cost. Consistently design application scaling capacities for different assets over numerous administrations nearly immediately. Auto-scaling includes a feature-rich and direct client interface that empowers you to construct scaling plans for different resources. These assets incorporate Amazon EC2 occasions and Spot Armadas, EC2 assignments, Dynamo DB tables and files, and Amazon Aurora Copies.

9. Amazon IAM (Identity and Access Management)

AWS Identity and Access Management provides secure access and management of resources in a secure and compliant manner. By leveraging IAM, you can create and manage users and groups by allowing and denying their permissions for individual resources.

There are no additional costs, people only get charged for the use of other services by their users.

10. Dynamo DB

DynamoDB is a document database with key-value structuring that delivers single-digit millisecond performance at scale. Dynamo has built-in security with a fully managed, multi-master, multi-region, durable database, backup and restore, and in-memory archiving for web-scale applications. DynamoDB can manage upward of 10 trillion requests daily and can support thresholds of more than 20 million requests per second.

11. Amazon SQS (Simple Queue Service)

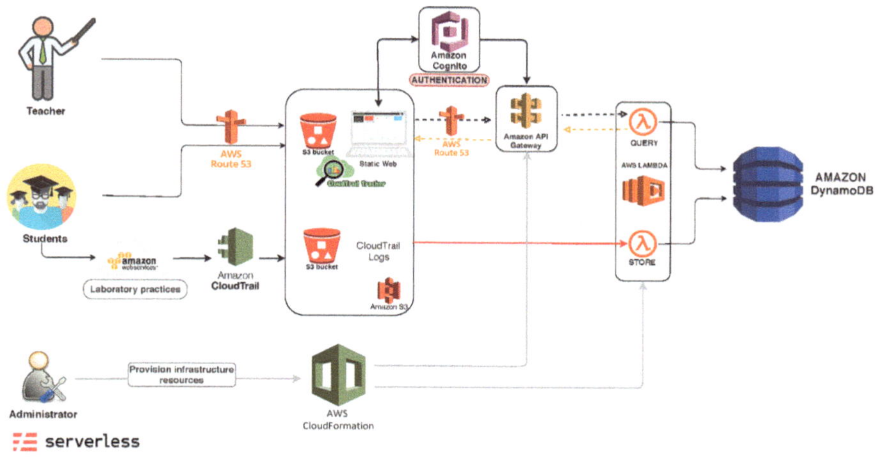

Fig. 2 Distributed platform related educational analytics for cloud training in Amazon Web Services (AWS) [8]

AWS SQS is a fully managed message queuing facility enabling you to decouple and scale microservices, distributed systems, and serverless apps. SQS purges the intricacies and overhead associated with managing and operating message-oriented middleware and permits developers to focus on diverse workloads.

12. Amazon ElastiCache

ElastiCache is an AWS service that effortlessly sets up, runs, and scales popular open-source, in-memory data storages in the cloud. Operate data-intensive apps or enhance the performance of existing databases by evaluating data from high throughput and low latency in-memory data stores.

13. AWS Athena

Amazon Athena facilitates straightforward analysis of data stored in Amazon S3 by utilizing the standard Structured Query Language (SQL), Since Athena is serverless, there is no infrastructure to maintain, and you only pay for the queries you execute (Fig. 2).

2.2 AWS Infrastructure and Data Centers

AWS operates in regions worldwide, with each region having multiple Availability Zones (data centers). AWS's global infrastructure is designed to provide high availability and low-latency access to services. AWS also invests heavily in data center sustainability and efficiency. AWS has the concept of a Region, which are physical locations around the world where we cluster data centers. We call each group

Fig. 3 Google cloud and some applications (*Source* Google)

of logical data centers an Availability Zone (AZ). Each AWS Region consists of multiple, isolated, and physically separate AZs within a geographic area. Each AZ consists of one or more physical data centers, and we design each AZ to be completely isolated from the other AZs in terms of location, power, and water supply.

Unlike other cloud providers, who often define a region as a single data center, the multiple AZ design of every AWS Region offers additional advantages for our customers, such as reliability, scalability, and the lowest possible latency. For example:

AZs allow for partitioning applications for high availability. If an application is partitioned across AZs, companies are better isolated and protected from issues such as power outages, lightning strikes, tornadoes, earthquakes, and more.

AZs in an AWS Region are interconnected with high-bandwidth, low-latency networking, over fully redundant, dedicated metro fiber, providing high-throughput, low-latency networking between AZs.

Traffic between AZs is encrypted. The network performance is sufficient to accomplish synchronous replication between AZs.

Additionally, AWS provides data center security in four layers or levels.

Perimeter layer

Perimeter security measures provide access control of the physical equipment by using:

- Security guards
- Fencing
- Security feeds
- Intrusion detection technology
- Entry control and monitoring

Infrastructure layer

Infrastructure layer security protects the equipment from damage and overheating. It includes measures such as:

- World-class cooling systems and fire suppression equipment
- Backup power equipment
- Routine machine maintenance and diagnostics
- Water, power, telecommunications, and internet connectivity backups

Data layer

Data layer security protects the data itself from unauthorized access and loss. Typical measures in this layer are:

- Threat and electronic intrusion detection systems in the data center
- Electronic control devices at server room access points
- External auditing of more than 2,600 requirements throughout the year

Environmental layer

The environmental layer is dedicated to environmental control measures that support sustainability. These are some of its measures:

- Sensors and responsive equipment that automatically detect flooding, fire, and other natural disasters
- An operations process guide outlining how to avoid and lessen disruptions due to natural disasters

100% renewable energy and environmental economies of scale. AWS has been extensively used in research across various domains. Researchers can leverage AWS for genomics research, climate modeling, high-performance computing (HPC), and machine learning. Notable research projects include the 1000 Genomes Project and the Square Kilometre Array radio telescope project.

2.3 *Google Cloud Platform (GCP)*

Google cloud platform is a medium with the help of which people can easily access the cloud systems and other computing services which are developed by Google. The platform includes a wide range of services that can be used in different sectors of cloud computing, such as storage and application development. Anyone can access the Google cloud platform and use it according to their needs.

The Google Cloud Platform, first established on October 6, 2011, has a market share of 13%, giving stiff competition to Amazon AWS cloud. It has turned out to be one of the best and most successful cloud computing services.

GCP has also added a number of cloud functionality and features including cloud storage, data analytics, developer options, and advanced machine learning, in addition to the various management tools that are available on Google Cloud Platform. The Google Cloud Platform's numerous optimization options and other benefits are what have made it so well-liked.

2.3.1 Overview of GCP Services and Offerings

Google Cloud Platform (GCP) provides a comprehensive suite of cloud services that encompass computing, storage, databases, machine learning, data analytics, and more. Key GCP services include Google Compute Engine, Google Cloud Storage, BigQuery, TensorFlow, and Google AI Platform. Google Compute Engine: This computing platform was offered together with the IaaS service by Google, effectively offering virtual machines (VMs) that are comparable to Amazon EC2.

Google Cloud App Engine: This app engine offers a PaaS service for directly hosting the right applications. This is a really potent and significant platform that aids in the creation of various mobile and online applications.

The Google Cloud Container Engine is a useful component since it enables users to run the Docker containers that are already existing on the Google Cloud Platform and are in fact started by Kubernetes.

Google Cloud Storage: It's crucial to have the ability to store data and other crucial resources on the cloud platform. The Google cloud platform is well-liked by storage facilities and enables customers to back up or store data on the cloud servers which can be accessed from anywhere at any time.

Google BigQuery Service: The Google BigQuery Service is a powerful data analysis tool that gives users the ability to examine big data in their organizations. Additionally, it boasts a sophisticated storage system with terabytes of storage capacity.

Google Cloud Dataflow: With this tool, customers may control reliable parallel data-processing pipelines. It aids in managing the pipelines' processing lifecycles on Google Compute servers.

Google Cloud Job Discovery: The Google Cloud Platform is a fantastic resource for career alternatives, job searches, etc. It is possible to learn about various methods of locating jobs and business prospects thanks to the sophisticated search engine and machine learning skills.

Google Cloud Job Discovery: The Google Cloud Platform is a fantastic resource for career alternatives, job searches, etc. It is possible to learn about various methods of locating jobs and business prospects thanks to the sophisticated search engine and machine learning skills.

Google Cloud Test Lab: This service from Google enables users to test their apps using real-world and virtual cloud-based hardware. Users can learn more about their applications through the numerous instrumentation testing and robotic tests.

The Google Cloud Endpoints functionality enables users to create and manage secure application program interfaces that are operating on the Google Cloud Platform.

Google Cloud Machine Learning Engine: As its name implies, this component of Google Cloud assists in developing models and structures which enables the users to concentrate on Machine learning abilities and framework.

2.3.2 GCP Infrastructure and Data Centers

GCP has a global presence with data centers in regions and zones. Google has been at the forefront of energy-efficient data center design, and GCP's infrastructure is known for its reliability. Google Cloud Platform (GCP), the cloud computing service of Alphabet Inc, provides compute, storage, and networking services through its data centers in over 20 countries and 35 locations around the world.

By the end of 2024, Google Cloud will have a total of 44 regions available, with 35 now operational and 9 more in the development phase. Three to four deployment areas, referred to by Google as zones and by other cloud service providers as availability zones, correspond to groups of data centers with different physical infrastructure (such as electricity, cooling, and networking) inside each Google Cloud region.

Presently, Google Cloud has 106 zones in operation and a further 27 under development, meaning that the company will have a total of 133 zones existing by the end of 2024. Customers of Google Cloud can distribute resources and workloads over several zones to assist safeguard against unforeseen failures, which is not always possible through a single, private data center.

GCP has played a significant role in research, particularly in data analytics and machine learning. It is used for genomics research, earthquake prediction, and climate modeling. GCP has supported projects such as the Broad Institute's data analysis for genomics and the LIGO project, which detected gravitational waves. One of the example of applications of Google cloud platform is the mapping of flood affected areas using Landsat satellite with Google Earth Engine Cloud Platform as shown in Fig. 4 [9]

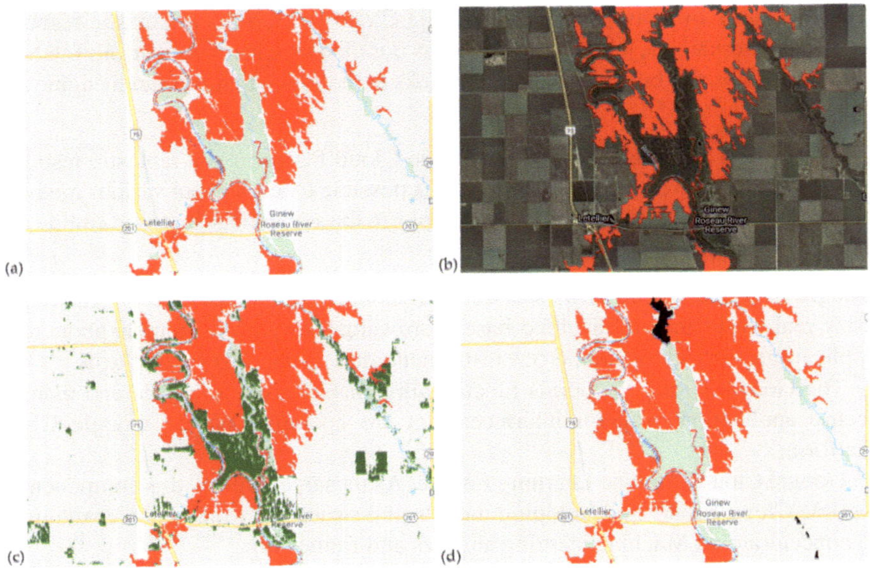

Fig. 4 NDVI and HAND mask for 2019 Red River flood [9]

2.4 *Microsoft Azure*

Microsoft Azure, often referred to as Azure, is a Microsoft-operated cloud computing platform. Through international data centers, it provides access, management, and the creation of applications and services. Azure offers a large collection of services, which includes platform as a service (PaaS), infrastructure as a service (IaaS), and managed database service capabilities.

Azure debuted on February 1, 2010, much later than its primary rival, AWS. It is free to start using and uses a pay-per-use business model, so you only pay for the services you use. It's interesting to note that Azure services are used by 80% of Fortune 500 organizations for their cloud computing requirements. Java, Node Js, and C# are just a few of the programming languages supported by Azure.

2.4.1 Overview of Azure Services and Offerings

Microsoft Azure provides a comprehensive set of cloud services, including virtual machines, databases, AI and machine learning tools, IoT, and more. Notable Azure services include Azure Virtual Machines, Azure Blob Storage, Azure SQL Database, Azure Machine Learning, and Azure IoT Hub. Some of the key services offered by Azure are:

Azure Active Directory One of the most popular cloud computing services from Microsoft Azure. Belonging to the Identity section, it is a universal identity platform to ensure the management and security of identities.

Azure Active Directory provides an enterprise identity service that offers single sign-on and multi-factor authentication to safeguard against cybersecurity threats. Identity-based security ensures comprehensive protection for users against cyberattacks. Authenticated login enables access to software from any location worldwide. The creation of a single identity platform by Azure Active Directory facilitates secure engagement with both internal and external users.

Azure CDN This particular service has been specifically designed to operate in conjunction with a variety of other services, such as web applications, Azure cloud services, and storage, all while ensuring secure delivery worldwide. It offers the added advantage of advanced analytics, which can assist in gaining valuable insights into customer workflows and business needs.

Azure Data Factory Azure Data Factory is responsible for the ingestion of data from multiple sources in order to automate the transmission and movement of data. In order to carry out its tasks, Azure Data Factory makes use of various Azure services for computing, such as Azure Machine Learning, Azure HDInsight Hadoop, and Azure Data Lake Analytics. Azure Data Lake can be perceived as an extensive data repository that stores data in its original format, specifically designed for the purpose of conducting Big data analytics.

Azure Cosmos DB Azure Cosmos DB provides a globally distributed and meticulously managed NoSQL database service. This distribution encompasses seamless multi-master replication, meticulously designed to deliver response times in the single-digit millisecond range. This ensures exceptional speed regardless of the scale of operations. Being fully managed, it relieves users from the burden of database administration through automated management, updates, and patching. Additionally, Cosmos DB is a multi-model database, offering wire protocol-compatible API endpoints.

DevOps When commencing the utilization of Microsoft Azure services, the adoption of the software as a service (SaaS) platform of DevOps becomes imperative for software development and deployment. This platform provides seamless integration capabilities with renowned industry tools and facilitates the orchestration of a DevOps toolchain. DevOps services effectively demonstrate the agility of these tools through the tracking, planning, and collaborative discussion of work among various teams. DevOps is highly advantageous for the majority of users, regardless of the platform, programming language, or cloud utilized for their applications.

Azure Backup Human error is an unfortunate and undeniable fact, and Azure Backup offers straightforward data protection solutions for Azure Web app services, ensuring the safeguarding of your data against ransomware attacks or any form of loss. The cost of implementing backups is highly affordable, and it can be utilized for securing SQL workloads and data from virtual machines as well.

Logic Apps This technology facilitates the expeditious development of robust integration solutions. It is capable of linking data, applications, and devices across diverse locations. Additionally, the Logic Apps' business-to-business (B2B) functionalities streamline collaboration with commercial associates by adhering to Electronic Data Interchange (EDI) and Enterprise Application Integration (EAI) standards.

Virtual Machine A virtual machine is commonly referred to as an image, which is a file that emulates the behaviour of a physical computer. Microsoft Azure provides the option of incorporating virtual machines in its Compute category, thereby enabling the creation of Windows or Linux systems within a matter of seconds on a tangible computer. The virtual machine remains distinct from the remainder of the computer, thereby furnishing an ideal environment for beta application testing, accessing virus-infected data, generating system backups, and executing applications that were not originally intended to operate on the designated operating system.

2.4.2 Azure Infrastructure and Data Centers

Azure operates in regions and Availability Zones globally, ensuring redundancy and availability. Microsoft is committed to sustainability and has been investing in renewable energy for its data centers. The Azure global infrastructure comprises two fundamental elements, namely the physical infrastructure and connective network components. The physical component encompasses over 200 physical datacenters, organized into regions, and interconnected by one of the most extensive networks on the planet.

The global Azure network's connectivity ensures that each datacenter offers high availability, low latency, scalability, and cutting-edge cloud infrastructure, all operating on the Azure platform. These components work in tandem to ensure that data remains entirely within the secure Microsoft network, with IP traffic never traversing the public internet.

Azure datacenters are unique physical buildings—located all over the globe—that house a group of networked computer servers. Azure has been instrumental in various research areas. It has been used for genomics research, climate modeling, and IoT applications. Projects like the Cancer Genome Atlas and the Square Kilometre Array rely on Azure for data storage and processing.

2.5 Other Cloud Computing Providers and Their Research Capabilities

There are numerous other cloud computing providers like IBM Cloud, Oracle Cloud, and Alibaba Cloud, each offering a range of services and infrastructure. These providers also have their own research capabilities and use cases. To learn more about their offerings and research involvement, you can visit their respective websites and explore their case studies and research partnerships.

Azure region An Azure region comprises a collection of datacenters that are strategically positioned within a defined perimeter to ensure optimal latency. These datacenters are interconnected via a specialized regional low-latency network. Azure, as a cloud provider, offers a greater number of global regions than any other provider, thereby affording customers the freedom to deploy their applications in locations that best suit their needs. It is worth noting that each Azure region has its own distinct pricing and service availability.

Azure Availability Zones The Azure region comprises of distinct physical locations that are exclusive and provide exceptional availability to safeguard applications and data from datacenter outages. Each zone is composed of one or more datacenters that are equipped with autonomous power, cooling, and networking facilities. The segregation of availability zones within a region ensures the protection of applications and data from facility-level challenges. Zone-redundant services replicate applications and data across Azure Availability Zones to mitigate the risk of single points of failure.

Azure Edge Zones Azure Edge Zones are geographical expansions of Azure, strategically located in densely populated regions. They facilitate the operation of virtual machines (VMs), containers, and a limited range of Azure services, enabling the execution of latency-sensitive and throughput-intensive applications in close proximity to end-users.

As an integral component of the Microsoft global network, Azure Edge Zones provide dependable, secure, and high-bandwidth connectivity between applications hosted at the Azure Edge Zone (in close proximity to the user) and the complete suite of Azure services operating across the broader Azure regions.

Azure Space Azure Space is a strategic endeavor aimed at expanding the functionality of Azure capabilities by leveraging space infrastructure. Its primary objective is to serve as the preferred platform and ecosystem for fulfilling the mission requirements of the space community. Azure Space facilitates enhanced accessibility to connectivity and computational resources across various sectors, such as agriculture, energy, telecommunications, and government.

Azure Orbital Azure Orbital is an integral component of the Azure Space strategic initiative. It offers a comprehensive ground station service that is fully managed, allowing customers to effectively communicate, download, and analyze data from their satellites or spacecrafts. This service operates on a flexible pay-as-you-go model, eliminating the necessity for customers to construct their own satellite ground stations.

Azure point of presence An Azure point of presence, often abbreviated as PoP, is an access point or physical location where traffic can enter or exit the Microsoft global network.

Regional network gateways Regional network gateways are highly parallel and hyperscale datacenter interconnects that facilitate seamless communication between datacenters within a specific region, eliminating the necessity of establishing individual network connections between each datacenter in the same region.

3 Cloud Computing Platforms for Research

3.1 Virtual Machines (VMs) and Infrastructure as a Service (IaaS)

Virtual Machines (VMs) enable the creation and utilization of virtual machines within the cloud environment. They offer Infrastructure as a Service (IaaS) in the form of a virtualized server and can be employed in various capacities. Similar to a physical computer, the software running on the VM can be fully customized. VMs are an optimal choice when one requires:

- Complete control over the operating system (OS)
- The capability to execute customized software
- The utilization of personalized hosting configurations

VM provides the flexibility of virtualization without the necessity of purchasing and managing the physical hardware that supports the VM. However, the VM still requires maintenance, including configuration, updates, and software upkeep.

IaaS is a cloud computing service model that makes on-demand compute, storage, and networking functionality available via an internet connection, on a pay-as-you-go basis.

In the IaaS model, a cloud service provider (CSP) manages large data centers, typically located around the world, with physical machines and virtualized resources that can make servers, virtual machines (VMs), storage, and networking services available to customers over the web. Customers rent access to these cloud infrastructure resources on a pay-as-you-go basis, using as much or as few services as they need at any given moment. The CSP is responsible for managing and maintaining the cloud infrastructure, minimizing the burden on in-house IT teams. With IaaS, businesses can also avoid the cost of building, maintaining, securing, and providing heating and cooling for data centers that would normally host these computing resources on site.

What types of infrastructure are available in IaaS offerings?

IaaS offerings fall into three categories:

- **Compute resources**. With IaaS offerings, businesses can access the essential hardware that every computer requires for processing: central processing units (CPUs), graphical processing units (GPUs), and random access memory (RAM).
- **Data storage**. IaaS providers offer access to block storage, file storage, and object storage technologies.

- **Networking**. These resources include virtualized routers, switches, and load balancers.

What are the benefits of IaaS?

- **Reduce capital expense**. With IaaS, there is no upfront cost required to purchase and install equipment in a physical data center.
- **Manage and optimize costs**. The pay-per-service pricing offered by IaaS providers allows businesses to pay only for the infrastructure services they need, reducing operational costs and optimizing IT budgets.
- **Scale easily**. IaaS solutions give businesses heightened scalability, allow them to add or minimize resources quickly to meet business needs, and accelerate speed to market.
- **Increase reliability**. An IaaS platform eliminates the single point of failure—if one component within a cloud environment fails, the redundant nature of the cloud means that IaaS services will still be available.
- **Improve security**. Most IaaS vendors are able to offer stronger and more advanced security for computing infrastructure than businesses can achieve in-house.
- **Gain agility**. The IaaS model allows businesses to provision the resources they need within minutes or hours, rather than days or weeks, increasing their ability to respond quickly to market conditions and business opportunities.

What are the challenges of IaaS?

For all its benefits, IaaS offers a variety of challenges that may prevent businesses from adopting this cloud technology.

- **Security risks**. Relying on a third-party provider to manage infrastructure and the data associated with it represents a certain loss of control, requiring IT teams to trust the security controls of the cloud service provider.
- **Lack of customization**. IaaS solutions may be less customizable than in-house technologies.
- **Vendor lock-in**. As businesses become reliant on IaaS providers, changing vendors can be costly and time-consuming, leading to a certain amount of vendor lock-in.
- **Connectivity issues**. As with any cloud computing solution, poor connectivity or internet outages can impact the performance of processes dependent on IaaS infrastructure.
- **Lack of transparency**. Because IT teams do not have access to infrastructure in IaaS solutions, gaining visibility into performance and security can be more difficult, making systems management more complex.
- **Competition for resources**. Because virtualization enables IaaS providers to provide infrastructure for multiple customers from the same physical server, bandwidth for one customer may be impacted by compute-intensive activity from another customer.

3.2 *Containers and Container Orchestration (E.G., Kubernetes)*

Containers are software units that are capable of executing applications. They package the application code, along with its libraries and dependencies, in a standardized manner, enabling the code to be executed on various platforms such as desktops, traditional IT systems, or the cloud.

To achieve this, containers utilize a type of operating system (OS) virtualization. This virtualization technique makes use of specific features of the OS kernel, such as Linux namespaces and cgroups, or Windows silos and job objects. These features allow for the isolation of processes and enable control over the resources, such as CPU, memory, and disk access, that these processes can utilize.

One of the key advantages of containers is their small size, high speed, and portability. Unlike virtual machines, containers do not require a separate guest OS for each instance. Instead, they can leverage the existing features and resources of the host OS, resulting in more efficient and lightweight execution.

While containers have been around for several decades, with earlier versions like FreeBSD Jails and AIX Workload Partitions, the modern container era is commonly associated with the introduction of Docker in 2013.

Benefits of containers

The primary advantage of containers, particularly when compared to virtual machines (VMs), lies in their ability to provide a level of abstraction that renders them lightweight and portable. Their main benefits encompass the following:

1. Lightweight: Containers share the operating system (OS) kernel of the host machine, eliminating the necessity for a complete OS instance per application. Consequently, container files are small in size and consume fewer resources. Their reduced dimensions, especially when contrasted with VMs, enable containers to be rapidly deployed and better support cloud-native applications that scale horizontally.
2. Portable and platform-independent: Containers encapsulate all their dependencies, allowing software to be written once and executed without the need for reconfiguration across various computing environments, including laptops, cloud platforms, and on-premises systems.
3. Supports modern development and architecture: Containers, owing to their deployment portability and consistency across platforms, as well as their compact size, are an ideal fit for contemporary development and application patterns. These patterns include DevOps, serverless computing, and microservices, which are built using regular code deployments in small increments.
4. Improves utilization: Similar to VMs, containers empower developers and operators to enhance the utilization of CPU and memory resources on physical machines. However, containers go a step further by enabling the deployment and scaling of application components at a more granular level. This capability is

particularly attractive as it offers an alternative to scaling up an entire monolithic application when a single component is struggling with its workload.

What is Container Orchestration?

Container orchestration is an automated process that provisions, deploys, scales, and manages containerized applications without the need to consider the underlying infrastructure. This allows developers to automate the life cycle management of containers wherever they are implemented.

How Does Container Orchestration Work?

Container orchestration tools, such as Google Kubernetes Engine (GKE), simplify the deployment and operation of containerized applications and microservices. While container orchestrators may differ in their methodologies and capabilities, they all enable organizations to automatically coordinate, manage, and monitor containerized applications.

Declarative programming is used in container orchestration, where the desired output is defined instead of the steps required to achieve it. Developers create a configuration file that specifies the location of container images, network establishment and security between containers, and container storage and resources. Container orchestration tools use this file to automatically achieve the requested end state.

When a new container is deployed, the tool or platform schedules the containers and selects the most suitable host based on predetermined constraints or requirements defined in the configuration file, such as CPU, memory, proximity to other hosts, or metadata.

Container orchestration tools automate life cycle management and operational tasks based on the container definition file, including provisioning and deployment, scaling containers up or down, load balancing, resource allocation between containers, moving containers to another host to ensure availability in case of resource shortage or unexpected outage, performance and health monitoring of the application, and service discovery.

3.3 Platform as a Service (PaaS) and Serverless Computing

Platform as a Service (PaaS) pertains to the services rendered by a cloud platform, wherein computing and software resources are provided with minimal or no obligations for infrastructure management. PaaS represents the organic progression from Infrastructure as a Service (IaaS).

In a PaaS offering, the cloud service provider manages the OS, underlying servers, network infrastructure, and most software configurations, leaving users free to develop and deploy applications rapidly.

Let us consider a straightforward deployment of a web application. In this scenario, we acquire a virtual server through an Infrastructure as a Service (IaaS) provider such as AWS EC2. However, before we can proceed with deploying the application, there are several preliminary steps that need to be taken:

1. Setting up the web server.
2. Installing the necessary dependencies and other required software.
3. Configuring the network and storage.

Alternatively, with a Platform as a Service (PaaS) offering like AWS Elastic Beanstalk, AWS takes care of various aspects such as EC2 instances, operating system, web server installation, and other resource configurations. Users can simply create a beanstalk environment with the appropriate configurations, and then deploy the application.

In addition to the fundamental server, storage, and networking services, PaaS also provides middleware, Business Intelligence (BI) services, database systems, and development tools. This eliminates the need for managing software licenses, as they are handled by the cloud provider.

All of these services are offered under a pay-as-you-go model, which helps to reduce expenses.

Advantages of PaaS include:

1. Simplified and cost-effective management of resources.
2. Ability to easily create scalable and highly available environments.
3. Reduction in infrastructure management and monitoring requirements, resulting in time and cost savings.
4. Support for automation, which reduces workload in the software development lifecycle.
5. Increased security.
6. More flexible development and deployment pipeline.

Some popular PaaS offerings include:

1. AWS Elastic Beanstalk.
2. Azure App Service.
3. Azure Cognitive Search.
4. Google App Engine.
5. RedHat OpenShift.
6. IBM Cloud Pak for Applications.
7. BitNami by VMWare.

Serverless overview

Serverless offerings aim to eliminate all management and configuration requirements from the software development and delivery process. By utilizing a serverless solution, users have the ability to write individual functions or services and deploy them directly on a serverless platform, without the need for any infrastructure or software configuration. All aspects of managing functions or services, such as scaling, availability needs, and managing bandwidth, are handled by the cloud provider. Serverless operates on a usage-based monetization model, where users only pay for the usage of the service. The serverless approach allows developers to focus more on creating functions, resulting in a simplified development and deployment experience. Serverless offerings are particularly well-suited for microservices-based architectures or event-driven architecture.

For instance, consider a web function that is used to monitor a data stream. In traditional development, the following steps are required:

1. Provision and configure the necessary resources.
2. Set up the webserver.
3. Deploy the function, which requires ongoing maintenance and cost, even when the function is not in use.

However, with a serverless solution like AWS Lambda, users only need to:

1. Select the programming language.
2. Create the function and specify the frequency of function execution or the triggering event.
3. The function will then be automatically executed based on the configured parameters. The cloud provider fully manages the function, and users are only billed for the number of executions.

There are two main serverless offerings: Function as a Service (FaaS) and Backend as a Service (BaaS). Let's briefly examine each one.

Function as a Service (FaaS) is the most widely used serverless solution. It allows users to execute code in stateless containers based on predefined events. This provides developers with a platform to easily develop and deploy code as functions, without the need to worry about infrastructure configurations.

Backend as a Service (BaaS): The BaaS model lets developers manage only the frontend and instead outsources most of the backend functionality to third-party services. BaaS provides services such as:

- Authentication
- Database management
- Hosting
- Notifications

Advantages of Serverless

- Eliminates any infrastructure configuration or management requirements
- Faster and simplified development and deployments due to the focus on individual functions or services
- Cost savings due to the usage-based payment model
- Highly scalable
- Reduced cloud provider lock-in

3.4 Big Data and Analytics Platforms in the Cloud

The term "Big Data" refers to data that is vast in size and presents challenges in terms of storage, management, and analysis using traditional databases. To efficiently handle this large volume of data, a scalable architecture is required. This data originates from various sources, including smartphones, social media posts, sensors (such as traffic signals and utility meters), point-of-sale terminals, and consumer wearables like fitness trackers and electronic health records. To extract valuable insights, improve decision-making, and gain a competitive advantage, a combination of technologies is integrated to uncover hidden values within this complex and diverse data.

The characteristics of big data can be summarized as follows:

1. Volume: This refers to the enormous amount of data generated every second from sources such as social media, cell phones, cars, credit cards, M2M sensors, photographs, and videos. The ability to mine this data allows users to discover hidden information and patterns.
2. Velocity: This refers to the speed at which data is generated, transferred, collected, and analyzed. As data is generated at an ever-accelerating pace, it must be analyzed in real-time to enable instant access to applications that rely on this data.
3. Variety: This refers to the different formats in which data is generated, including structured and unstructured formats. Structured data, such as names, phone numbers, addresses, and financials, can be easily organized within a database. On the other hand, unstructured data, which accounts for 80% of today's data, is more challenging to sort and extract value from. Examples of unstructured data include text messages, audio, blogs, photos, video sequences, social media updates, log files, machine data, and sensor data.
4. Variability—Refers to the significant inconsistency in the flow of data and its fluctuations during peak periods. This variability arises from the multitude of dimensions in the data, resulting from various types and sources of disparate data. Variability can also pertain to the inconsistent speed at which large amounts of data are incorporated into the data repositories.
5. Value—Refers to the concealed value that is discovered through data analysis for the purpose of decision making. Substantial value can be derived from big data, including gaining a better understanding of customers, targeting them effectively, optimizing processes, and enhancing machine or business performance.
6. Veracity—Refers to the quality and reliability of the data source. Its significance lies in the context and the meaning it contributes to the analysis. Knowledge of the veracity of the data aids in comprehending the risks associated with analysis and making business decisions based on the dataset.
7. Validity—Refers to the accuracy of the data collected for its intended use. It is essential to adopt proper data governance practices to ensure consistent data quality, standardized definitions, and metadata.

8. Vulnerability—Refers to the security aspects of the data that is collected and stored.
9. Volatility—Refers to the duration for which data remains valid and needs to be stored historically before it becomes irrelevant for current analysis.
10. Visualization—Refers to the process of presenting data in a manner that is understandable to non-technical stakeholders and decision makers. Visualization involves creating intricate graphs that transform data into information, information into insights, insights into knowledge, and knowledge into an advantage for decision making.

Classification of Big Data

1. Analysis Type—The classification of data analysis based on whether it is conducted in real-time or through a batch process. Real-time analysis is commonly used by banks for fraud detection, while batch processing is often utilized for making strategic business decisions.
2. Processing Methodology—The methodology chosen for data processing, such as predictive analysis, ad-hoc analysis, or reporting analysis, is determined by the specific requirements of the business.
3. Data Frequency—The frequency at which data is ingested and the rate at which it arrives. Data can be continuous, as seen in real-time feeds, or based on time series.
4. Data Type—Data can be classified as historical, transactional, or real-time, such as streams.
5. Data Format—Structured data, such as transactional data, can be stored in relational databases. Unstructured and semi-structured data, on the other hand, can be stored in NoSQL data stores. The format of the data determines the appropriate type of data store for storage and processing.
6. Data Source—The origin of the data, whether it is generated from social media, machines, or human input, is a determining factor in its classification.
7. Data Consumers—A comprehensive list of all users and applications that utilize the processed data.

Cloud Analytics Platforms

The realm of business software is in a constant state of flux, particularly with regard to cloud-based business intelligence tools. The most exceptional cloud analytics platforms, which possess the ability to scale responsively and support a company's evolving business models, are rapidly replacing on-premise reporting solutions. This trend is easily explicable, as legacy systems lack the user-friendly interfaces, prototyping tools, and self-service BI that cloud platforms offer. Businesses of all sizes are facing the challenge of scaling their business models while simultaneously maintaining responsive and efficient operations. As a result, cloud-based analytics platforms are narrowing the gap between the capabilities of legacy systems and the requirements of modern enterprises in order to compete and expand.

Cloud-based business intelligence (BI) tools that can be implemented using agile methodology will persist in achieving the quickest outcomes. Cloud platforms that can be implemented incrementally, while considering the needs of internal and external stakeholders, including customers, IT, marketing, sales, service, and senior management, yield the most favourable results. In contemporary enterprise software development, agile methodologies are prevalent. Progressive cloud platform providers are affording their clients the adaptability to adopt an agile-based approach to deploying their analytics platforms.

3.5 AI and Machine Learning Platforms in the Cloud

The domain of artificial intelligence (AI) and machine learning is expanding at an impressive pace. As per the findings of Tractica, a research firm, the revenues associated with AI are projected to escalate to $36 billion by 2025, as compared to $1.5 billion in 2016. However, given the plethora of platforms available, it can be arduous to identify the most suitable cloud computing services that align with your requirements and financial constraints. There are several alternatives accessible, each with its own advantages and disadvantages. This article aims to elucidate some of the leading AI cloud providers, enabling you to make an informed decision regarding the service that best suits your machine learning or AI project.

Let's take a look at five top cloud providers offering dedicated AI solutions today!

IBM Cloud is widely recognized as a highly sought-after cloud computing platform currently in extensive use. It boasts an extensive array of services encompassing analytics, security, IoT, mobile, Watson (AI), blockchain, and various others. Developers are empowered to construct applications utilizing a diverse range of programming languages on IBM Cloud's PaaS (Platform as a Service) or SaaS (Software as a Service) platforms. Furthermore, IBM Cloud extends its offerings to encompass private cloud solutions tailored to meet the needs of enterprise clients who prefer to retain their data on-premises.

Microsoft Azure is a robust cloud computing platform that has been specifically designed to assist companies in constructing, deploying, and managing applications on their own infrastructure. It offers a diverse array of services, encompassing data storage, server management, analytics tools, and various other functionalities. In recent years, Microsoft has made substantial investments in the field of artificial intelligence (AI) and machine learning, leading many experts to anticipate its emergence as one of the primary beneficiaries as AI continues to permeate various industries. In fact, Gartner's Magic Quadrant report for 2018 ranked Microsoft as the second-best cloud provider overall, considering both the comprehensiveness of its vision and its ability to execute. Nevertheless, it is important to note that there are certain limitations associated with utilizing Azure for the development of custom algorithms or models. Azure is most suitable for scenarios where pre-trained models are employed or when constructing applications that can operate as web services or mobile apps.

AWS, an abbreviation for Amazon Web Services, is a prominent cloud computing platform renowned for its proficiency in machine learning and artificial intelligence. With a history spanning over a decade, it stands as one of the most well-established cloud computing platforms currently available. Its extensive array of services encompasses analytics, machine learning, artificial intelligence, databases, networking tools, and various other functionalities. The platform's popularity stems from its user-friendly interface, comprehensive documentation, and tutorials. Furthermore, it caters to both large-scale and small-scale data applications, showcasing its exceptional versatility. Among its notable offerings, Amazon Elastic Compute Cloud (EC2) stands out, enabling users to effortlessly execute workloads on-demand, eliminating the need for server and software management. Additionally, the platform's resource scalability feature allows users to adjust their resources according to their requirements, resulting in a cost-effective payment structure where users only pay for the resources they utilize.

Google Cloud is renowned for its machine learning tools, particularly TensorFlow, which was created by Google. It provides a diverse array of cloud services that can be accessed through an API. Should one desire to construct a machine learning model independently, they may utilize one of its pre-trained models or commence from the beginning with one of its APIs. Additionally, it features a tool named AutoML, which assists users in constructing their own neural networks, even if they lack coding expertise. Other tools, such as BigQuery for querying vast datasets and Pub/Sub for publishing messages to subscribers in real-time, are also available.

3.6 *High Performance Computing (HPC) and Cloud-Based Clusters*

High-Performance Computing (HPC) is a technological advancement that utilizes clusters of powerful processors operating in parallel to efficiently process vast multi-dimensional datasets, commonly referred to as big data. This cutting-edge technology enables the resolution of complex problems at exceptionally high speeds. HPC systems typically exhibit performance levels exceeding one million times that of the fastest commodity desktop, laptop, or server systems.

Traditionally, the HPC system paradigm revolved around supercomputers, purpose-built computers incorporating millions of processors or processor cores. Supercomputers continue to be prevalent today, with the current fastest supercomputer being the US-based Frontier, boasting a processing speed of 1.102 exaflops, or quintillion floating point operations per second (flops). However, there is a growing trend among organizations to adopt HPC solutions that operate on clusters of high-speed computer servers, either hosted on premises or in the cloud.

HPC workloads play a pivotal role in unearthing valuable insights that advance human knowledge and provide significant competitive advantages. For instance, HPC is instrumental in DNA sequencing, automating stock trading, and executing artificial intelligence (AI) algorithms and simulations. These simulations, such as those employed in self-driving automobiles, analyze terabytes of real-time data streaming from IoT sensors, radar, and GPS systems, enabling split-second decision-making.

A conventional computing system primarily solves problems through serial computing, where the workload is divided into a sequence of tasks and executed one after the other on the same processor.

In contrast, HPC utilizes massively parallel computing. Parallel computing allows for the simultaneous execution of multiple tasks on multiple computer servers or processors. Massively parallel computing takes this a step further by employing tens of thousands to millions of processors or processor cores.

HPC clusters, also known as computer clusters, play a crucial role in this process. These clusters consist of multiple high-speed computer servers that are interconnected. They are managed by a centralized scheduler, which handles the parallel computing workload. The nodes within these clusters utilize high-performance multi-core CPUs or, more commonly today, GPUs (graphical processing units). GPUs are particularly well-suited for demanding mathematical calculations, machine learning models, and graphics-intensive tasks. It is not uncommon for a single HPC cluster to include 100,000 or more nodes.

Furthermore, all other computing resources within an HPC cluster, such as networking, memory, storage, and file systems, are high-speed, high-throughput, and low-latency components. These components are designed to keep up with the nodes and optimize the computing power and performance of the cluster.

At an elevated level, a computer cluster refers to a collection of two or more computers, or nodes, that operate in parallel to accomplish a shared objective. This facilitates the distribution of workloads comprising a substantial number of individual, parallelizable tasks among the nodes in the cluster. Consequently, these tasks can exploit the combined memory and processing capabilities of each computer to enhance overall performance.

To construct a computer cluster, the individual nodes must be interconnected in a network to enable internode communication. Computer cluster software can then be utilized to unite the nodes and establish a cluster. It may feature a shared storage device and/or local storage on each node. Typically, at least one node is designated as the leader node, which serves as the entry point to the cluster. The leader node may be responsible for delegating incoming work to the other nodes and, if necessary, aggregating the results and returning a response to the user.

Ideally, a cluster should operate as if it were a single system. A user accessing the cluster should not be required to discern whether the system is a cluster or an individual machine. Furthermore, a cluster should be designed to minimize latency and prevent bottlenecks in node-to-node communication.

Types of Cluster Computing

Computer clusters can generally be categorized as three types:

1. Highly available or fail-over
2. Load balancing
3. High performance computing

Four advantages to cluster computing

Cluster computing provides a number of benefits: high availability through fault tolerance and resilience, load balancing and scaling capabilities, and performance improvements. Let's expand upon each of these features and examine how clusters enable them.

1. High availability

An application operating on a single machine possesses a singular point of failure, resulting in inadequate system reliability. In the event that the machine hosting the application experiences a shutdown, there will typically be a period of downtime while the infrastructure recovers. To enhance reliability, it is advisable to maintain a level of redundancy, which can reduce the duration of application unavailability. This can be accomplished by proactively running the application on a secondary system (which may or may not be receiving traffic) or by configuring a dormant system with the application. These configurations are commonly referred to as active-active and active–passive configurations, respectively. In the case of a failure, an active-active system can promptly switch over to the second machine, whereas an active–passive system will only switch over once the second machine is operational.

Computer clusters consist of multiple nodes concurrently running the same process, making them active-active systems. Active-active systems are typically fault-tolerant since the system is inherently designed to handle the loss of a node. If a node fails, the remaining node(s) are prepared to assume the workload of the failed node. However, in the case of a cluster that necessitates a leader node, it is recommended to operate a minimum of two leader nodes in an active-active configuration. This precautionary measure can prevent the cluster from becoming unavailable in the event of a leader node failure.

2. Load Balancing

Load balancing refers to the practice of evenly distributing traffic among the nodes of a cluster in order to optimize performance and prevent any single node from being overwhelmed with excessive workload. A load balancer can be installed on the leader node(s) or provisioned separately from the cluster. By conducting regular health checks on each node within the cluster, the load balancer is capable of identifying any node failures and redirecting incoming traffic to the other nodes in the cluster.

While a computer cluster does not inherently possess load balancing capabilities, it facilitates the implementation of load balancing across its nodes. This particular configuration is commonly known as a "load balancing" cluster and often serves as a highly available cluster simultaneously.

3. Scaling

Scaling can be classified into two categories: vertical scaling and horizontal scaling. Vertical scaling, also known as scaling up/down, involves adjusting the allocated resources for a process, such as memory capacity, number of processor cores, or available storage. On the other hand, horizontal scaling, referred to as scaling out/in, entails running additional parallel tasks on the system.

When managing a cluster, it is crucial to monitor resource usage and scale accordingly to ensure optimal utilization of cluster resources. Fortunately, the inherent nature of a cluster simplifies horizontal scaling, as the administrator can easily add or remove nodes as needed, while considering the minimum redundancy level required to maintain high availability of the cluster.

4. Performance

When it comes to parallelization, clusters can achieve higher performance levels than a single machine. This is because they're not limited by a certain number of processor cores or other hardware. Additionally, horizontal scaling can maximize performance by preventing the system from running out of resources.

"High performance computing" (HPC) clusters leverage the parallelizability of computer clusters to reach the highest possible level of performance. A supercomputer is a common example of an HPC cluster.

Clustering challenges

One of the primary challenges posed by clustering is the heightened intricacy associated with installation and maintenance. In order to establish a cluster, it is imperative to install and update the operating system, application, and their respective dependencies on each individual node. This task becomes even more intricate if the nodes within the cluster are not uniform. Furthermore, it is essential to closely monitor resource utilization on each node and aggregate logs to guarantee proper software functionality. Moreover, managing storage becomes more arduous as a shared storage device must prevent nodes from overwriting one another, and distributed data stores must be synchronized.

4 Cloud Computing Software for Research

4.1 Data Storage and Database Services

Cloud computing has revolutionized data storage and database services for research by offering scalable, secure, and cost-effective solutions. Researchers can now store vast amounts of data in the cloud, eliminating the need for on-premises hardware and infrastructure. Cloud providers like AWS, Azure, and Google Cloud offer a range of storage options, from object storage for unstructured data to managed relational databases for structured information.

These services enable researchers to efficiently manage and access their data, ensuring data integrity and availability. Researchers can easily scale storage capacity up or down as their needs change, reducing costs and optimizing resource usage. The cloud also facilitates data replication and backup, safeguarding research data from loss or corruption. Furthermore, cloud-based databases support advanced features like automatic backups, high availability, and security compliance, making them a powerful asset for research projects. Researchers can focus on their work without the burden of managing database infrastructure, leading to increased productivity and improved data-driven insights.

4.2 Data Processing and Analytics Tools

Cloud computing offers a diverse array of data processing and analytics tools tailored to the unique needs of researchers. These tools allow researchers to manipulate, analyze, and derive insights from data at a scale and speed that was previously unattainable. Services like AWS Lambda, Google Cloud Dataflow, and Apache Spark on Azure enable researchers to process large datasets efficiently. They offer parallel and distributed processing capabilities, which are essential for handling big data in fields such as genomics, climate modeling, and social sciences.

Cloud-based analytics platforms, including Google BigQuery and Amazon Redshift, empower researchers to perform complex queries and gain deeper insights into their data. Machine learning and artificial intelligence tools, like TensorFlow and PyTorch, enable predictive modeling and pattern recognition, which are valuable for various scientific applications. These cloud services provide researchers with the computational power to accelerate their research, reduce time-to-insight, and make more informed decisions based on data-driven analysis. Collaboration and Workflow Management Tools for Research.

4.3 Collaboration and Workflow Management Tools

Cloud computing has transformed collaboration and workflow management in research by offering integrated tools that enhance teamwork, streamline processes, and facilitate effective project management. These cloud-based solutions promote efficient knowledge sharing and accelerate research outcomes. Collaboration platforms like Microsoft Teams and Slack enable real-time communication, file sharing, and project coordination. Researchers can easily collaborate with peers across different geographical locations, fostering a global research community.

Workflow management tools such as Trello, Asana, and Jira help researchers organize tasks, set priorities, and track progress. These platforms allow for agile project management, providing researchers with flexibility and adaptability in the face of changing research needs. Cloud-based project management systems, like Basecamp and Monday.com, provide centralized hubs for research projects. They offer features such as document sharing, task assignments, and milestone tracking, ensuring that research teams stay organized and on schedule. Overall, cloud-based collaboration and workflow management tools empower researchers to work more efficiently, foster innovation, and optimize project outcomes by providing a structured and collaborative environment for research teams.

4.4 Data Visualization and Reporting Tools

Cloud computing has revolutionized data visualization and reporting in research, making it easier to communicate findings and insights through compelling visuals and reports. Cloud-based tools provide researchers with powerful resources to transform data into meaningful presentations. Tools like Tableau and Power BI offer interactive and dynamic data visualization capabilities, enabling researchers to create charts, graphs, and dashboards that convey complex information in a user-friendly manner. These visualizations enhance data comprehension and support better decision-making.

Cloud-based reporting tools like Google Data Studio and Microsoft Power BI enable researchers to generate customized reports and share them securely with collaborators or stakeholders. These reports can incorporate real-time data, making it easy to keep all interested parties informed. Furthermore, cloud-based data visualization and reporting tools are often integrated with data storage and analytics platforms, allowing researchers to streamline the entire data-to-insight process. This integration ensures that data visualization and reporting become an integral part of the research workflow, enhancing the overall research experience.

In summary, cloud-based data visualization and reporting tools empower researchers to present their findings effectively, improving data communication and enhancing the impact of research outcomes.

4.5 Machine Learning and AI Frameworks

Cloud computing has democratized access to machine learning (ML) and artificial intelligence (AI) frameworks, providing researchers with the tools to solve complex problems and unlock new insights across various domains. Cloud-based ML and AI platforms offer scalability, ease of use, and a wealth of pre-built models, making them invaluable in research. Popular cloud ML platforms like Google Cloud AI, Amazon SageMaker, and Microsoft Azure Machine Learning provide researchers with environments for training, deploying, and managing ML models. These platforms offer a wide range of ML algorithms and deep learning frameworks like TensorFlow and PyTorch, simplifying the development of predictive models.

Researchers can leverage pre-built AI services for tasks such as natural language processing, computer vision, and speech recognition. These cloud services, including Amazon Comprehend and Google Cloud Vision, expedite research by eliminating the need to develop custom AI models from scratch. Moreover, cloud-based ML and AI frameworks support collaborative research by providing shared workspaces, version control, and integration with other research tools. This promotes knowledge exchange and accelerates innovation in various fields, from healthcare and finance to climate science and astronomy. Cloud computing has transformed the research landscape, enabling researchers to harness the power of machine learning and AI, thereby advancing the boundaries of human knowledge.

4.6 Simulation and Modeling Software

Cloud computing has had a profound impact on simulation and modeling software in research by offering scalable, cost-effective, and accessible resources for complex computational tasks. Researchers can now perform simulations and model various phenomena with greater efficiency and accuracy. Cloud-based High-Performance Computing (HPC) clusters, such as AWS EC2 instances, Azure Virtual Machines, and Google Cloud Compute Engine, provide researchers with the computational horsepower required for intricate simulations. These resources can be provisioned on-demand, eliminating the need for researchers to invest in expensive on-premises HPC infrastructure.

Specialized simulation software, like ANSYS and COMSOL, are available on cloud platforms, allowing researchers to design experiments and analyze results remotely. Cloud-based simulations enable collaboration among geographically dispersed teams, fostering a global exchange of ideas and expertise. Additionally, the cloud supports big data-driven simulations and modeling, enhancing research in fields such as climate science, computational chemistry, and aerospace engineering. Researchers can process and analyze massive datasets alongside their simulations, leading to more comprehensive and data-driven insights. Overall, cloud computing has democratized access to simulation and modeling resources, making it easier for researchers to conduct experiments and explore complex phenomena across a wide range of scientific disciplines.

4.7 Cloud Based IOT Applications like for Example Smart Cities

Cloud computing has played a pivotal role in the development of IoT (Internet of Things) applications for smart cities. These applications leverage cloud resources to collect, process, and analyze data from a multitude of interconnected devices, ultimately enhancing urban infrastructure and quality of life. In smart cities, cloud-based IoT platforms like AWS IoT Core and Microsoft Azure IoT Hub act as central hubs for data ingestion and device management. They allow cities to seamlessly integrate various sensors, cameras, and other IoT devices into a unified network. The cloud ensures that data is transmitted securely and reliably from these devices to centralized data repositories.

Cloud computing supports real-time data processing and analytics, enabling smart cities to monitor and respond to changing conditions efficiently. For example, traffic management systems can use IoT data to optimize traffic flow and reduce congestion, while environmental monitoring systems can provide air quality alerts and help combat pollution. Machine learning and AI algorithms, hosted in the cloud, enable predictive maintenance for critical infrastructure like bridges, water supply networks, and public transportation. These systems can detect anomalies and predict when maintenance is required, helping to prevent breakdowns and improve overall infrastructure reliability. Overall, cloud-based IoT applications empower smart cities to enhance urban living through data-driven decision-making, increased efficiency, and improved services for residents and visitors. They represent a promising approach to solving many of the challenges faced by modern urban environments.

5 Security, Privacy, and Compliance in Cloud Computing for Research

5.1 Cloud Security Best Practices

Ensuring robust security in cloud computing for research is paramount, given the sensitive and valuable data often involved. Cloud security best practices encompass a range of measures to protect data and systems. Encryption is fundamental, both in transit and at rest, to safeguard data from unauthorized access. Access controls and identity management systems should be rigorously implemented to limit who can access what information.

Multi-factor authentication (MFA) is another key practice, bolstering login security. Regular security audits and vulnerability assessments help identify and address weaknesses in the cloud infrastructure. Monitoring and logging all activities in the cloud environment can aid in detecting and responding to security incidents promptly. Implementing a well-defined incident response plan is crucial to minimize the impact of breaches, while regular staff training on security best practices helps create a security-aware organizational culture. Selecting a cloud provider with a strong security track record and compliance certifications can also enhance overall security.

5.2 Data Privacy and Protection Considerations

Data privacy is of utmost importance in cloud computing for research, particularly when dealing with personal or sensitive information. Researchers must adhere to data protection laws and best practices. Data should be anonymized or pseudonymized whenever possible, and clear policies for handling sensitive data should be in place. Privacy impact assessments (PIAs) can help identify potential privacy risks and mitigation strategies. Researchers should minimize data collection to what is strictly necessary and avoid data retention beyond its purpose. Consent mechanisms should be transparent and unambiguous. Secure data transfer and storage, coupled with strong access controls, are vital for data privacy. Regular data audits can help ensure compliance and identify potential breaches. Collaboration with legal experts can help ensure that research activities conform to local and international privacy regulations.

5.3 Compliance with Regulatory Requirements (E.G., GDPR, HIPAA)

Compliance with regulatory requirements is critical in cloud computing for research, as non-compliance can result in legal consequences and damage to reputation. Researchers must be well-versed in relevant regulations, such as the General Data Protection Regulation (GDPR) in Europe or the Health Insurance Portability and Accountability Act (HIPAA) in the United States When dealing with personal health information, HIPAA requires strict data protection and access controls, and non-compliance can lead to severe penalties. GDPR mandates the protection of personal data with stringent consent requirements and the right to erasure.

To comply with these regulations, researchers should implement strong data protection measures, including encryption, access controls, and regular audits. They should also appoint data protection officers and establish clear procedures for data breaches and incident reporting. Researchers should work closely with their cloud providers to ensure that they also adhere to these regulations.

5.4 Data Governance and Access Control

Data governance and access control are essential aspects of cloud computing for research, particularly when managing large datasets and collaborating with multiple stakeholders. Researchers should establish clear data governance policies that define ownership, accountability, and data lifecycle management. Access control mechanisms are crucial to restrict data access to authorized personnel only. Role-based access control (RBAC) ensures that users have appropriate permissions based on their roles. Encryption and tokenization of data protect it from unauthorized access even in the event of security breaches.

Auditing and monitoring tools help keep track of data access and usage, ensuring compliance with data governance policies. Researchers should regularly review access permissions and revoke unnecessary access rights. Data governance and access control not only protect data but also facilitate data sharing and collaboration, which is often essential in research. Properly managed, these aspects enhance data security, privacy, and compliance in cloud computing for research.

6 Challenges and Future Directions

6.1 *Data Transfer and Bandwidth Limitations*

Data transfer and bandwidth limitations are persistent challenges in cloud computing for research. Researchers often deal with large volumes of data, and transferring this data to and from the cloud can be time-consuming and costly. Bandwidth constraints can lead to slow data uploads and downloads, hindering research productivity.

One solution is to employ data compression techniques and efficient data transfer protocols to reduce the amount of data that needs to be moved. Researchers can also consider utilizing dedicated high-speed networks or leveraging cloud-based data transfer services to optimize data migration. These services are specifically designed to overcome bandwidth limitations and facilitate the rapid transfer of data.

Future directions in this area may involve the development of more efficient data transfer technologies and network infrastructure. Enhanced data compression algorithms and novel methods for data synchronization can help mitigate these challenges. Furthermore, advancements in 5G and 6G networks may significantly improve data transfer speeds, making cloud-based research more accessible and efficient.

6.2 *Interoperability and Vendor Lock-In*

Interoperability and vendor lock-in are critical concerns in cloud computing for research. Researchers often use a mix of cloud services and applications, and ensuring seamless interoperability between these services can be challenging. Vendor lock-in occurs when research data and processes become tightly integrated with a particular cloud provider's technology stack, making it difficult to switch providers or use on-premises resources. To address these challenges, research institutions can establish cloud-agnostic strategies that prioritize the use of open standards and application programming interfaces (APIs). This approach promotes interoperability and ensures that research assets remain flexible and independent of any specific vendor.

Future directions may involve the development of industry standards and best practices for cloud interoperability, making it easier for researchers to transition between cloud providers or integrate multiple cloud services. Additionally, the adoption of multi-cloud and hybrid cloud architectures can help mitigate vendor lock-in by allowing researchers to distribute their workloads across different cloud platforms or on-premises infrastructure.

6.3 Integration of Cloud Computing with On-Premises Infrastructure

Integrating cloud computing with on-premises infrastructure is a crucial challenge and future direction in research computing. Many research organizations maintain legacy systems and hardware that need to coexist with cloud resources. Achieving seamless integration between these environments is essential for efficient resource utilization. Hybrid cloud architectures and containerization technologies like Docker and Kubernetes play a pivotal role in addressing this challenge. They enable researchers to deploy applications and workloads across both on-premises and cloud environments, ensuring compatibility and efficient resource allocation.

Future directions may involve the development of more sophisticated orchestration tools and management platforms that simplify hybrid cloud deployment and management. Researchers should also consider optimizing data workflows to ensure that data can flow seamlessly between on-premises and cloud environments, minimizing latency and data transfer bottlenecks.

6.4 Advances in Cloud Computing Technologies for Research

Advances in cloud computing technologies are continuously shaping the landscape of research. Cutting-edge innovations include serverless computing, edge computing, and quantum computing, all of which have the potential to transform how research is conducted. Serverless computing simplifies resource management by allowing researchers to focus solely on code development. Edge computing extends the cloud's reach to the network's edge, enabling real-time data processing and analysis. Quantum computing, while still in its early stages, offers the potential to revolutionize complex computations in fields such as cryptography, materials science, and optimization problems.

Moreover, cloud providers are continually improving their services to cater to research needs, offering specialized research clouds and dedicated research networks. These advancements lead to enhanced scalability, reduced costs, and improved performance, driving further adoption of cloud computing in research. As research evolves, cloud computing technologies will continue to adapt to meet the increasingly sophisticated demands of researchers, enabling breakthroughs in a wide range of disciplines.

6.5 Ethical Considerations in Cloud-Based Research:

Ethical considerations are gaining prominence in cloud-based research as the technology becomes more deeply integrated into scientific investigations. Key ethical concerns include data privacy, data ownership, and the responsible use of AI and machine learning. Researchers must grapple with questions about data sovereignty and the handling of sensitive information in the cloud. Clear data protection policies and consent mechanisms are essential to ensure that research is conducted ethically and in compliance with regulations like GDPR and HIPAA.

Moreover, the responsible use of AI and machine learning models in research raises ethical issues regarding bias, fairness, and transparency. Researchers must be mindful of potential algorithmic biases and take steps to mitigate them to ensure fairness and equity in research outcomes. Future directions in cloud-based research ethics may involve the development of industry-specific ethical guidelines and frameworks. Ethical review boards for research projects should adapt to the nuances of cloud-based research, and researchers should receive training in ethical data handling and AI model development.

As cloud computing continues to play a pivotal role in research, ethical considerations will be an integral part of the research process, and researchers must address them diligently to ensure the responsible and ethical conduct of research in the cloud.

7 Conclusions

In conclusion, cloud computing infrastructure, platforms, and software have fundamentally reshaped the landscape of scientific research. The flexibility and accessibility offered by cloud-based resources have enabled researchers to tackle complex problems with unprecedented ease and efficiency. The shift towards cloud-based solutions has opened up new horizons for collaboration, allowing researchers from diverse geographical locations to work together seamlessly. The scalability of cloud computing resources ensures that scientific experiments and data analysis can be performed at a speed and scale previously unattainable, while the pay-as-you-go model helps control costs and allocate resources efficiently. This adaptability is particularly important for research that demands varying levels of computational power and data storage. While the advantages are clear, there are also challenges associated with cloud adoption, including data security, compliance, and potential vendor lock-in. Researchers must carefully consider these aspects and implement best practices to mitigate risks. Additionally, ensuring equitable access to cloud resources is an ongoing concern in the scientific community, as not all researchers or institutions have equal access to these powerful tools. In summary, cloud computing infrastructure, platforms, and software represent a powerful ally for scientific research, enhancing productivity and accelerating the pace of discovery. As cloud technology continues to evolve, researchers should remain vigilant in their adoption strategies, and policymakers should work towards equitable access and data privacy regulations. With the right balance of innovation and diligence, cloud computing will remain an essential driver of scientific progress in the years to come.

References

1. Al Masarweh, M., Alwada'n, T., Afandi, W.: Fog computing, cloud computing and IoT environment: advanced broker management system. J. Sens. Actuator Netw. **11**(4), 84 (2022). https://doi.org/10.3390/jsan11040084
2. Alouffi, B., Hasnain, M., Alharbi, A., Alosaimi, W., Alyami, H., Ayaz, M.: A systematic literature review on cloud computing security: threats and mitigation strategies. IEEE Access **9**, 57792–57807 (2021)
3. Alwada'n, T., Al-Zitawi, O., Khawaldeh, S., Almasarweh, M.: Privacy and control in mobile cloud systems. Int. J. Comput. Appl. **113**, 12–15 (2015)
4. Stergiou, C., Psannis, K.E., Gupta, B.B., Ishibashi, Y.: Security, privacy & efficiency of sustainable cloud computing for big data & IoT. Sustain. Comput. Inform. Syst. **19**, 174–184 (2018)
5. Mbarek, N.: Service level management in the cloud. Serv. Level Manag. Emerg. Environ. **20**, 45–81 (2021)
6. Washizaki, H., Xia, T., Kamata, N., Fukazawa, Y., Kanuka, H., Kato, T., Yoshino, M., Okubo, T., Ogata, S., Kaiya, H., et al.: Systematic literature review of security pattern research. Information **12**, 36 (2021)
7. Nayyar, A.: Handbook of Cloud Computing: Basic to Advance Research on the Concepts and Design of Cloud Computing. BPB Publications, Noida, India (2019)
8. Moltó, G., Naranjo, D.M., Damian Segrelles, J.: Insights from learning analytics for hands-on cloud computing labs in AWS. Appl. Sci. **10**(24), 9148 (2020). https://doi.org/10.3390/app10249148
9. Mehmood, H., Conway, C., Perera, D.: Mapping of flood areas using Landsat with google earth engine cloud platform. Atmosphere **12**(7), 866 (2021). https://doi.org/10.3390/atmos12070866

Expansion of AI and ML Breakthroughs in HPC with Shift to Edge Computing in Remote Environments

Kumud Darshan Yadav

Abstract This chapter explores the dynamic intersection of High-Performance Computing (HPC), Artificial Intelligence (AI), and Machine Learning (ML) while emphasizing the shift towards edge computing solutions in remote and challenging environments. HPC, traditionally centralized and known for solving complex computational problems, has converged with AI and ML, unleashing unprecedented capabilities. However, the challenges posed by remote settings, such as resource constraints, harsh conditions, latency, and data transfer issues, have necessitated innovative solutions. The concept of edge computing, which involves deploying computation closer to data sources, emerges as a key solution for these challenges. Edge computing minimizes latency, reduces bandwidth usage, enhances scalability, and offers robustness, making it ideal for real-time applications in remote environments. This chapter further delves into the integration of AI and ML with edge computing, highlighting the importance of customized hardware, distributed AI/ML models, and anomaly detection systems. Through case studies in deep-sea exploration and precision agriculture, the practical applications of this convergence in remote settings come to light. The future prospects of this transformative shift are promising, driven by advancements in AI hardware, algorithms, and edge computing technologies. This evolution promises to unlock innovative possibilities in scientific research, autonomous systems, and various domains operating in challenging and previously inaccessible locations.

Keywords Artificial intelligence · Machine learning · Edge computing · High performance computing · Computational environment · Multi frameworks

In the rapidly evolving landscape of High-Performance Computing (HPC), the integration of Artificial Intelligence (AI) and Machine Learning (ML) has ushered in a

K. D. Yadav (✉)
GVRC Robotics Laboratory, Universidad de Seville, Camino de los Descubrimientos s/n, 41092 Seville, Spain
e-mail: ykumuddarshan@us.es

K. A. Ahmad et al. (eds.), *High Performance Computing in Biomimetics*, Series in BioEngineering, https://doi.org/10.1007/978-981-97-1017-1_5

new era of innovation and efficiency. This chapter delves into the significant developments in AI and ML within HPC, emphasizing the transition towards edge computing in remote and challenging environments.

1 Artificial Intelligence—A Comprehensive Approach

Artificial intelligence basically refers to replicate the human intelligence by machines especially by modern computing devices. The problem-solving abilities, logical reasoning, language processing are some of the main criteria considered for artificial intelligence [1] without any human input. An AI is a self-sufficient system designed to reduce the human workload and sometimes eliminate the need for human input.

Primarily, the AI is classified into following sets [2]: (1) Narrow AI, whose operation of work is limited to handle one narrow task such as facial recognitions, chatbot. (2) General AI, where domain of operation is more intellectual and usually takes over the need for human intelligence. (3) Artificial super intelligence generally refers to the epitome of creating an AI beyond the cognition of human intellect.

Artificial Intelligence techniques encompasses the machine learning, natural language processing, computer vision and many other fields which depicts the human intelligence. Machine learning is the subset of Artificial Intelligence which has the ability to learn from data and improve with time without being programmed specifically for the task, and is a subset of AI as depicted in the Fig. 1. Whereas Deep Learning is the subset of Machine Learning which trains over a huge amount of data and could recognise pattern and features, it imitates the neural networks present inside the brain to forecast the intelligence possessed by the humans.

2 Machine Learning

The term Machine learning coined in 1959 by Arthur [4] quoted "Programming computers to learn from experience should eventually eliminate the need for much of this detailed programming effort". It is a subset of AI and generally refers to field of study which enables the computing devices the capability to learn from the data gathered over the course of time. The name coined is alluding to similar learning experience as humans. The computing device detects the pattern present inside the data by identifying the hidden correlations between the data gathered at various time instances. Machine learning algorithms are categorized into different types based on the learning type [5]:

Supervised Learning: The model is trained on the labelled dataset which means the data has properly defined input and output parameters. The supervised learning is further divided into two different sets of learning namely Classification and

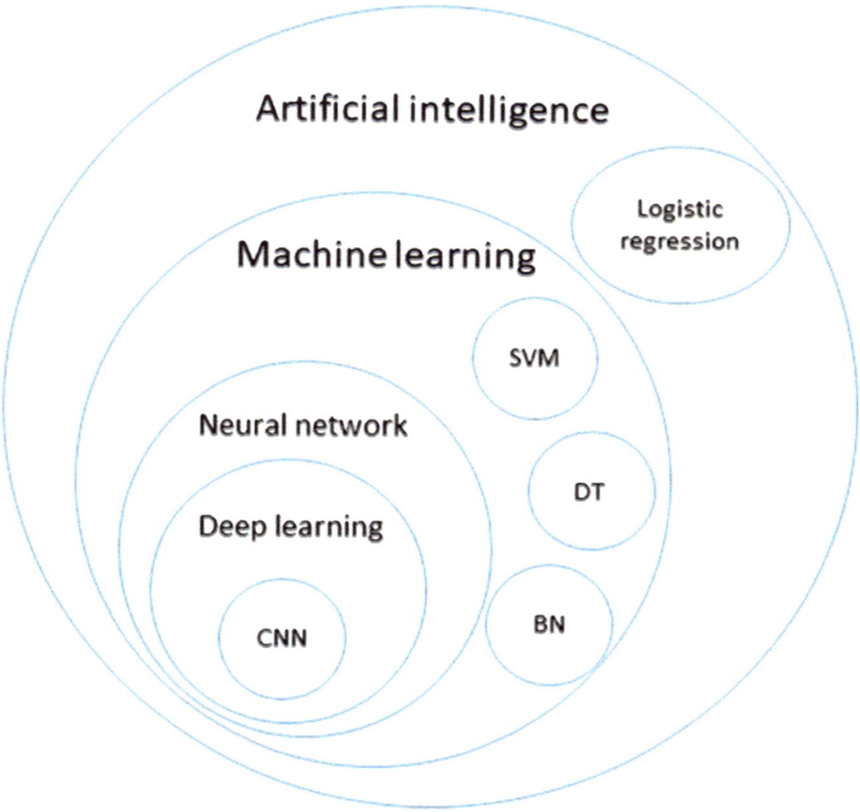

Fig. 1 Venn diagram—AI, ML and DL approaches [3]: https://doi.org/10.3390/cancers14061370

Regression problems where different algorithms are used based on the problem defined.

Unsupervised Learning: These algorithms find the underlying pattern in the data without any human intervention. The unsupervised learning is also divided into two sections Clustering and Association. The data type used for such learning technique is commonly Unstructured and Unlabelled data.

Semi—Supervised Learning: It lies between the Unsupervised and Supervised Learning techniques where some of the data is labelled, and rest is unlabelled. Generally, the data is first labelled using the unsupervised learning and then fed into the supervised algorithms. It is useful in the case of image dataset where the images are not properly labelled.

Reinforcement Learning: This kind of learning is based on the reward feedback mechanism. The system keeps on increasing the performance as the learning progresses based on the rewards set up during the training. The network learns each time the data is fed into the system. So, in theory the more network is trained the better it gets. The real-life example of using reinforcement learning is the google self-driving car.

3 Neural Networks

Neural Networks has evolved as a promising tool in the field of AI and ML. It tries to imitate the behaviour of the neural networks present inside human brain [6]. It has evolved now revolutionized technology in the field of feature detection, pattern recognition, Natural Language Processing (NLP), image processing and speech recognition.

A NN consists of multiple neurons connected to each other producing a sequence of activation functions. Initial neurons get activated with the input from the environment and remaining neurons gets activated through the previous neurons by weighted connections. Some of the neurons inside the network can also influence the environment. Learning is a process of finding weights of the connection to exhibit the desired behaviour [7]. Activation functions provides the non-linearity to the model providing the tools to model complex behaviour between the input to the network to the output.

Neurons are generally organised into layers, the first layer called Input Layer, it can be a sequence input layer or input layer for feature determination, and the last layer of the network is the output layer, which depends on the task which is being performed for example for regression problems, classification problem or any other problem. The layers between the input layer and the output layer are called the hidden layers. The connectivity inside the network often determines the flow of information from beginning to end. Fully connected layer connects all the neurons from the previous layer to the coming layer while other network architecture such as Recurrent Neural Networks (RNNs), Convolutional Neural Networks (CNNs) introduce specific connectivity inside the network [8] (see Fig. 2).

While training, neural networks learns to adjust the internal parameters, weights, and biases, depending upon the desired output to the input of the system. Feedforward propagation generates the predictions by computing the intermediate outputs while feeding input data through the network. The error between the predicted output through the network and the desired output is used to further update the network parameters using the backpropagation algorithm [8] which calculates the gradient of the error wrt weights and biases.

Feed Forward Network [9] is the elementary type of NN in which the information flows from the input to output layer in one direction. These are mainly used for the tasks of classification, regression, and pattern recognition. On the other hand, CNNs [8] are specifically designed to process the data in a grid like pattern such as a image can be divided into a grid of datapoints. CNNs deploy several layers for specialised tasks such as convolutional layers for feature extraction, fully connected layer for classification, and pooling layer for down sampling. CNNs are widely used in the field of perception, attitude determination using images.

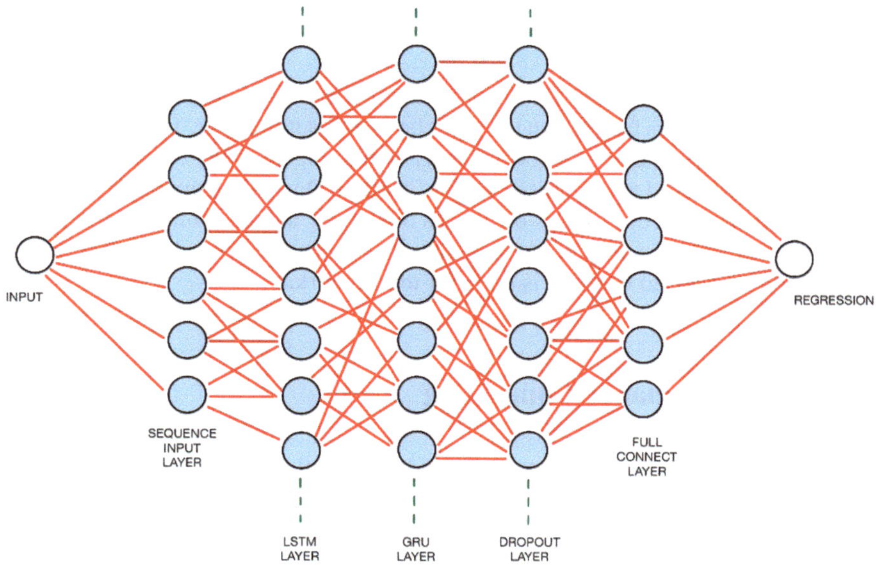

Fig. 2 A custom made RNN for regression

Recurrent Neural networks also knows as RNNs are specially designed for sequential data where the information is passed from one step to the other. They have a feedback mechanism which enables them to capture the temporal dependencies in the input data through maintaining hidden states in the network. RNNs [10] consists of multiple internal activation functions one for each time step. The hidden state in the network contains the past information of the previous time step and it gets updates every time step. Their ability to retain information about the data makes them the ideal candidate for language modelling, speech recognition, regression problems, time series forecasting, and machine translation.

Though RNNs have a lot of advantages but they suffer from the problem of vanishing gradient [10]. This specific problem arises in the long sequential data while training through back propagation. The gradients either vanish or explode during back propagation making it incapable of learning long term dependencies in the data. Long term short Memory (LSTM) network are specifically designed to address the problem of vanishing gradients which makes them the perfect candidate for learning long term dependencies in the data. LSTM uses gating mechanism and memory cells to retain information over time which makes them highly effective in NLP, time series analysis, trajectory designing.

4 Deep Neural Network

A deep neural network (DNNs) [10] is an Artificial Neural network (ANNs) with multiple hidden layers in between the input layer and output layer. The hidden layers in DNNs can be as large as 1000 layers. For training of DNNs gradient descent method is widely used to minimise the loss function. One such example is ImageNet which is a repository of millions of images to classify a dataset into categories like cats and dogs. They usually produce better results than the normal ML. The DNNs shine when supervised learning is done for the network.

5 High Performance Computing

The domain of computing has seen a tremendous growth in past few decades through advancement in technology, system architecture and usage of the systems. The second half of 1970s marks the beginning of modern Supercomputing when the vector computing systems were introduced [11]. With the introduction of Cray-1 in 1976 by cray research a significant milestone was achieved, which could perform calculations up to 160 million FLOPS. Every passing decade since then has seen a rise in performance of the computers exponentially, also known as Moore's Law [12]. 1980s marks the rise of parallel computing and subsequent decade 1990s witnessed the parallel supercomputers breaking the gigaflop barrier [13].

HPC delivers exceptional computing power which is being used in various fields i.e. solving highly complex non-linear problems to managing big chunks of data, from predicting weather to solving the mysteries of cosmos. It is being seen as the main tool in the advancement of the new technology. The training of complex network system generally requires huge amount of data to process for training which usually takes a lot of computational power. HPC shines when dealing with big data, NLP (Natural Language Processing), Image processing, pattern recognition (see Fig. 3).

The report about parallel computing [15] explains "the shift towards increasing parallelism is not a triumphant stride forward based on break throughs in novel software and architectures for parallelism; instead, this plunge into parallelism is actually a retreat from even greater challenges that thwart efficient silicon implementation of traditional single processor architectures".

The next big advancement in the field of big data and HPC is the introduction of Apache Spark [16] which presents the unified programming model and engine for big data. Spark not only enables the intermediate computation by providing in memory storage but also encompasses the libraries with composable APIs made for machine learning, SQL for interactive environments which makes it faster than the Hadoop MapReduce [17].

The convergence of Machine learning and High Performance computing provides a positive outlook in boosting performance efficiency of the system. ML driven HPC also referred to as ML around HPC presents a viable option for enhancing the

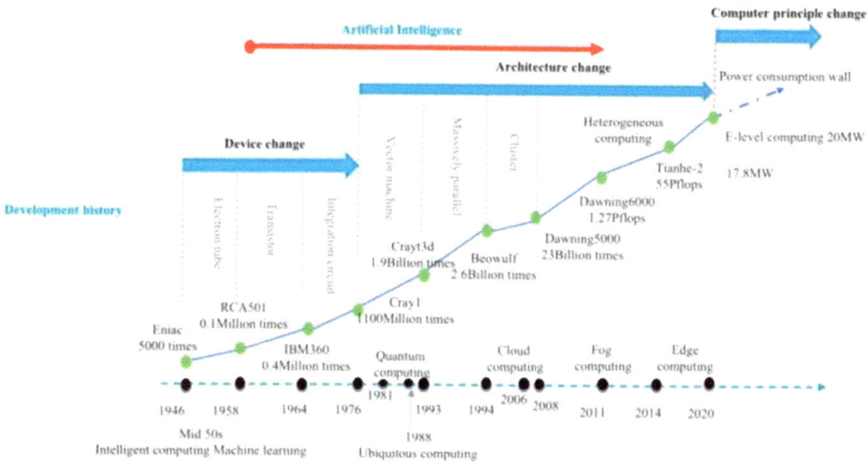

Fig. 3 Details of timeline of advanced computing [14] https://doi.org/10.2991/icitme-18.2018.43

efficiency and has been widely discussed in [18]. It can be achieved in several ways by executing simulations to directly train the AI system instead of adding an AI system to learn the system, also called as learning outputs from inputs. There are several strategies discussed in the [19]. The article also discusses about the possible limitation on the performance gains from hardware and thereby increasing the performance of the system. There exists a possibility that we might have reached the epitome of current configuration and there is a substantiate need for major functionality and performance gain independent from the hardware.

The domain of computational fluid dynamics relies on huge amount to be processed to make any prediction in the behaviour of the fluid. One such study has been presented in the article [20] which shows the importance of architecture implementing the problem while keeping the same computational resource. The improvement was just made because of the change in the architecture of the solving the problem.

HPC with ML has its own set of benefits but it does come with its own challenges. There are several areas which need to be properly defined before implementing the ML algorithm. Due to the shear size of datasets and memory requirement for training it becomes very crucial to ensure the scalability to address the challenge of data partitioning, load balancing. Coherent storage, movement mechanism is important to minimise the issue of bottlenecks on performance. HPC and ML algorithms must be designed together to maximise the performance since traditional ML are designed without considering the architecture of HPC. The memory requirement for training the deep networks is very enormous, to cater for this requirement both HPC system and memory optimisation (data parallelism, distributed training, and memory efficient algorithms) should be considered to get the maximum performance from the system. Overall HPC system should also be energy efficient, and it should be able to reproduce or replicate the research.

6 Edge Computing

An Edge computing can be defined as a distributed computing architecture at the end receiving of the network i.e., user or closer to the data thereby reducing the latency while processing the data and saving the network cost (see Table 1).

The Edge Computing roots date back to 1990s when the CDNs (Content Delivery Networks) was introduced to increase the web performance [22]. The CDNs used nodes at the end of network to cache web content to boost the performance. The Edge computing uses the same principal of CDN and extending it further by leveraging the cloud computing infrastructure. It provides several benefits by reducing the latency which is very beneficial in the field of autonomous systems such as autonomous cars. When the data is processed locally, it saves the bandwidth of the network which would have been used to transmit big chunks of data from user to the server. It also improved the reliability of the system because of the local processing of the data with the added benefit of enhanced data privacy and security.

The combination of Edge computing and Artificial intelligence provides a promising solution to the challenges of AI. This new amalgamation is called Edge intelligence [23], also called Mobile intelligence [24]. It is a set of system which collects the data from the source and process it locally without sending it to the cloud based server, which reduces the latency and saves the bandwidth of the network [25]. Moreover, it enables the user to train the ML/DL model with the self-generated data [26], one such application is the G-board by Google which is trained with the input from the user and predicts the suggestion for the coming sentences. Intelligence can be used at the edge with two ways, one is using the DL (Deep Learning) at the edge which means some optimization being performed at the edge with DL while the other is edge computing for DL which focuses on applying the DL in context of edge computing [27] (see Fig. 4).

Simple Neural networks at the edge can also be implemented on the edge [28], however the performance of the system might be compromised and if DL are used it requires a lot of computing, energy requirement. Such configuration might not be beneficial for the devices where a high computational source is not viable such as UAVs [28]. Potential solutions to the resource constraints would be training smaller

Table 1 Computing evolution [21]

Early computing	Runs on isolated computing machines
Personal computing	Runs locally at user's end or data center
Cloud computing	Runs at the data center, processing via cloud
Edge computing	Runs closer to the user

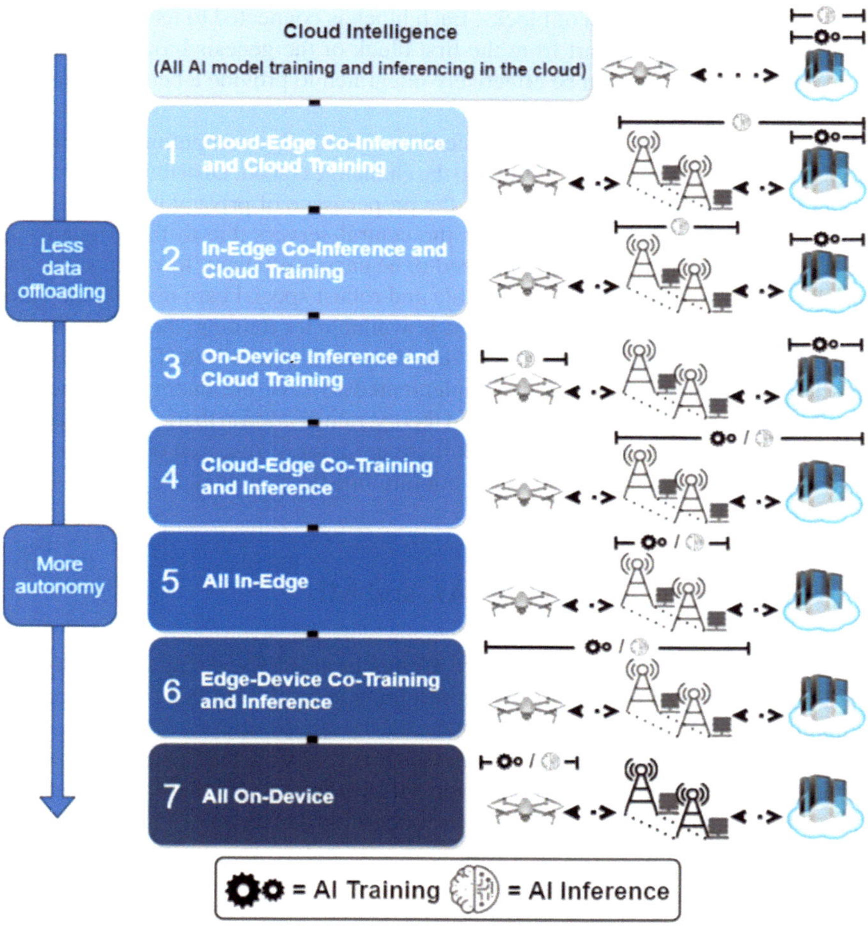

Fig. 4 Edge AI categorization [28]

networks, pruning, low-rank factorisation, quantization, hardware, and software co-design, blockchain, federation learning [29, 30].

Federation Learning and Blockchain are some of the promising solutions for the challenges faced sur to resource constraints. Federation Learning is a decentralized ML technique where instead of collecting data and training using combined data at one place, each device at the edge trains a model based on the local data. These trained models are further sent to the server (global) using encryption and a global model is created using multiple trained models at the edge. The global model is sent back to the edge and this process continues till a sufficient accuracy is achieved.

Blockchain in another technique which can mitigate the challenges of resource constraints at the edge with Federation Learning. It is a secure ledger which records all the transactions into a chain of blocks. Each block is connected to its previous block by storing a hash value apart from the first block or the genesis block. Blockchain and Federation learning can be effectively integrated to provide a better privacy and security to the system.

The report which focuses on the implementation of edge computing in the field of UAVs [28] provides possible solutions to the challenges faced. Distributed algorithms are a reliable solution to the to address the proper issue of privacy when mission is critical, and training cannot happen at the central servers due to the criticality of the missions. Distributed algorithms need to be energy efficient, low—capacity and simple. To make the system more reliable and robust special care needs to be taken care of data scarcity, when not much data is available for training, the system should exchange the data with the central servers through a secure link. Data and model parallelization techniques must be implemented to save the energy and memory requirements, along with the mean—field theory, rate distortion theory tools to reduce the latency in the system. Even though there are a lot of research questions which needs to be answered to evolve the system into full-fledged model.

7 The Convergence of HPC, AI, and ML

The convergence of High-Performance Computing (HPC), Artificial Intelligence (AI), and Machine Learning (ML) represents a pivotal moment in the world of computing. This convergence has the potential to reshape scientific research, industrial processes, and decision-making across numerous domains. Understanding the historical significance of HPC, the AI and ML revolution, and the synergy between these fields is crucial to appreciate the impact of this convergence.

7.1 HPC's Historical Significance

HPC has a rich history dating back to the mid-twentieth century when scientists and engineers first began using computers to solve complex problems in fields like physics, weather modeling, and aerospace engineering. These early HPC systems were characterized by their exceptional computational power and ability to tackle large-scale simulations and numerical analyses. Supercomputers, with their massive processing capabilities, became instrumental in scientific breakthroughs and technological advancements.

However, HPC's historical significance goes beyond scientific research. It has played a critical role in diverse applications, including national security, financial modeling, and drug discovery. These centralized computing environments provided the foundation for addressing complex problems that were previously insurmountable.

7.2 The AI and ML Revolution

The AI and ML revolution has been a game-changer, introducing a new era of computing. AI, powered by deep learning neural networks and other ML techniques, has transformed industries by enabling machines to learn from data, recognize patterns, and make autonomous decisions. This shift has given rise to applications like voice assistants, autonomous vehicles, recommendation systems, and fraud detection.

AI and ML have also gained prominence in research, healthcare, and finance, with applications ranging from genomics and drug discovery to predictive analytics and algorithmic trading. The ability to process vast datasets and extract valuable insights has made AI and ML indispensable in our data-driven world.

7.3 HPC Meets AI/ML

The convergence of HPC and AI/ML is a natural progression in computing. HPC's immense computational power is highly complementary to AI and ML workloads, which often involve training complex neural networks on massive datasets. HPC can accelerate AI model training, making it faster and more efficient. Researchers and data scientists can leverage HPC clusters and supercomputers to experiment with larger neural networks, optimize hyperparameters, and improve the performance of AI models.

Moreover, AI and ML techniques can enhance HPC in return. AI algorithms can be employed to monitor HPC systems, predict and prevent system failures, and optimize resource allocation. Machine learning can help researchers make sense of the massive amounts of data generated by HPC simulations and experiments, leading to more insightful discoveries.

The convergence of HPC, AI, and ML not only enhances their individual capabilities but also opens up new possibilities. It allows us to tackle more complex problems, from climate modeling and drug discovery to autonomous robotics and advanced scientific simulations. This synergy is reshaping the landscape of research, industry, and technology, and its full potential is yet to be realized.

8 Challenges in Remote Environments

Remote environments present a unique set of challenges when it comes to deploying and maintaining high-performance computing (HPC) and artificial intelligence/machine learning (AI/ML) systems. These challenges are diverse and multifaceted, requiring innovative solutions to enable computational capabilities in settings where traditional infrastructure is limited or unavailable.

8.1 Remote Environments Defined

Remote environments encompass a broad spectrum of locations where access to conventional computing infrastructure is either restricted or altogether absent. These settings can include outer space, deep-sea research vessels, arid deserts, isolated rural areas, and even disaster-stricken regions. The common thread among these diverse environments is the absence of readily available resources, particularly in terms of reliable power sources and high-speed network connections.

8.2 Challenges Faced

a. **Resource Constraints**:

In many remote environments, one of the primary challenges is the scarcity of essential resources. Reliable power sources are often limited, if available at all. Traditional HPC and AI/ML systems are power-hungry and require a consistent and ample energy supply. Overcoming this limitation involves developing energy-efficient computing solutions, utilizing renewable energy sources, or relying on power-saving hardware components.

High-speed network connections, which are fundamental for data transfer and remote collaboration, are often lacking. These limitations can impede real-time data analysis and require the development of strategies for efficient data transmission over low-bandwidth or unreliable networks.

b. **Harsh Conditions**:

Many remote environments expose computing equipment to extreme conditions. Harsh temperatures, high humidity, and exposure to dust, moisture, or radiation can lead to hardware damage or malfunctions. Specialized equipment, such as ruggedized servers and protective enclosures, is necessary to safeguard technology in such conditions. Additionally, efficient cooling solutions are essential to prevent overheating in extreme climates.

c. **Latency**:

Low-latency processing is imperative in certain remote applications, especially in scenarios where timely decision-making is crucial. For example, autonomous vehicles, drones, and remote robotic systems must make split-second decisions without relying on cloud-based solutions that introduce network latency. To address this challenge, edge computing solutions, which enable data processing closer to the data source, are often employed.

d. **Data Transfer**:

Transferring large datasets from remote locations to centralized data centers can be a cumbersome and costly endeavor due to bandwidth limitations and network latency.

In some cases, it may be more practical to develop in-situ data processing capabilities in remote environments, allowing data to be processed locally without the need for frequent and extensive data transfers.

Meeting the computational demands of remote environments while addressing these challenges requires a combination of innovative technologies, energy-efficient solutions, and adaptation of existing HPC and AI/ML systems. Researchers and engineers are continually developing strategies and technologies to expand the reach of computing capabilities to previously inaccessible or resource-constrained areas, pushing the boundaries of what is achievable in remote computing.

9 Integration of AI and ML with Edge Computing in Remote Environments

The integration of Artificial Intelligence (AI) and Machine Learning (ML) with edge computing in remote environments represents a transformative synergy that enables real-time decision-making, data analysis, and automation in locations with limited access to traditional computing infrastructure. Customized hardware, distributed AI/ML models, and anomaly detection systems are key components of this integration, facilitating a range of applications in challenging remote settings.

9.1 Customized Hardware

In remote environments, where conditions can be harsh and traditional hardware may not withstand extreme temperatures, humidity, or other environmental factors, the development of specialized hardware is crucial. This can include:

- Ruggedized GPUs (Graphics Processing Units): Custom-designed GPUs that are built to operate reliably in harsh conditions. These ruggedized GPUs can withstand temperature variations, moisture, and dust, making them suitable for remote locations like deserts, jungles, or deep-sea vessels.
- TPUs (Tensor Processing Units): Customized TPUs optimized for AI workloads, tailored to function efficiently in remote environments. TPUs are designed for high-performance AI inference and can enhance the capabilities of edge devices in these challenging conditions.
- AI Accelerators: Edge devices in remote environments often require specialized AI accelerators to speed up machine learning inference tasks while minimizing power consumption. These accelerators may be purpose-built for specific AI workloads encountered in remote applications, such as image analysis or natural language processing.

Customized hardware is critical for ensuring that edge devices can operate reliably in remote environments, where standard equipment may not survive or perform optimally.

9.2 Distributed AI/ML Models

Distributed AI/ML models play a pivotal role in enabling edge devices to operate efficiently in remote environments. These models allow edge devices to perform AI/ML inference and decision-making locally without relying on centralized servers, which may introduce latency and are not always feasible in resource-constrained remote locations. Key aspects of this integration include:

- Lightweight ML Models: Edge devices in remote environments often employ lightweight ML models optimized for their specific tasks. These models require fewer computational resources and can run on hardware with limited processing capabilities. They are ideal for applications like object detection, environmental monitoring, and equipment health analysis in challenging settings.
- Federated Learning: In scenarios where data privacy and bandwidth constraints are critical concerns, federated learning techniques enable edge devices to collaboratively train and update ML models without centralizing data. This distributed approach is particularly valuable in remote environments where the transfer of large datasets to a central location may not be practical.
- Offline Inference: Edge devices can conduct offline AI/ML inference, making decisions based on pre-trained models and locally collected data. This approach minimizes the need for constant connectivity to central servers, enhancing the autonomy and responsiveness of edge devices.

Distributed AI/ML models empower edge devices in remote environments to process data and make decisions at the point of data generation, improving efficiency and reducing reliance on external infrastructure.

9.3 Anomaly Detection

AI-driven anomaly detection systems are instrumental in monitoring equipment health and environmental conditions in remote environments. These systems use machine learning algorithms to identify deviations from expected behavior, providing early warnings and insights that can help predict maintenance needs and prevent equipment failures. Key components of this integration include:

- Sensor Data Analysis: Anomaly detection systems process data from various sensors deployed in remote environments, such as temperature sensors, vibration sensors, and environmental sensors. By analyzing this sensor data, these systems can identify irregular patterns or conditions that could indicate potential issues.

- Predictive Maintenance: The ability to predict when equipment or machinery is likely to require maintenance is invaluable in remote locations. AI-driven anomaly detection can provide actionable insights, enabling maintenance teams to plan interventions before equipment failures occur, reducing downtime and associated costs.
- Environmental Monitoring: In remote environments like deep-sea research vessels or desert installations, monitoring environmental conditions is crucial for safety and equipment longevity. AI-based anomaly detection can help detect and respond to adverse environmental changes, such as extreme temperatures, humidity fluctuations, or pressure anomalies.

Integration of AI and ML with edge computing in remote environments transforms these settings into data-driven, smart ecosystems. Customized hardware, distributed AI/ML models, and anomaly detection systems empower edge devices to function autonomously, making real-time decisions, enhancing equipment reliability, and ensuring the effective operation of remote infrastructure. This convergence is vital for applications across various domains, including scientific research, industrial operations, and environmental monitoring in remote and challenging locations.

Future Prospects, conclusions and prospective usage in the field of Bio-pneumatics:

The expansion of AI and ML breakthroughs in HPC with a shift to edge computing in remote environments is poised for significant growth. Advancements in AI hardware, algorithms, and edge computing technologies will continue to unlock new possibilities for applications in challenging and previously inaccessible locations. The fusion of HPC, AI, and ML with edge computing in remote environments represents a transformative shift in the way we harness computational power. As technology continues to evolve, the potential for innovation in fields ranging from scientific research to autonomous systems in remote locations is boundless. This chapter has explored the pivotal role of AI and ML in this journey, emphasizing their vital contributions to addressing the unique challenges posed by remote environments. The amalgamation of High performance computing with Machine Learning and Deep Learning and its application in Edge Computing provides real time decision-making, enhanced security and privacy, offline capabilities when there is some issue with the network, better co-ordination between the edge's nodes and the main server. The Edge intelligence has opened a new frontier of research and development which a lot of potential to address the challenges of the new age of advanced computing. Its main application will not be just limited to autonomous system but also towards developing smart cities and building advanced and robust Health care systems. The field of Bio-Pneumatics is the application of pneumatics system in medical context. Edge intelligence will help in real time monitoring and diagnosis, example: by monitoring the real time vital signs such as oxygen level, blood pressure etc. and would detect the abnormalities as soon as they occur and thereby reducing the risk of patient. The continuous data can also be utilised to determine the maintainability of the pneumatic system based on the historical data which would help in preventing the sudden breakdown of critical subsystems. The optimisation of the resources allocated to a certain area and reducing the need for high bandwidth to transmit data to the main server is also one of the main added advantages of edge intelligence. It can further be adapted to personalised needs based on the requirements such as a patient would require certain amount of oxygen based on the oxygen levels inside the body. The system would then automatically adapt to the situation and transfer a certain amount of oxygen based on the normal levels of oxygen inside the body. These are such few examples of application of Edge intelligence in the field of Bio-pneumatics. The door to this new frontier has just opened and there are endless possibilities.

References

1. Russell, S.J. (Stuart Jonathan), 1962: Artificial Intelligence: A Modern Approach. Prentice Hall, Upper Saddle River, N.J. (2010)
2. Bostrom, N.: Superintelligence: Paths, Dangers, Strategies (2014)
3. The Tech Advocate (n.d.)/A tech portal. https://www.thetechedvocate.org/the-two-main-barriers-against-deep-learning/
4. Samuel, A.L.: Some studies in machine learning using the game of checkers. IBM J. Res. Dev. **44**(1.2), 206–226 (2000). https://doi.org/10.1147/rd.441.0206
5. GeeksforGeeks (n.d.): A computer science portal for geeks. https://www.geeksforgeeks.org/
6. Bain, A.: Mind and Body: The Theories of Their Relation. D. Appleton and Company, New York (1873)
7. Bishop, C.M.: Pattern Recognition and Machine Learning. Chapter 5: Neural Networks (2006)
8. Schmidhuber, J.: Deep learning in neural networks: an overview. Neural Netw. **61**, 85–117 (2015)
9. Bengio, Y., LeCun, Y., Hinton, G.: Deep learning. Nature **521**, 436–444 (2015)
10. Goodfellow, I., Bengio, Y., Courville, A.: Deep Learning. MIT Press (2016)
11. Gustafson, J.L.: Moore's law. In: Padua, D. (ed.) Encyclopedia of Parallel Computing. Springer, Boston, MA (2011). https://doi.org/10.1007/978-0-387-09766-4_81
12. Spataro, W., Trunfio, G.A., Sirakoulis, G.C.: High performance computing in modelling and simulation. Int. J. High Perform. Comput. Appl. **31**(2), 117–118 (2015)
13. Asanovic, K., Bodik, R., Demmel, J., Keaveny, T., Keutzer, K., Kubiatowicz, J., Morgan, N., Patterson, D., Sen, K., Wawrzynek, J., Wessel, D., Yelick, K.: A view of the parallel computing landscape. Commun. ACM. **52**, 56–67 (2009). https://doi.org/10.1145/1562764.1562783
14. Zaharia, M., Xin, R.S., Wendell, P., Das, T., Armbrust, M., Dave, A., Meng, X., Rosen, J., Venkataraman, S., Franklin, M.J., Ghodsi, A., Gonzalez, J., Shenker, S., Stoica, I.: Apache Spark: a unified engine for big data processing. Commun. ACM **59**(11), 56–65 (2016). https://doi.org/10.1145/2934664
15. Reed, D., Gannon, D., Dongarra, J.: Reinventing High Performance Computing: Challenges and Opportunities (2022)
16. Dean, J., Ghemawat, S.: MapReduce: simplified data processing on large clusters. Commun. ACM. **51**, 137–150 (2004). https://doi.org/10.1145/1327452.1327492
17. Fox, G.C., Glazier, J.A., Kadupitiya, J.C.S., Jadhao, V., Kim, M., Qiu, J., Sluka, J.P., Somogy, E., Marathe, M., Adiga, A., Chen, J., Beckstein, O., Jha, S.: Learning everywhere: pervasive machine learning for effective high-performance computation. In: IEEE International Parallel and Distributed Processing Symposium Workshops, IPDPSW 2019, Rio de Janeiro, Brazil, May 20–24, 2019, pp. 422–429 (2019)

18. Fox, G., Jha, S.: Understanding ML Driven HPC: Applications and Infrastructure (2019). https://doi.org/10.1109/eScience.2019.00054
19. Garcia-Gasulla, M., Mantovani, F., Josep-Fabrego, M., Eguzkitza, B., Houzeaux, G.: Runtime mechanisms to survive new HPC architectures: a use case in human respiratory simulations. Int. J. High Perform. Comput. Appl. **34**(1), 42–56 (2020). https://doi.org/10.1177/1094342019842919
20. Simplilearn (n.d.): https://www.simplilearn.com/
21. Satyanarayanan, M.: The emergence of edge computing. Computer **50**(1), 30–39 (2017). https://doi.org/10.1109/MC.2017.9
22. Wang, X., Han, Y., Wang, C., Zhao, Q., Chen, X., Chen, M.: In-edge ai: Intelligentizing mobile edge computing, caching and communication by federated learning. IEEE Netw. **33**(5), 156–165 (2019)
23. Wang, Z., Cui, Y., Lai, Z.: A first look at mobile intelligence: architecture, experimentation and challenges. IEEE Netw. (2019)
24. Khelifi, H., Luo, S., Nour, B., Sellami, A., Moungla, H., Ahmed, S.H., Guizani, M.: Bringing deep learning at the edge of information-centric internet of things. IEEE Commun. Lett. **23**(1), 52–55 (2018)
25. Chen, F., Dong, Z., Li, Z., He, X.: Federated meta-learning for recommendation (2018). arXiv:1802.07876
26. Wang, X., Han, Y., Leung, V.C., Niyato, D., Yan, X., Chen, X.: Convergence of edge computing and deep learning: a comprehensive survey. IEEE Commun. Surv. Tutor. (2020)
27. Anwar, M.A.: Enabling edge-intelligence in resource-constrained autonomous systems. Ph.D. dissertation, Georgia Institute of Technology (2021)
28. McEnroe, P., Wang, S., Liyanage, M.: A survey on the convergence of edge computing and AI for UAVs: opportunities and challenges. IEEE Internet Things J. 9 (2022)https://doi.org/10.1109/JIOT.2022.3176400
29. Faniadis, E., Amanatiadis, A.: Deep learning inference at the edge for mobile and aerial robotics. In: 2020 IEEE International Symposium on Safety, Security, and Rescue Robotics (SSRR), pp. 334–340. IEEE (2020)
30. Ansari, M.S., Alsamhi, S.H., Qiao, Y., Ye, Y., Lee, B.: Security of distributed intelligence in edge computing: threats and countermeasures. In: The Cloud-to-Thing Continuum, pp. 95–122. Palgrave Macmillan, Cham (2020)

Role of Distributed Computing in Biology Research Field and Its Challenges

Bahiyah Azli and Nurulfiza Mat Isa

Abstract Experimental biology and bioinformatics have amazingly allowed us to understand nature, its living organisms, and its environments. In this book chapter, we explained the role of distributed computing applications in solving biology research questions and their challenges during application. The use of high-performance computing, specifically distributed computing into biological protocols, reduces the runtime, captures discrete biological interactions, increases collaborative teamwork initiatives, and speeds up the process of bridging the gap of biological knowledge. Although the integration of computer science and the biology research field is elucidated with promising advantages, researchers who adopted this system into their experiments are still faced with several challenges, such as cost-to-demand issues, lack of expertise, compatibility issue, and more. Nevertheless, many intensive interventions from biologists, mathematicians, statisticians, and computer scientists have started collectively utilizing advanced distributed computing environments in these biological research techniques. This backbreaking move hopes to be the driving force toward the progress of biology research and findings. This book chapter will summarise some recent applications of distributed computing in experimental biology and bioinformatics. It will further discuss its advantages, challenges, and limitations, as well as future directions for integrating both knowledge.

Keywords Bioinformatics · Biology · Distributed computing · Experimental biology · High-throughput sequencing · Omics

B. Azli · N. Mat Isa (✉)
Laboratory of Vaccine and Biomolecules, Institute of Bioscience, Universiti Putra Malaysia, 43400 Serdang, Malaysia
e-mail: nurulfiza@upm.edu.my

N. Mat Isa
Faculty of Biotechnology and Biomolecular Sciences, Department of Cell and Molecular Biology, Universiti Putra Malaysia, 43400 Serdang, Malaysia

K. A. Ahmad et al. (eds.), *High Performance Computing in Biomimetics*, Series in BioEngineering, https://doi.org/10.1007/978-981-97-1017-1_6

1 Introduction

This book chapter aims to reveal a comprehensive picture of the role of a distributed computing environment in the biology research field and its challenges during applications. The discussion throughout this chapter may provide a better understanding of different sub-biological fields with recent examples. The hope is that the elucidation and illustration explained in this chapter will guide researchers around the globe to point their research scope direction or redirection and set a valuable basis for future advancement and utilisation of high-performance computing- specifically distributed computing in answering biological research objectives of interest. The potential benefits of adopting distributed computing in experimental biology research are huge, significant, and worthwhile.

This book chapter will cover the following:

1. What is the difference between experimental biology and bioinformatics?
2. How are conventional and 'modern' biology related?
3. How is high-performance, parallel, and distributed computing applied to the biology research field protocols?
4. What are the advantages and challenges of adopting distributed computing into the biology research field protocols?
5. What are the future directions and researchers' hopes in applying distributed computing in the biology research field?

1.1 Biology: Experimental Biology Versus Bioinformatics

Biology is the scientific study of living organisms, chemical processes, molecular interactions, physiological manner and mechanisms, their physical structures, and interactions with the surrounding environments. This research field branch encompasses a wide range of sub-fields, such as genetics, molecular biology, biochemistry, ecology, evolution, and physiology. The interdisciplinary nature of biology research makes it a constantly evolving and dynamic research field, with discoveries and breakthroughs being made all the time. Biologists have used various techniques and advanced tools, including microscopy, gene editing via genetic engineering, computational modelling, and molecular techniques, intending to study living organisms at different levels of complexity, from molecules to cells to ecosystems. Like any other field of study, biology has also undergone some advancement, progress, and growth. After years of ongoing research, current biologists have established a more detailed sub-category of biology-related research investigation- experimental biology and bioinformatics [1].

Experimental biology is the set of protocols, methods, and approaches in the biology field that investigates and answers biological research questions by conducting several designated experiments. Experimental biology is the opposed term of theoretical biology which concerns the abstract explanations of the biological system [2]. In experimental biology, laboratory-focused experimentation retrieves valuable quantitative and qualitative data to validate a biology theory. Techniques such as tissue culturing, microscopy, plaque assay, genetic engineering via kits and reagents, biochemical assay, and more are a few common protocols applied in the laboratory setting. Millions of laboratory setting experiments have successfully retrieved important data, allowing global initiatives to construct proven biology mechanisms and processes. Nevertheless, experiment-based techniques and data still have limitations during the data retrieval or processing stage [3].

Meanwhile, bioinformatics is a relatively new interdisciplinary field that emerged in early 1980 at the intersection of biology and computer science [4]. It is expected that the bioinformatics research field's progress will exponentially grow alongside the advancement of computing and information technology, as hypothesised by Moore's Law [5, 6]. This field involves the application of computational tools and techniques to analyse biological data, particularly large-scale genomic data. The start of the bioinformatic field is famously known from the Human Genome Project initiative to sequence the inter-population genome into a large central database [7, 8]. This project generated vast amounts of data that required advanced computational power for storage, processing, and analysing. Presently, bioinformatics is the only discipline that has the potential to handle those sequence data and extract valuable significant insights from it. This newly emerged subfield is a powerful tool that has revolutionised the field of biology, further enabling researchers around the world to analyse genomic data efficiently with promising excellent efficiency and high accuracy discovery.

The biology research field is vital to the global understanding of life and has numerous practical applications in essential practices such as medicine, agriculture, and biotechnology. Both experimental biology and bioinformatics are indeed interrelated, despite having different focuses, techniques of data retrieval, and data analysis proses. Only integrating both findings allows us quickly solve biological questions with high confidence towards the established novel insights.

1.2 From Conventional to the 'Modern' Experimental Biology

The conventional and 'traditional' biology experiment protocols have contributed immensely to our understanding of living organisms and environments. Nevertheless, it also has several limitations. Here are some of the limitations of conventional laboratory setting biology research protocols:

1. **Reductionism**: Conventional biology research often focuses on individual components of a system, such as a single protein or gene. This approach potentially results in an oversimplified understanding of complex biological systems. Although various successful previous published experiments have managed to construct detailed biological mechanisms, these approaches normally took years of research from various groups of biologists. Also, this research approach tends

to ignore the interactions between different components as designed experiments are intended to be highly specific towards listed experiment variables.

2. **Laboratory experiment limitations**: Many biological phenomena are difficult or impossible to study in vivo or in vitro due to ethical or practical limitations. Unless the designated laboratory experiments are endorsed and approved by certain research ethics committees (such as IACUC for animals and IRB for humans), it may not be possible to perform certain experiments on humans or endangered species.
3. **Sample size**: In conventional biology research, sample sizes are often limited due to cost or practical considerations. This lack of suitable sample size can lead to a bias or incomplete result, as small sample sizes may not represent larger populations. Additionally, current experimentations require an approved and statistically significant sample size to deem statistically appropriate [9].
4. **Time and resource constraints**: Biological experiments can be time-consuming and resource intensive. Experimental constraints limit the number of experimental trials that can be performed and lead to slower progress in research.
5. **Lack of integration**: Conventional biology research often focuses on a single aspect of biology, such as genetics or physiology, and may not integrate multiple disciplines. The lack of integration between outputs can lead to a fragmented understanding of the whole concept of complex biological systems.
6. **Limited generalisation**: Many biological experiments are performed in highly controlled and artificial laboratory conditions, which may not accurately reflect real-world environments. This can limit the generalisations of results against the natural world settings.

Due to the key limitations of experimental biology mentioned above, many biologists and scientists have started to integrate their research with the newly emerged bioinformatics protocols. In this field, bioinformatics caters to the large biological 'big data' retrieved from workbench biological samples, which are then translated into a suitable computer-friendly binary format [5, 10, 11]. However, despite the availability of bioinformatics application software on the net for utility, bioinformaticians have slowly recognised the limitations of these tools in analysing their ever-growing big data [12, 13].

The top two most common problems during bioinformatics data processing are the time taken to complete a task and the cost of setting up research-personalised workstations. These limitations limit the researchers from conducting in-depth analyses, especially in the heavy human DNA sequence data [14]. Hence, this field has slowly adopted and utilised high-performance computing, such as parallel and distributed computing, to solve problems efficiently [15].

2 High-Performance Computing, Parallel, and Distributed Computing

High-performance computing (HPC), parallel computing, and distributed computing are all related concepts under computational that involve using multiple computing resources, aiming to solve computationally intensive problems [16, 17]. HPC refers to using powerful computers or a set of clusters of computers utilised to perform computationally exhaustive tasks. This system typically involves specialised hardware, such as high-speed interconnects and Graphical User Interface (GUI), as well as specialised software to manage and facilitate the HPC system. Meanwhile, parallel computing involves breaking down a single large problem or task into smaller, independent sub-task that can be solved simultaneously on multiple processors or computing cores. This computing can speed up input processing with large datasets, dynamic simulations, or other complex analysis tasks.

In contrast, distributed computing involves using multiple computers connected over a network for a better sub-individual task. Each computer connected to the network will perform a portion of the task, and the results of each completed sub-task to produce the final result. Each job is divided into several explicitly defined, independent work units set in the environment. Distributed computing can solve large-scale problems that cannot be solved on a single computer due to memory or processing constraints. Distributed computing uses the strategy of dividing a large workload among multiple computers to reduce processing time or to make use of resources such as programs and databases that are not available on all computers.

Distributed computing is the process of connecting multiple computers via a local area network (LAN) or wide area network (WAN) so that these computers can act together as a single ultra-powerful computer capable of performing computations that no single computer within the network would be able to perform on its own. A few advantages nature of distributed computing is (1) easy scalability, (2) redundancy, (3) memory power, (4) synchronisation, and (5) usage. This type of computing has allowed the system to add connected computers to expand the system for more efficient usage. Besides that, the distributed computing service can keep running despite having one (or more) failed computers going down within the same connected system since many machines provide the same role and services. Additionally, in distributed computing system style, each computer has its processors and memory, allowing the service to have more memory capacity and not utilise a single shared master clock, hence adopting a synchronisation algorithm. Lastly, the distributed computing style technology enhances scalability and promotes resource sharing. Distributed computing is the best application for building and deploying powerful applications that run across many users and geographies. The known cloud-based computing, edge computing, and software as a service (SaaS) are a few examples of applications built using the distributed computing system architecture [18].

These three computing systems are advanced, necessary, and beneficial in solving computationally demanding experimental biology and bioinformatics research questions, leveraging multiple computing resources and systems.

3 Role of Distributed Computing Application in the Biology Research Field

Previously, HPC and parallel computing have been increasingly applied and used in performing experimental biology and bioinformatics research. Parallelisation of biology tools such as protein modelling software PyMOL [19], multiple sequence alignment software MEGA-X [20], and genome editing software BioEdit [21] successfully reduced the execution time at a fraction of the setup cost. However, with increased genomic data, parallelisation has begun to slacken.

Presently, bioinformaticians and pipeline developers have come forth with the adoption of distributed computing into the analysis software. These computing systems allow the processing and analysing of a large amount of sequence data generated via high-throughput sequencing, imaging, and other experimental techniques with great performance [22]. However, it is important to note that not all bioinformatics analysis pipelines have utilised the distributability capacity of distributed computing. According to consensus, data storage and retrieval, genome analysis, and molecular modelling are the only top three bioinformatics analyses mentioned by researchers which require intensive and very good distributability of multiple computers, as these analyses tend to cater to a large amount of data, up to Mbs and Gbs per file [23, 24].

Below are several of identified role of distributed computing in solving biology research questions;

1. **Genomics**: High-throughput sequencing technologies generate vast amounts of data that can take hours, weeks, or months, up to a year, to analyse on a single computer. The application of distributed computing facilitates the analysis of the magnitude of data retrieved from omics experiments [25]. Likewise, the assessment of genome sequence alignment against an available database such as BLAST [26, 27], HMMER [28], RGI [29], and else involves a lot of local and global nucleotide calculation amongst the vast sequences archived in bioinformatics databases. This process overwhelms the computational capacity of a computer. Hence, several genetic sequence alignment pipelines have adopted the distributed computing environment to allow alignment tasks executed in multiple machines [30].

The MultiPhyl project: This project integrates distributed computing into the MultiPhyl software to study the evolution of species [31] and to create a map of the evolutionary relationships between all living organisms [32]. This project is the first high throughput implementation of distributed phylogenetic platforms capable of benefiting semi-idle-connected computers. The program is set to utilise the most accurate phylogenetic calculation algorithm, the maximum-likelihood (ML) models [33], and various statistical models for end-users to choose from during analysis. The developer has deployed the distributed system across the National University of Ireland via the idle computing resources of several hundred desktop computers.

DPRml project: Like MultiPhyl, DPRml is another distributed computing-bioinformatics pipeline project built to execute phylogenetic tree construction and analysis [13]. The group of bioinformaticians had identified the suitability of phylogenetic analysis to the heterogeneous distributed computing nature and successfully developed DPRml, a fully cross-platform distributed application for other users. It was reported that multiple DPRml had run simultaneously while achieving a near linear speed up upon execution [34].

Genome Analysis Toolkit (GATK) project: The GATK project is constructed to help researchers to analyse large genome sequences efficiently [35]. The GATK is mentioned to be built based on MapReduce and has shown great efficiency in calculating sequence depth coverage and calling out variants [36].

2. **Structural biology**: Structural biology experiments often involve analysing large datasets of protein structures, which can be computationally demanding.

The Genome@home project: The Genome@home project uses a special algorithm to design new genes that can form proteins for various purposes [37]. This pipeline allows researchers to virtually design an amino acid sequence of interest against a match existing proteins in the Protein Data Bank [38], generating a homology protein prediction structure. Through the three-dimensional structure of interest protein, researchers can annotate its potential functions and evolution from different species.

3. **Image analysis**: Experimental biology researches often involve imaging large samples at high resolution, generating large amounts of image data. Distributed computing can process and analyse these images, enabling researchers to identify patterns, quantify features, and perform statistical analysis.
4. **Simulation and modelling**: Computational modelling and simulation are increasingly used in experimental biology research to understand complex biological simulations. HPC and distributed computing can be used to accelerate the processing of simulations, and modelling of mathematical equations, further enabling researchers to illustrate large-scale biological systems [39, 40]. The valuable insights from simulation and in silico modelling help scientists basis information for the experiment designing stage more efficiently.

The Folding@home project: The Folding@home project uses a client-server model, where the client software is installed on the volunteers' computers, and the server distributes simulation tasks to them [41]. This project has been instrumented to understand how proteins fold and are misfolded, which sheds light on the pathology of hereditary diseases such as Alzheimer's [42] and Cystic fibrosis [43].

5. **System biology**: System biology is a field that aims to understand biological systems as a whole rather than an individual component in isolation [44]. It involves the integration of experimental and computational approaches to model and thoroughly analyse complex biological systems, such as metabolic pathways [45] or gene regulatory networks [46]. System biology researchers use mathematical models and simulations to predict the behavior of biological systems. They also often rely on high-throughput experimental techniques such as microarray

analysis or mass spectrophotometry that capture a tenth to a hundredth of data in a single setting [47].

Gene Expression Omnibus (GEO) project: Gene expression analysis involves studying how genes are turned on or off in response to different stimuli. The data generated is large; hence distributed computing environment is required during the processing and analysing stage. Ever since established in the year 2000, the GEO database has started to support analyses from genome methylation, genome-protein interactions, and even chromatin structure. Currently, GEO allows annotation and visualisation of archived metadata on their homepage [48].

6. **Accelerated drug discovery and personalised medicine**: Analyses of structure and functions of molecules of interest via distributed computing-adopted software provide a great medium for researchers to discover new drugs or execute optimisation of existing ones.

The Gromacs@home project is a distributed computing project that uses the Gromacs molecular dynamics simulation software to study the behavior of proteins and other biomolecules. Initially, the project aim is to develop new drugs and therapies for diseases such as cancer, Alzheimer's, and Human immunodeficiency virus (HIV) [49, 50]. Nowadays, research groups that study Multiple sclerosis [51], old-age diseases [52], and autoimmune [53], to name a few, have integrated Gromacs into their research methods, resulting in the identification of potential drug structure prediction for therapy purposes.

7. **Inspiration for biomimicry**: Despite having different goals and applications, experimental biology, bioinformatics, and biomimicry can complement each other. Both experimental biology and bioinformatics findings can enlighten and provide a vital understanding of the structure, functions, and mechanism information of biological systems of interest, inspiring biomimicry. Bioinformatics data can also inform biomimicry and help identify key biological features to mimic, optimise and prioritise. Despite the indirect outcome, bio-inspiration and biomimicry have been listed as benefits of distributed-computing applications in biological research protocols [54, 55].

4 Challenges and Limitations of Distributed Computing Application in a Biology Research Field

Without a doubt, HPC and distributed computing system plays a huge role in the current biological research field's novel findings, pace, and ability. The wide range of resource sharing and collaborations over wide and open networks permits researchers to maximise their research methods to the fullest to answer every possible biological research question in their research scope. The huge volume of biological data from experimental biology and bioinformatics sequencing, which is heterogeneous, autonomous, and dynamic, is fully analysed and not left unprocessed.

Implementing HPC and distributed computing onto biological data research pipelines is considered the stepping stone of progress and direction of the current biology research field. Nevertheless, researchers have elucidated several challenges, both when applying distributed computing systems to relevant analyse software and applying a distributed computing-adopted analysis pipeline into their research methods to solve experimental biology research questions. For ease of understanding, these challenges will be sorted into two points of view; end-users and pipeline developers.

From the point-of-view of end-users, such as biologists, chemists, biostatisticians, computational biologists, and end-user bioinformaticians:

1. **Interpretation of knowledge**: Acknowledging the gaps between the real setting and the 'calculate' setting during system biology's mathematical modelling is crucial. Due to the specificity of expertise and integration of research fields, there is a need for constant communication between biologists and mathematicians to ensure parameters set during analysis via supercomputers are not overestimated or underestimated [56]. These two expert groups need to speak the same language to interpret the system biology output significantly and thoroughly, to bridge the gap, and reduce the limitations in solving real biological phenomena. Previously, a list of common bioinformatics tools had inaccurately overestimated the detrimental effect of a few tested gene mutations [54], which, if not rectified soon, could lead to incorrect biology understanding.
2. **Technical expertise**: Setting up and managing distributed computing environment requires technical experts in software engineering, system administration, and networking. Researchers who lack this expertise may need to collaborate with IT professionals or seek external support [57, 58]. Moreover, using existing hardware infrastructure such as networks and semi-idle work station computers to set up the distributed computing environment system will only increase the complexity of managing such a system, compared to an easier setup of a single computer or parallel computing [23]. Besides, this supercomputing environment is highly suggested to be maintained by a full-time system administrator to upgrade and update the system in the long term [13].
3. **Cost**: Distributed computing is indeed one of the most chosen systems as applications with efficient running time and sensitive output. However, distributed computing can also be cost-effective compared to maintaining local computing resources. It can still be expensive to use or purchase access to computing clusters if the user application demand exceeds the configuration cost. Next, it is important to note that not all biological data is suitable for employment in the distributed computing environment. Hence, most of the mentioned pipelines which adopted the distributed computing system often cater to specific groups of researchers online. To make supercomputing worthwhile initiatives, the application of distributed computing to biological data needs to exhibit a high 'compute-to-data' ratio to make the costing application worthwhile, especially in building a connected computer rather than just parts of a local system [13].

4. **Biological noise**: Biological samples have much noise that is hard to omit and can only be canceled if the researchers are well aware of the type of noise factor presence. Despite the ability of HPC or advanced distributed computing prowess, biological noise cancelation retrieved together during the experimental data retrieval process are occasionally highly technical and requires manual consideration of cancellation- depending on each factor. This biological trait raises the potential of producing a systematic error during model construction [59] and imaging [60].
5. **Unknown mechanisms and open boundaries**: Not all mechanisms for cellular or metabolic activities are studied in great detail nor one experiment setting. Different organisms are elucidated to have species-specific metabolic mechanisms carried out internally that may mimic similarly between paralogues groups but differ from the orthologues groups. This highlights the challenges in interpreting system biology insights. Moreover, most large biological data retrieved disregard the biological boundary's presence, as external conditions and factors are usually unknown, cannot be controlled, and are hard to measure. Due to this, to an extent, any mathematical modelling constructed is safe to only called a snapshot view of the current understanding derived from currently available datasets. Nevertheless, the prediction made via the biological model is still significant and valuable as of the systematically designed new informative experiments in the lab [61].

From the point-of-view of pipeline developers, such as computer scientists, mathematicians, statisticians, and back-end bioinformaticians:

1. **Network connectivity issue**: The manager or worker architecture of distributed computing requires reliable and constant network connections between components inside the environment. In brief, high bandwidths are needed to ensure the downloading process of any large databases file required for analyses in a reasonable time.
2. **Suitability of biology research software and its complexity**: Setting up and configuring a distributed computing network may not be very complex for a specialised expert. Additionally, there is a need to ensure that algorithm and configuration settings that allow the speed up of the rate of running time do not interfere with the sensitivity of searches nor the replicability of output results [13]. Experimental biology protocols are not just defined as significant when retrieving statistically significant results but also with the rate of replicability between experiments. Few researchers have raised concern about the potential of lowering the sensitivity of data searching or analysing due to the greedy heuristic algorithm that will take the best immediate and quick solutions, which can result in output that is far than optimal against the mimic 'natural' settings [23, 24, 61].
3. **Data transfer limitations**: Transferring large datasets between computers can be time-consuming and may require high-speed network connections. No applications can be executed well without a secure network connectivity issue [24]. The data sent for processing is dynamically transferred to computing nodes attached to the distributed computing system at a certain runtime. Currently,

sequencing machines such as Illumina, PacBio, and Nanopore have been updated and improved yearly with a low-cost fee per sample [62]. So, we are seen exponentially increasing genomic sequence uploading and strong activities in related databases [63]. Without an efficient data file transfer system, the cost of data transfer will be the limiting factor counteracting the benefits of HPC and distributed computing.
4. **Interoperability and compatibility issues**: The hardware and software used on different computers in a distributed network may not be compatible, leading to issues with data transfer or analysis [61].
5. **Data management**: Distributed computing involves processing and analysing vast amounts of biologically sequenced data, which requires careful data management. Researchers must ensure that the data is properly labelled, annotated, and stored to avoid errors and confusion.

Addressing these challenges requires careful planning, coordination, and technical expertise. Researchers must ensure that their workflows are optimised for distributed computing environments and that their data is properly managed and protected.

5 Future Directions of Distributed Computing Application

Distributed computing has already had a significant impact on the biology research field, and it is expected to grow tremendously. Here are some of the potential future directions of distributed computing applications in the biology field:

1. **Edge computing**: Edge computing involves processing data at the edge of the network, close to the source of data. In biology, edge computing could be used for real-time data analysis from sensors or wearable devices, allowing researchers to monitor biological processes in real time.
2. **Machine learning**: Machine learning algorithms require large data and computational power to train and optimise models. Distributed computing can speed up the training process and develop more accurate and powerful biological models and system biology insights [64, 65].
3. **Data sharing and collaboration**: Distributed computing can facilitate data sharing and collaboration between researchers, making it easier for resource pooling and exchanges to tackle more complex biological questions that demand numerous data.

6 Conclusion

Distributed computing application in experimental biology and bioinformatics is a computing strategy that helps solve many biological research problems. Larger bodies of scientific application have proven the benefits of HPC during processing and analysing large biological data, such as genomic sequencing, image analysis, system biology, biomimicry, and more. The utilisation of the distributed computing prowess in the biology field allows the mathematical model to be constructed and genomic findings to be tested, albeit to an extent. Most experimental biology and bioinformatics pipelines are still in the stage of utilising the parallel computing environment. However, only certain software adopts the loosely coupled distributed network, despite its proven role and benefits. Indeed, aggregating distributed computing power, storage capacity, relevant software tools, and data can solve research problems and push forward in the global progress of findings novel insights and findings, especially involving biotechnology applications such as drug discovery for personalised medicine and bio-inspired biomimicry.

Although the surge of popularity and demand of these two mentioned computational powers within the biologists and scientists' communities, the implementation of distributed computing is still relatively slow. In this chapter, it has been mentioned that the application of distributed computing into the biology research field has challenges from both spectrums of users. The application of this type of computing is still seen as a risk by many from the biology field due to

1. The lack of detailed understanding of the computational prowess, capacity, and potential of distributed computing have on generating significant biological outputs,
2. The lack of collaboration between biologists, computer scientists, mathematicians, and statisticians in constructing significant novel findings with integrated interpretations, implementations, and applications,
3. The beneficial trajectories towards the researchers' groups (especially in developing countries) do not interest most researchers as available grants budget are highly limited, and oftentimes, constructing a biological pipeline that only caters to a small group of end-users does not benefit back even a fraction of the cost, to the ongoing research groups.

Nevertheless, a small group of research teams is all we need to start this newly emerging biology-computer science knowledge integration. The number of literature which explained elucidated and discussed these advancing ideas of knowledge integration is still underwhelming. Nevertheless, the increasing number of published articles is promising evidence that this integrated study is getting into the spotlight, despite the slow and steady manner. Finding the current framework and then mimicking the real-life biological system phenomena might take time and a lot of computational capacity to overcome all the challenges of conventional experimental biology protocols and limited bioinformatics software potential. Currently, an increasing number of organisations offer more sophisticated experimental biology protocols with custom solutions that enable better data analysis processes in automation and resource management.

The idea of the computer science field, such as distributed computing, in relation to the biology field- bound or not, has been rigorously tested and subjected to various research fields. With the observed exponential growth of the computational field, specifically HPC and distributed computing system, it is natural to expect more biological research problems to be solved robustly and rapidly in due time. The goal of researchers, biologists, computer scientists, mathematicians, and statisticians is to provide an excellent environment for answering biological research questions via computationally effective and biologically accurate methods.

References

1. Lapatas, V., Stefanidakis, M., Jimenez, R.C., et al.: Data integration in biological research: an overview. J. Biol. Res. **22**, 9 (2015). https://doi.org/10.1186/s40709-015-0032-5
2. Turney, J.: Life in the laboratory: public responses to experimental biology. Public Underst. Sci. **4**, 153–176 (1995). https://doi.org/10.1088/0963-6625/4/2/004
3. Ilzins, O., Isea, R., Hoebeke, J.: Can bioinformatics be considered as an experimental biological science? Open Sci. J. Biosci. Bioeng. **2**, 60–62 (2015)
4. Trifonov, E.N.: Earliest pages of bioinformatics. Bioinformatics **16**, 5–9 (2000). https://doi.org/10.1093/bioinformatics/16.1.5
5. Gough, E.S., Kane. M.D.: Evaluating parallel computing systems in bioinformatics. In: Third International Conference on Information Technology: New Generations (ITNG'06), pp. 233–238. IEEE (2006)
6. November, J.: More than moore's mores: computers, genomics, and the embrace of innovation. J. Hist. Biol. **51**, 807–840 (2018). https://doi.org/10.1007/s10739-018-9539-6
7. Sawicki, M.P., Samara, G., Hurwitz, M., Passaro, E.: Human genome project. Am. J. Surg. **165**, 258–264 (1993). https://doi.org/10.1016/S0002-9610(05)80522-7
8. Hood, L., Rowen, L.: The human genome project: big science transforms biology and medicine. Genome. Med. **5**, 79 (2013). https://doi.org/10.1186/gm483
9. Kang, H.: Sample size determination and power analysis using the G*Power software. J. Educ. Eval. Health Prof. **18**, 17 (2021). https://doi.org/10.3352/jeehp.2021.18.17
10. Sayers, E.W., Bolton, E.E., Brister, J.R., et al.: Database resources of the national center for biotechnology information. Nucleic. Acids. Res. **50**, D20–D26 (2022). https://doi.org/10.1093/nar/gkab1112
11. Tower, S., Spine, C., Nagar, V.: Bio-Informatics. Bonding Genes. With IT **302023**, 1–8 (2010)
12. Sheng, H.-F., Zhou, H.-W.: Methods, challenges and opportunities for big data analyses of microbiome. Nan Fang Yi Ke Da Xue Xue Bao **35**, 931–934 (2015)
13. Keane, T.M., Page, A.J., McInerney, J.O., Naughton, T.J.: A high-throughput bioinformatics distributed computing platform. In: 18th IEEE Symposium on Computer-Based Medical Systems (CBMS'05), pp. 377–382. IEEE (2005)
14. Goldfeder, R.L., Wall, D.P., Khoury, M.J., et al.: Human genome sequencing at the population scale: a primer on high-throughput DNA sequencing and analysis. Am. J. Epidemiol. **186**, 1000–1009 (2017). https://doi.org/10.1093/aje/kww224
15. Ostrovsky, B., Smith, M.A., Bar-Yam, Y.: Applications of parallel computing to biological problems. Annu. Rev. Biophys. Biomol. Struct. **24**, 239–267 (1995). https://doi.org/10.1146/annurev.bb.24.060195.001323
16. Nobile, M.S., Cazzaniga, P., Tangherloni, A., Besozzi, D.: Graphics processing units in bioinformatics, computational biology and systems biology. Brief Bioinform. bbw058 (2016). https://doi.org/10.1093/bib/bbw058
17. Sanbonmatsu, K.Y., Tung, C.-S.: High performance computing in biology: multimillion atom simulations of nanoscale systems. J. Struct. Biol. **157**, 470–480 (2007). https://doi.org/10.1016/j.jsb.2006.10.023
18. Dowell, R.D., Jokerst, R.M., Day, A., et al.: The distributed annotation system. BMC Bioinformatics **2**, 7 (2001). https://doi.org/10.1186/1471-2105-2-7

19. Rigsby, R.E., Parker, A.B.: Using the PyMOL application to reinforce visual understanding of protein structure. Biochem. Mol. Biol. Educ. **44**, 433–437 (2016). https://doi.org/10.1002/bmb.20966
20. Kumar, S., Stecher, G., Li, M., et al.: MEGA X: Molecular evolutionary genetics analysis across computing platform. Mol. Biol. Evol. **35**, 1547–1549 (2018). https://doi.org/10.1093/molbev/msy096
21. Hall, T., Biosciences, I., Carlsbad, C.: BioEdit: an important software for molecular biology. GERF Bull. Biosci. **2**, 60–61 (2011)
22. Siepel, A.C., Tolopko, A.N., Farmer, A.D., et al.: An integration platform for heterogeneous bioinformatics software components. IBM Syst. J. **40**, 570–591 (2001). https://doi.org/10.1147/sj.402.0570
23. Jain, E.: Distributed computing in bioinformatics. Appl. Bioinformatics **1**, 13–20 (2002)
24. Xu, G., Lu, F., Yu, H., Xu, Z.: A Distributed parallel computing environment for bioinformatics problems. In: Sixth International Conference on Grid and Cooperative Computing (GCC 2007), pp. 593–599. IEEE (2007)
25. Courneya, J.-P., Mayo, A.: High-performance computing service for bioinformatics and data science. J. Med. Libr. Assoc., 106 (2018).https://doi.org/10.5195/jmla.2018.512
26. Yim, W.C., Cushman, J.C.: Divide and Conquer (DC) BLAST: fast and easy BLAST execution within HPC environments. PeerJ. **5**, e3486 (2017). https://doi.org/10.7717/peerj.3486
27. McGinnis, S., Madden, T.L.: BLAST: at the core of a powerful and diverse set of sequence analysis tools. Nucleic Acids Res. **32**, W20–W25 (2004). https://doi.org/10.1093/nar/gkh435
28. Finn, R.D., Clements, J., Eddy, S.R.: HMMER web server: interactive sequence similarity searching. Nucleic Acids Res. **39**, W29–W37 (2011). https://doi.org/10.1093/nar/gkr367
29. Alcock, B.P., Raphenya, A.R., Lau, T.T.Y., et al.: CARD 2020: antibiotic resistome surveillance with the comprehensive antibiotic resistance database. Nucleic Acids Res. (2019). https://doi.org/10.1093/nar/gkz935
30. Kumar, R., Kumar, A., Agarwal, S.: A distributed bioinformatics computing system for analysis of DNA sequences. In: Proceedings 2007 IEEE SoutheastCon, pp. 358–363. IEEE (2007)
31. Keane, T.M., Naughton, T.J., McInerney, J.O.: MultiPhyl: a high-throughput phylogenomics webserver using distributed computing. Nucleic Acids Res. **35**, W33–W37 (2007). https://doi.org/10.1093/nar/gkm359
32. Delsuc, F., Brinkmann, H., Philippe, H.: Phylogenomics and the reconstruction of the tree of life. Nat. Rev. Genet. **6**, 361–375 (2005). https://doi.org/10.1038/nrg1603
33. Guindon, S., Gascuel, O.: A simple, fast, and accurate algorithm to estimate large phylogenies by maximum likelihood. Syst. Biol. **52**, 696–704 (2003). https://doi.org/10.1080/10635150390235520
34. Keane, T.M., Travers, S.A.A., McInerney, J.O., et al.: DPRml: distributed phylogeny reconstruction by maximum likelihood. Bioinformatics **21**, 969–974 (2005). https://doi.org/10.1093/bioinformatics/bti100
35. McKenna, A., Hanna, M., Banks, E., et al.: The genome analysis toolkit: a MapReduce framework for analyzing next-generation DNA sequencing data. Genome Res. **20**, 1297–1303 (2010). https://doi.org/10.1101/gr.107524.110
36. Mohammed, E.A., Far, B.H., Naugler, C.: Applications of the MapReduce programming framework to clinical big data analysis: current landscape and future trends. BioData Min. **7**, 22 (2014). https://doi.org/10.1186/1756-0381-7-22
37. Raha, K., Wollacott, A.M., Italia, M.J., Desjarlais, J.R.: Prediction of amino acid sequence from structure. Protein Sci. **9**, 1106–1119 (2000). https://doi.org/10.1110/ps.9.6.1106
38. Berman, H.M.: The protein data bank. Nucleic Acids Res. **28**, 235–242 (2000). https://doi.org/10.1093/nar/28.1.235
39. Segel, L.A.: Modeling Dynamic Phenomena in Molecular and Cellular Biology. Cambridge University Press (1984)
40. Schuster, S., Fell, D.A., Dandekar, T.: A general definition of metabolic pathways useful for systematic organization and analysis of complex metabolic networks. Nat. Biotechnol. **18**, 326–332 (2000). https://doi.org/10.1038/73786

41. Beberg, A.L., Ensign, D.L., Jayachandran, G. et al.: Folding@home: lessons from eight years of volunteer distributed computing. In: 2009 IEEE International Symposium on Parallel & Distributed Processing, pp. 1–8. IEEE (2009)
42. Polychronidou, E., Avramouli, A., Vlamos, P.: Alzheimer's Disease: The Role of Mutations in Protein Folding, pp. 227–236 (2020)
43. van Willigen, M., Vonk, A.M., Yeoh, H.Y. et al.: Folding–function relationship of the most common cystic fibrosis–causing CFTR conductance mutants. Life Sci. Alliance **2**, e201800172 (2019). https://doi.org/10.26508/lsa.201800172
44. Hucka, M., Finney, A., Sauro, H.M., et al.: The systems biology markup language (SBML): a medium for representation and exchange of biochemical network models. Bioinformatics **19**, 524–531 (2003). https://doi.org/10.1093/bioinformatics/btg015
45. Gaasterland, T., Selkov, E.: Reconstruction of metabolic networks using incomplete information. Proc. Int. Conf. Intell. Syst. Mol. Biol. **3**, 127–135 (1995)
46. Qu, L., Wang, Z., Huo, Y. et al.: Distributed local bayesian network for gene regulatory network reconstruction. In: 2020 6th International Conference on Big Data Computing and Communications (BIGCOM), pp. 131–139. IEEE (2020)
47. Kitano, H.: Perspectives on systems biology. New Gener. Comput. **18**, 199–216 (2000). https://doi.org/10.1007/BF03037529
48. Clough, E., Barrett, T.: The Gene Expression Omnibus Database, pp. 93–110 (2016)
49. Bekker, H., Berendsen, H., Dijkstra, E.J., et al.: Gromacs: a parallel computer for molecular dynamics simulations—ScienceOpen. Phys. Comput. **92**, 252–256 (1993)
50. Van Der Spoel, D., Lindahl, E., Hess, B., et al.: GROMACS: Fast, flexible, and free. J. Comput. Chem. **26**, 1701–1718 (2005). https://doi.org/10.1002/jcc.20291
51. Payab, N., Mahnam, K., Shakhsi-Niaei, M.: Computational comparison of two new fusion proteins for multiple sclerosis. Res. Pharm. Sci. **13**, 394 (2018). https://doi.org/10.4103/1735-5362.236832
52. Pereira, G.R.C., Da Silva, A.N.R., Do Nascimento, S.S., De Mesquita, J.F.: In silico analysis and molecular dynamics simulation of human superoxide dismutase 3 (SOD3) genetic variants. J. Cell. Biochem. **120**, 3583–3598 (2019). https://doi.org/10.1002/jcb.27636
53. Taidi, L., Maurady, A., Britel, M.R.: Molecular docking study and molecular dynamic simulation of human cyclooxygenase-2 (COX-2) with selected eutypoids. J. Biomol. Struct. Dyn. **40**, 1189–1204 (2022). https://doi.org/10.1080/07391102.2020.1823884
54. Graeff, E., Maranzana, N., Aoussat, A.: Biological practices and fields, missing pieces of the biomimetics' methodological puzzle. Biomimetics **5**, 62 (2020). https://doi.org/10.3390/biomimetics5040062
55. Snell-Rood, E.: Interdisciplinarity: bring biologists into biomimetics. Nature **529**, 277–278 (2016). https://doi.org/10.1038/529277a
56. Cechova, M.: Ten simple rules for biologists initiating a collaboration with computer scientists. PLOS Comput. Biol. **16**, e1008281 (2020). https://doi.org/10.1371/journal.pcbi.1008281
57. Condon, A., Kirchner, H., Larivière, D., et al.: Will biologists become computer scientists? EMBO Rep., 19 (2018). https://doi.org/10.15252/embr.201846628
58. Linshiz, G., Goldberg, A., Konry, T., Hillson, N.J.: The fusion of biology, computer science, and engineering: towards efficient and successful synthetic biology. Perspect. Biol. Med. **55**, 503–520 (2012). https://doi.org/10.1353/pbm.2012.0044
59. Tsimring, L.S.: Noise in biology. Reports Prog. Phys. **77**, 026601 (2014). https://doi.org/10.1088/0034-4885/77/2/026601
60. Wang, Y.-L.: Noise-induced systematic errors in ratio imaging: serious artefacts and correction with multi-resolution denoising. J. Microsc. **228**, 123–131 (2007).https://doi.org/10.1111/j.1365-2818.2007.01834.x
61. Disz, T., Kubal, M., Olson, R. et al.: Challenges in large scale distributed computing: bioinformatics. In: CLADE 2005. Proceedings Challenges of Large Applications in Distributed Environments, pp. 57–65. IEEE (2005)
62. Bansal, V., Boucher, C.: Sequencing technologies and analyses: where have we been and where are we going? iScience **18**, 37–41 (2019). https://doi.org/10.1016/j.isci.2019.06.035

63. Slatko, B.E., Gardner, A.F., Ausubel, F.M.: Overview of next-generation sequencing technologies. Curr. Protoc. Mol. Biol., 122 (2018).https://doi.org/10.1002/cpmb.59
64. Greener, J.G., Kandathil, S.M., Moffat, L., Jones, D.T.: A guide to machine learning for biologists. Nat. Rev. Mol. Cell Biol. **23**, 40–55 (2022). https://doi.org/10.1038/s41580-021-00407-0
65. Xu, C., Jackson, S.A.: Machine learning and complex biological data. Genome Biol. **20**, 76 (2019). https://doi.org/10.1186/s13059-019-1689-0

HPC Based High-Speed Networks, ARM Processor Architecture and Their Configurations

Srikanth Prabhu, Richa Vishwanath Hinde, and Balbir Singh

Abstract This chapter looks into the critical aspects of High-Performance Computing (HPC) platforms, high-speed networks, ARM processor architecture, and their configurations for research. High-Performance Computing plays a pivotal role in modern research, enabling the rapid processing of vast datasets and complex simulations across various scientific domains. Key features of HPC platforms, including parallel processing, large memory, high throughput, scalability, and specialized hardware, are explored, highlighting their significance in accelerating scientific discoveries. High-speed networks are essential components of HPC platforms, facilitating efficient communication between nodes and data centers. These networks offer low latency, high bandwidth, fault tolerance, and support for various technologies like InfiniBand and Ethernet. Their role in data transfer, job scheduling, and overall system performance is discussed. The ARM processor architecture, historically associated with mobile devices, is gaining prominence in HPC environments due to its energy efficiency, scalability, vector processing capabilities, and customizability. ARM's increasing adoption in research computing is exemplified by supercomputers like Fugaku, AWS Graviton2 processors, and ARM-based clusters for AI and machine learning workloads. To harness the power of ARM in HPC, specific configurations are required. These configurations involve selecting an ARM-compatible operating system, compilers, libraries, toolchains, cluster management systems, application porting, benchmarking, and energy efficiency optimization. Careful consideration of these aspects is necessary to make the most of ARM-based

S. Prabhu
Department of Computer Science and Engineering, Manipal Institute of Technology, Manipal Academy of Higher Education, Manipal 576104, Karnataka, India

R. V. Hinde
Department of Online Education, Manipal Academy of Higher Education, Manipal 576104, Karnataka, India

B. Singh (✉)
Department of Aeronautical and Automobile Engineering, Manipal Institute of Technology, Manipal Academy of Higher Education, Manipal 576104, Karnataka, India
e-mail: balbir.s@manipal.edu

Department of Aerospace Engineering, Faculty of Engineering, Universiti Putra Malaysia, 43400 Serdang, Selangor, Malaysia

K. A. Ahmad et al. (eds.), *High Performance Computing in Biomimetics*, Series in BioEngineering, https://doi.org/10.1007/978-981-97-1017-1_7

HPC systems. In a rapidly evolving research landscape, understanding the interplay of HPC platforms, high-speed networks, and ARM processors is crucial. Researchers need to adapt to emerging technologies and make informed decisions about hardware and configurations to stay at the forefront of their fields.

Keywords ARM processor · Architecture · High-speed network · High-performance computing

1 Introduction

In the ever-evolving landscape of scientific research and computational studies, High-Performance Computing (HPC) platforms, high-speed networks, and processor architectures play a pivotal role. This chapter explores the importance of these components in research, with a focus on the ARM processor architecture, its configurations, and their relevance. In the pursuit of scientific knowledge and technological innovation, the tools and technologies at our disposal are paramount. Among these, High-Performance Computing (HPC) platforms, high-speed networks, and processor architectures hold a unique and indispensable place. In this chapter, we embark on a comprehensive exploration of the role these components play in the realm of research, with a particular focus on the rising star of processor architecture—ARM, and how its configurations are reshaping the landscape of scientific discovery. The relentless advance of science has led to an explosion of data and computational demands. Researchers are tasked with unraveling complex problems in fields as diverse as astrophysics, genomics, climate modeling, drug discovery, and artificial intelligence. In this context, High-Performance Computing has emerged as the linchpin of scientific progress. HPC platforms, comprised of supercomputers and high-speed clusters of interconnected machines, are designed to tackle these Herculean computational tasks.

2 High-Performance Computing (HPC) Platforms

High-Performance Computing (HPC) is a field of computing that involves the use of supercomputers and high-speed clusters of computers to solve complex computational problems. These platforms are indispensable for research across a wide range of domains, including physics, chemistry, biology, engineering, climate modeling, and data analysis. High-Performance Computing is not merely a niche within the world of computing; it is a powerful force that drives research forward. These platforms are characterized by several key features that make them a vital tool in the researcher's arsenal. First and foremost, HPC platforms harness the power of parallel processing. Instead of relying on a single processor to execute tasks sequentially, these systems utilize multiple processors or cores that work in tandem. This parallelism greatly accelerates the execution of computational tasks, enabling researchers to perform simulations and analyses that would be impossible with conventional computing resources.

Another defining feature of HPC platforms is their ability to accommodate vast amounts of memory. Whether it's processing immense genomic datasets, running climate simulations, or conducting large-scale molecular dynamics simulations, HPC platforms offer the memory capacity required to handle these data-intensive tasks. In addition to parallel processing and ample memory, HPC platforms are engineered for high throughput. This means that they can efficiently move data within the system and between nodes, ensuring that data transfer doesn't become a bottleneck in the research workflow.

Scalability is another hallmark of HPC systems. They are designed to grow with the increasing computational demands of researchers. Whether it's adding more nodes to a cluster or upgrading to faster processors, these systems can be scaled to meet the evolving needs of research projects. Furthermore, HPC platforms often integrate specialized hardware. Graphics Processing Units (GPUs) and other accelerators are commonly used to accelerate specific tasks, such as deep learning, molecular dynamics simulations, and weather modeling. The fusion of general-purpose processors and specialized accelerators enables HPC platforms to excel in a wide range of scientific applications. Figure 1 shows the UVM optimization of the CUDA model for the HPC platform.

As researchers strive to tackle more complex and data-intensive problems, the importance of HPC platforms becomes increasingly apparent. They have become the workhorses of modern scientific inquiry, powering the simulations and analyses that drive innovation and discovery.

2.1 Key Features of HPC Platforms

1. Parallel Processing: HPC platforms use multiple processors or cores to perform computations in parallel, significantly speeding up the execution of tasks.
2. Large Memory: HPC systems are equipped with ample memory to handle massive datasets and complex simulations.
3. High Throughput: HPC clusters are designed for high data throughput, enabling researchers to process vast amounts of data quickly.
4. Scalability: These platforms can be easily scaled to accommodate increasing computational demands.
5. Specialized Hardware: Many HPC systems employ GPUs (Graphics Processing Units) and specialized accelerators for tasks like machine learning and simulations [1, 2].

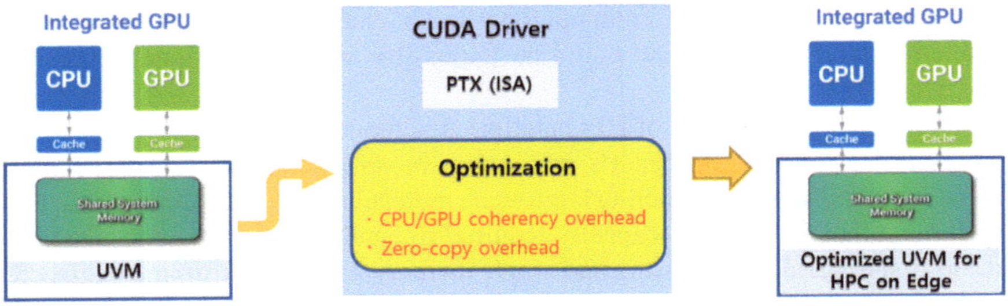

Fig. 1 UVM (Unified Virtual Memory) optimization of the CUDA model for HPC on edge [3]

3 High-Speed Networks

High-speed networks are the backbone of HPC platforms, enabling seamless communication between nodes and data centers. These networks are essential for data transfer, job scheduling, and overall system performance. While HPC platforms are pivotal, their capabilities can only be fully realized when complemented by high-speed networks. These networks form the backbone of the HPC ecosystem, facilitating seamless communication between nodes and data centers.

High-speed networks are distinguished by several key attributes. Low latency is a critical feature, as it minimizes the delay in transmitting data between nodes. In real-time applications, such as weather forecasting or high-frequency trading, low latency is essential for timely decision-making.

High bandwidth is another hallmark of high-speed networks. This attribute ensures that large volumes of data can flow smoothly within the HPC system. Researchers can efficiently transfer and process the vast datasets required for their work. One such example is shown in Fig. 2 with high speed flow control based HPC architechture called FOSquare and the CG latenscy comparence of both FOSquare and Leaf-Spine architectures in Fig. 3.

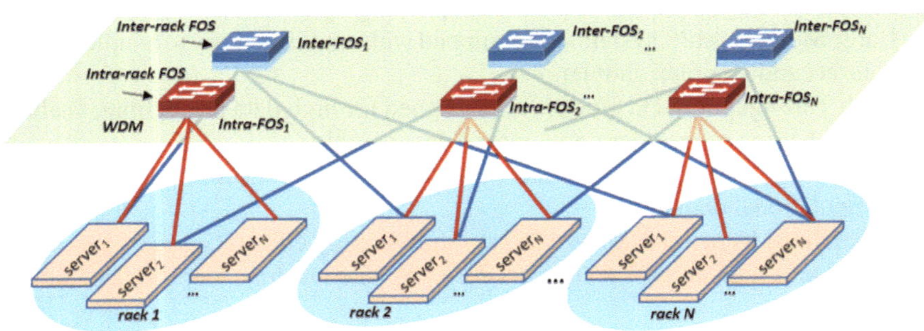

Fig. 2 The FOSquare High-Performance Computing (HPC) architecture based on fast optical switch with flow control [4]

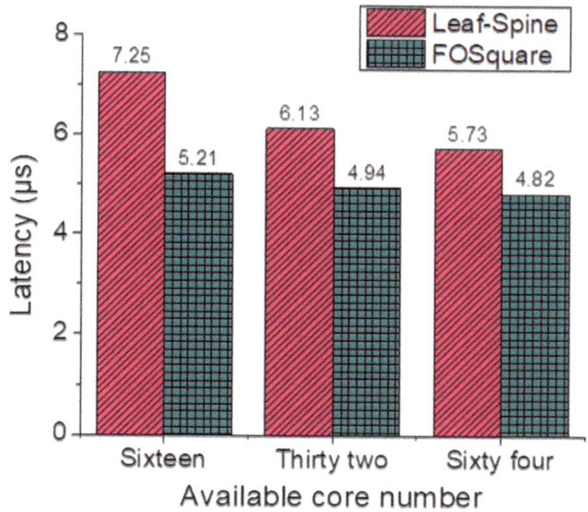

Fig. 3 CG application Latency of FOSquare and Leaf-Sine [4]

Moreover, high-speed networks are designed with fault tolerance in mind. Redundancy and fault-tolerant mechanisms are incorporated to enhance network reliability. In a high-stakes research environment, ensuring continuous access to computational resources and data is of paramount importance.

There are various technologies that underpin high-speed networks, with InfiniBand and Ethernet being the most prominent. The choice between these technologies often depends on the specific requirements of the research tasks at hand. Ethernet is widely used for its ubiquity and compatibility, while InfiniBand offers low-latency and high-bandwidth advantages.

High-speed networks play a critical role in data transfer, job scheduling, and overall system performance. Without efficient data exchange between nodes, HPC platforms would not be able to perform at their peak capacity. As we look deeper into the intricacies of HPC research, the interplay between high-speed networks and HPC platforms becomes ever more evident [5, 6].

3.1 Key Features of High-Speed Networks

1. Low Latency: High-speed networks have minimal delays in transmitting data, crucial for real-time applications.
2. High Bandwidth: These networks offer high data transfer rates, accommodating large data flows.
3. Fault Tolerance: Redundancy and fault-tolerant mechanisms ensure network reliability.
4. InfiniBand and Ethernet: Common high-speed network technologies used in HPC clusters.

3.2 ARM Processor Architecture

The ARM (Advanced RISC Machine) processor architecture has gained prominence in recent years due to its power-efficiency and adaptability. Historically, ARM was primarily associated with mobile devices, but it is increasingly being adopted in HPC environments. In a technology landscape where energy efficiency and adaptability are increasingly paramount, the ARM (Advanced RISC Machine) processor architecture has emerged as a compelling choice. Historically associated with mobile devices, ARM is now making inroads into the HPC domain, promising a fresh perspective on computational research. The ARM architecture is characterized by several key features that set it apart from traditional processors. Perhaps the most celebrated feature is its energy efficiency. ARM processors are renowned for their low power consumption, making them an attractive option for institutions and data centers aiming to reduce their carbon footprint and operating costs.

Scalability is another hallmark of ARM-based systems. Researchers can tailor these systems to meet the performance requirements of various workloads. ARM processors come in a range of configurations, from embedded systems to high-performance server-grade chips, allowing organizations to choose the right fit for their needs.

ARM processors often feature Single Instruction, Multiple Data (SIMD) vector processing units. This capability is particularly valuable for data-intensive applications, as it allows the processor to perform multiple operations on data in parallel. This vector processing power is vital for scientific applications, such as molecular dynamics simulations and data analysis. Figure 4 shows the typical ARM processor configuration.

The customizability of ARM architecture is a boon for research institutions with unique requirements. Whether it's adapting the system for specialized scientific tasks or integrating hardware accelerators, ARM's flexibility enables tailored solutions.

ARM's growing adoption in the HPC space is evidenced by notable examples. Fugaku, the world's fastest supercomputer as of 2020, is powered by ARM-based processors, marking a significant shift in the supercomputing landscape. In the cloud computing realm, Amazon's AWS Graviton2 processors, based on ARM architecture, have opened the door to ARM-powered HPC solutions. Additionally, ARM-based clusters are becoming increasingly popular for AI and machine learning workloads, leveraging the architecture's energy efficiency for large-scale data processing and training deep neural networks [7, 8].

3.3 Key Features of ARM Architecture

1. Energy Efficiency: ARM processors are known for their low power consumption, making them suitable for energy-conscious HPC deployments.

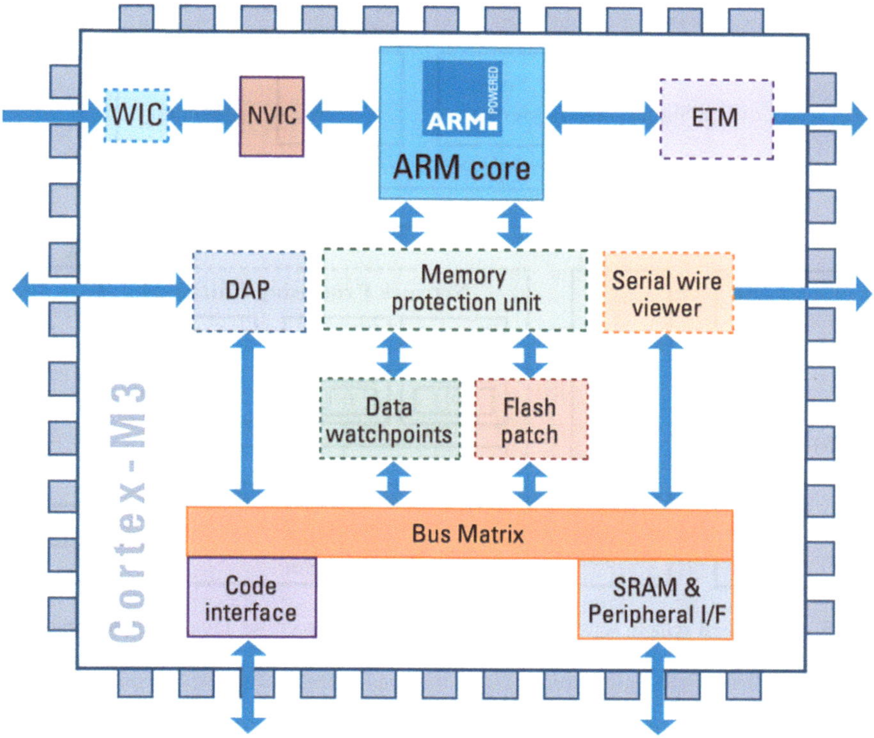

Fig. 4 Typical ARM processor configuration [*Source* https://doi.org/10.5815/ijisa.2017.07.08]

2. Scalability: ARM-based systems can be scaled to meet the performance requirements of various workloads.
3. Vector Processing:ARM-based processors often feature SIMD (Single Instruction, Multiple Data) vector processing units, ideal for data-intensive applications.
4. Customizability: ARM architecture allows customization to meet specific research needs.

3.4 ARM in HPC

Recent developments have seen ARM processors being integrated into HPC environments:

- Fugaku, the world's fastest supercomputer as of 2020, is powered by ARM-based processors.
- The ARM-based AWS Graviton2 processors are used in Amazon's cloud-based HPC solutions.

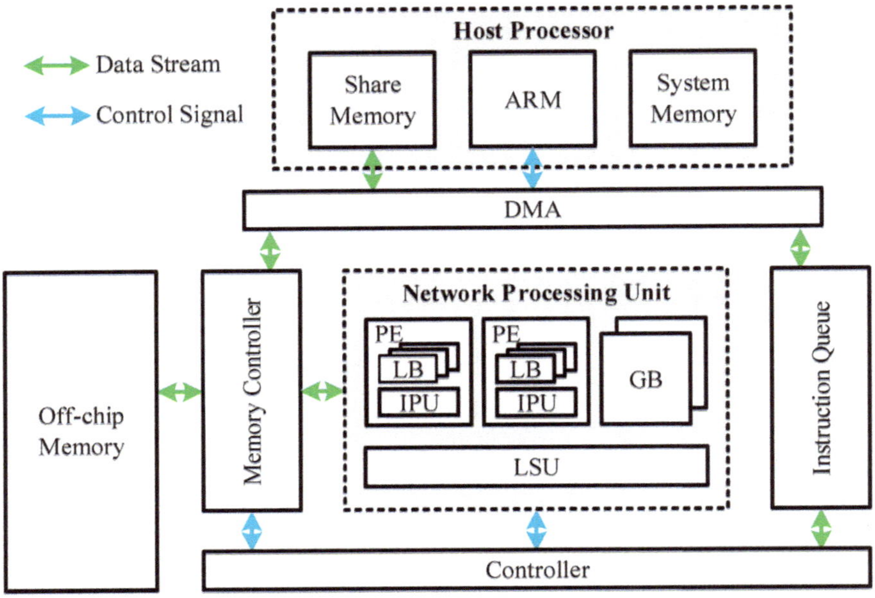

Fig. 5 CNN based remote sensing accelerator with ARM processor [*Source* https://doi.org/10.3390/rs15245784]

- ARM-based clusters are becoming popular for AI and machine learning workloads. One such example is shown in Fig. 5 where an ARM processor controls the complete system and performs post-processing tasks for CNN based remote sensing accelerator.

4 Configurations for ARM-Based HPC

While the adoption of ARM in HPC holds immense potential, effectively configuring ARM-based systems for research purposes is a nuanced task. It requires careful consideration of both hardware and software components. Here are the essential aspects that researchers and system administrators must address:

Operating System: Selecting an ARM-compatible operating system is the foundation of configuring ARM-based HPC systems. Popular choices include Linux distributions tailored for ARM, such as Ubuntu or CentOS.

Compiler and Libraries: Ensuring that the compilers and libraries used in the system are optimized for ARM architecture is crucial. This step allows researchers to fully leverage the computational power of ARM-based processors.

Toolchains: ARM toolchains, like the GNU Compiler Collection (GCC), are essential for cross-compilation. These tools enable the development of software on one platform for execution on another, allowing researchers to compile code specifically for ARM processors.

Cluster Management: Selecting an appropriate cluster management system that is compatible with ARM architecture is crucial. Systems like Slurm and Torque are well-suited to managing ARM-based clusters.

Application Porting: Many scientific applications are initially developed for ×86 or x86_64 architectures. To run these applications on ARM-based systems, porting may be necessary. This process involves adapting or recompiling the application code to ensure compatibility with ARM architecture.

Benchmarking: Assessing the performance of ARM-based HPC systems is essential. Benchmarking provides valuable insights into how well the system meets the needs of specific research workloads. Benchmarks like High-Performance Linpack (HPL) and High-Performance Conjugate Gradient (HPCG) are commonly used to evaluate system performance.

Energy Efficiency Optimization: One of the standout features of ARM architecture is its energy efficiency. Researchers and administrators should explore strategies for optimizing the system to maximize its green computing potential, reducing both operational costs and environmental impact.

The meticulous configuration of ARM-based HPC systems is essential for unlocking their full potential. The unique features of ARM architecture, coupled with the specific requirements of research workloads, demand a tailored approach to ensure optimal performance. Figure 6 shows the the Maxim Integrated MAX78000 ultra-low power microcontroller, which is a relatively new device specially designed for edge Artificial Intelligence (AI) applications. Integrates a dedicated Convolutional Neural Network (CNN) accelerator along with a low-power ARM Cortex-M4 core and a RISC-V core [9]

As we venture further into this chapter, it becomes apparent that understanding the interplay of HPC platforms, high-speed networks, and ARM processor architecture

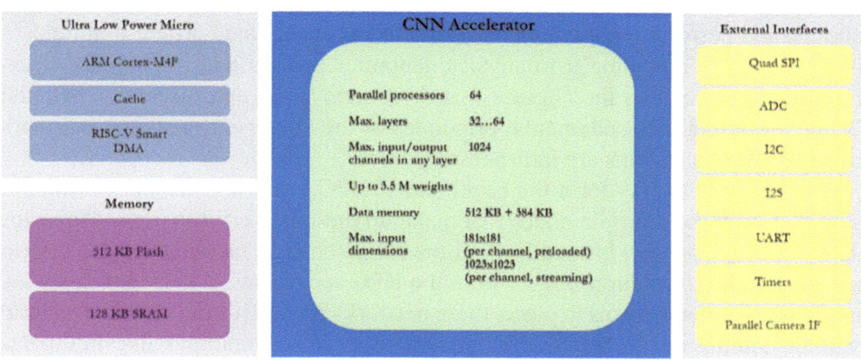

Fig. 6 The architecture of MAX78000 [9]

is crucial in the realm of modern research. Each of these components plays a distinct yet interdependent role, and their effective integration is central to the advancement of knowledge and innovation.

In the dynamic field of HPC and research, staying updated on the latest advancements is not just a luxury; it's a necessity. The tools and technologies available to researchers are in a constant state of flux, with new possibilities and paradigms emerging regularly. The references provided in this chapter offer a starting point for those seeking a deeper understanding of these topics and the evolving landscape of HPC research configurations. By exploring the convergence of HPC, high-speed networks, and ARM architecture, we are taking a significant step towards unlocking new frontiers in scientific exploration and discovery [10, 11].

5 Conclusion

HPC platforms, high-speed networks, and ARM processor architecture have become integral components of modern research. Researchers must make informed decisions about hardware and configurations to maximize the potential of these technologies. As ARM processors continue to gain ground in the HPC space, they offer an energy-efficient alternative that holds immense promise for diverse scientific applications. In the dynamic field of HPC and research, staying updated on the latest advancements is essential. The provided references offer a starting point for further exploration into these topics and the evolving landscape of HPC research configurations. **Conclusion (500 words):**

In the realm of scientific research, the utilization of High-Performance Computing (HPC) platforms, high-speed networks, and ARM processor architecture has become instrumental in pushing the boundaries of what is possible. This chapter has provided an in-depth exploration of these critical components, shedding light on their importance and configurations in the context of research.

HPC platforms have evolved into powerful tools for researchers across a wide array of disciplines. Their ability to execute complex simulations, process massive datasets, and perform computations in parallel has accelerated progress in fields as diverse as climate modeling, astrophysics, genomics, and artificial intelligence. These platforms have become the engines that drive innovation and discovery. Whether it's understanding the fundamental principles of the universe or solving real-world problems, HPC platforms are indispensable.

High-speed networks form the backbone of HPC, ensuring seamless communication between the multiple nodes that make up these powerful systems. Their low latency and high bandwidth enable real-time data exchange, making them the arteries through which information flows within the HPC ecosystem. With redundancy and fault tolerance mechanisms in place, these networks provide the reliability necessary for long-running computations. The choice between technologies like InfiniBand and Ethernet is dictated by the specific requirements of the research tasks at hand.

ARM processor architecture, originally confined to mobile devices, has begun to make a significant impact in the HPC landscape. ARM's appeal lies in its energy efficiency, scalability, vector processing capabilities, and customizability. These features make ARM-based systems ideal for a wide range of workloads, from numerical simulations to machine learning tasks. Notable examples of this trend include Fugaku, currently the world's fastest supercomputer, which is powered by ARM-based processors, and AWS Graviton2 processors used in cloud-based HPC solutions. ARM-based clusters are gaining traction for AI and machine learning workloads due to their energy efficiency, which is a critical factor in large-scale data processing and deep learning.

To effectively harness the power of ARM in HPC, researchers and system administrators must carefully configure these systems. This includes selecting an ARM-compatible operating system, compilers, libraries, and toolchains that are optimized for ARM architecture. Cluster management systems must also be chosen to match the platform's capabilities. Application porting is often required to ensure software compatibility with ARM-based systems. Extensive benchmarking helps assess the performance of these systems, enabling researchers to optimize their workloads effectively. Furthermore, energy efficiency optimization is paramount in the era of sustainability and green computing, making ARM an attractive choice for environmentally conscious research facilities.

In conclusion, HPC platforms, high-speed networks, and ARM processor architecture have transformed the landscape of scientific research. They empower researchers to tackle increasingly complex problems and dig deeper into the mysteries of the universe. As ARM-based systems continue to gain prominence in the HPC ecosystem, they offer a promising alternative that combines performance with energy efficiency, further enriching the toolkit available to the research community. Staying informed and adapting to these emerging technologies is essential for researchers and institutions looking to remain at the forefront of their respective fields, fostering innovation and advancing knowledge for the betterment of society.

References

1. Dongarra, J. et al.: The international exascale software project roadmap. Int. J. High-Performance Comput. Appl. (2020)
2. Sterling, T. et al.: High-Performance Computing: Modern Systems and Practices. Morgan Kaufmann (2018)
3. Kang, P.: Programming for high-performance computing on edge accelerators. Mathematics **11**(4), 1055 (2023). https://doi.org/10.3390/math11041055
4. Yan, F., Yuan, C., Li, C., Deng, X.: FOSquare: a novel optical HPC interconnect network architecture based on fast optical switches with distributed optical flow control. Photonics **8**(1), 11 (2021). https://doi.org/10.3390/photonics8010011
5. Snir, M. et al.: Addressing big data challenges in high-performance computing and scientific data centers. Int. J. High-Performance Comput. Appl. (2016)
6. Gupta, I. et al.: A survey of network fault tolerance in high-performance computing. J. Parallel Distrib. Comput. (2019)

7. Wulf, W. et al.: The ARM SVE vector architecture and its applicability to scientific computing. Concurr. Comput. Pract. Exp. (2021)
8. Sabry, A. et al.: ARM-based HPC systems: exploring the performance, energy efficiency, and architectural advancements. IEEE Comput. Archit. Lett. (2019)
9. Lucan Orăşan, I., Seiculescu, C., Căleanu, C.D.: A brief review of deep neural network implementations for ARM cortex-M processor. Electronics **11**(16), 2545 (2022). https://doi.org/10.3390/electronics11162545
10. Sur, S. et al.: Challenges in porting scientific applications to ARM-based HPC systems. J. Parallel Distrib. Comput. (2019)
11. Lee, T. et al.: An In-depth analysis of Fugaku's Arm A64FX vector architecture for HPC. ACM Trans. Archit. Code Optim. (2020)

High-Performance Computing Based Operating Systems, Software Dependencies and IoT Integration

Nor Asilah Wati Abdul Hamid and Balbir Singh

Abstract This chapter delivers the critical aspects of operating systems and software dependencies within the context of High-Performance Computing (HPC) using Nvidia Jetson devices, while seamlessly integrating them with the Internet of Things (IoT) ecosystem. High-performance computing has witnessed a paradigm shift towards edge computing, where the Nvidia Jetson platform plays a pivotal role due to its impressive computational power and energy efficiency. The chapter begins by providing an overview of the Nvidia Jetson platform and its relevance in the HPC and IoT domains. It explores the various operating systems that are compatible with Nvidia Jetson, highlighting their strengths and trade-offs. Special attention is given to Linux-based distributions, including Ubuntu, NVIDIA's JetPack, and custom-built OS images, discussing their configuration processes. A significant portion of the chapter is dedicated to dissecting the intricate web of software dependencies in HPC applications. It addresses the challenges of managing complex software stacks on edge devices, emphasizing the importance of package managers, containerization technologies like Docker, and virtual environments. Best practices for optimizing software performance on Nvidia Jetson devices are also elucidated. Furthermore, the chapter explores the integration of HPC capabilities with IoT, showcasing practical examples of how Nvidia Jetson can be used as a powerful edge device for data analysis, machine learning, and real-time decision-making. This integration is pivotal in domains such as autonomous robotics, smart surveillance, and industrial automation. By understanding these intricacies, researchers, developers, and practitioners

N. A. W. A. Hamid (✉)
Institute for Mathematical Research, Universiti Putra Malaysia, 43400 Serdang, Selangor, Malaysia
e-mail: asila@upm.edu.my

Faculty of Computer Science and Information Technology, Universiti Putra Malaysia, 43400 Serdang, Selangor, Malaysia

B. Singh
Department of Aerospace Engineering, Faculty of Engineering, Universiti Putra Malaysia, 43400 Serdang, Selangor, Malaysia

Department of Aeronautical and Automobile Engineering, Manipal Institute of Technology, Manipal Academy of Higher Education, Manipal, Karnataka 576104, India

K. A. Ahmad et al. (eds.), *High Performance Computing in Biomimetics*, Series in BioEngineering, https://doi.org/10.1007/978-981-97-1017-1_8

will be better equipped to harness the full potential of HPC and IoT integration for their specific applications, fostering innovation in edge computing environments.

Keywords NVIDIA Jetpack · HPC framework · GPU · Edge computing · IoT integration · Software dependencies

1 Introduction

1.1 Background

High-Performance Computing (HPC) is a rapidly evolving field that focuses on the use of advanced computing systems to solve computationally demanding problems. HPC applications span various domains, including scientific research, data analysis, simulations, and machine learning. Nvidia Jetson, specifically Jetson TX1, has emerged as a popular platform for HPC, thanks to its high-performance GPU architecture and power efficiency. Moreover, the integration of IoT with HPC allows for real-time data processing and decision-making at the edge. This chapter aims to provide a comprehensive understanding of the operating systems and software dependencies required for leveraging Nvidia Jetson and IoT in high-performance computing environments. It will cover the selection criteria for operating systems, the significance of software dependencies, and explore case studies that highlight the successful integration of Nvidia Jetson and IoT in HPC applications. High-Performance Computing (HPC) refers to the use of advanced computing systems and architectures to solve complex problems that require substantial computational power. HPC systems typically consist of multiple processors, parallel computing architectures, and specialized hardware accelerators like GPUs. The purpose of HPC is to enable faster and more accurate simulations, data analysis, and modeling, leading to advancements in various scientific, industrial, and research fields.

1.2 Key Components and Architecture

HPC systems comprise several key components, including processors, memory modules, interconnects, storage systems, and software frameworks/stacks. The architecture of an HPC system is designed to maximize computing power and minimize communication latency. This is achieved through the use of parallel computing techniques, such as distributed memory systems, shared memory systems, or hybrid combinations of both. HPC finds applications in numerous domains, such as weather forecasting, climate modeling, computational fluid dynamics, molecular dynamics simulations, financial modeling, genomics, and machine learning. HPC enables

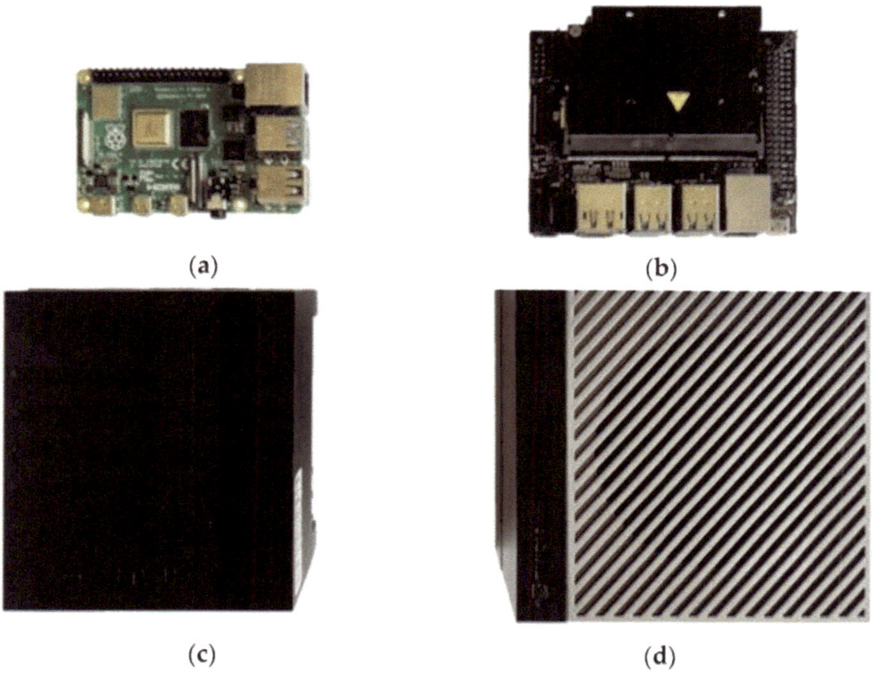

Fig. 1 Raspberry Pi 4B and members of NVIDIA Jetson series [*Source* https://doi.org/10.3390/en16186677]

researchers, scientists, and engineers to tackle complex problems that were previously computationally infeasible or time-consuming. Nvidia Jetson is a family of embedded computing platforms that provide high-performance GPU capabilities for edge computing applications. These platforms are designed to enable AI-powered applications in resource-constrained environments. Nvidia Jetson modules integrate powerful GPUs, CPUs, memory, and I/O interfaces in a compact form factor, making them ideal for HPC at the edge.

As far as Jetson hardware platforms are concerned, we have Jetson Nano: The Jetson Nano is an entry-level Jetson module, providing a cost-effective solution for AI and robotics projects. It features a Quad-core ARM Cortex-A57 CPU and a Maxwell GPU with 128 CUDA cores. Jetson TX2: The Jetson TX2 is a more powerful module with a Dual-core Denver 2 CPU and a Pascal GPU with 256 CUDA cores. Jetson Xavier NX: The Jetson Xavier NX is designed for AI workloads and features a hexa-core Carmel ARM CPU and a Volta GPU with 384 CUDA cores. Jetson AGX Xavier: The Jetson AGX Xavier is the most powerful module in the Jetson family, boasting an octa-core Carmel ARM CPU and a Volta GPU with 512 CUDA cores. It is ideal for high-performance edge computing applications. Jetson TX series is one of the popular members of the Nvidia Jetson family. It features an NVIDIA Maxwell architecture GPU with 256 CUDA cores, a quad-core ARM Cortex-A57

CPU, and 4 GB LPDDR4 memory. Jetson TX is known for its power efficiency and is suitable for applications that require both high-performance computing and low power consumption. Figure 1(a–d) depicts Raspberry Pi 4B and NVIDIA Jetson series members nano, Xavier AGX and Orin AGX in this series [1, 2].

2 Role of Jetson in High-Performance Computing

Nvidia Jetson, including Jetson TX2, plays a vital role in HPC by providing an embedded platform that can handle computationally intensive workloads efficiently. The combination of Jetson's GPU architecture and its support for AI frameworks like TensorFlow and PyTorch makes it well-suited for deep learning applications. Furthermore, the integration of Jetson with IoT enables real-time processing and decision-making at the edge, reducing latency and enhancing the overall performance of HPC applications. The role of Nvidia Jetson, including the Jetson TX2, in High-Performance Computing (HPC) is significant and continues to evolve as it brings the power of GPU-accelerated computing to the edge [3].

Nvidia Jetson modules, such as the Jetson TX2, offer powerful GPU and CPU capabilities in a compact and energy-efficient form factor. This makes them ideal for running computationally intensive workloads, including scientific simulations, data analytics, and image processing, which are common in HPC. Jetson platforms are equipped with Nvidia GPUs based on the Maxwell or Pascal architectures, depending on the model. These GPUs are designed for parallel processing and excel at accelerating tasks that can be parallelized, which is a core requirement in many HPC applications. Jetson modules come with comprehensive support for AI and deep learning frameworks such as TensorFlow, PyTorch, and Caffe. This support allows researchers and developers in the HPC domain to leverage GPU acceleration for their AI and machine learning workloads [4].

Jetson's integration with the Internet of Things (IoT) and edge computing is a game-changer for HPC. It enables real-time processing and decision-making at the edge of the network, reducing the need to send data to centralized data centers. This reduces latency and enhances the overall performance of HPC applications, especially in scenarios where low-latency processing is critical. Jetson modules are increasingly finding applications in HPC scenarios where edge computing is necessary. For example, they are used in autonomous vehicles for real-time object recognition, in scientific research for data analysis, and in robotics for high-performance control systems. Nvidia's Jetson lineup offers various models with different performance levels, enabling scalability and flexibility in HPC deployments. Developers and researchers can choose the Jetson module that best suits their computational requirements [5].

In summary, Nvidia Jetson, including the Jetson TX2, plays a crucial role in High-Performance Computing by providing embedded platforms that are capable of handling computationally intensive workloads efficiently. Its GPU architecture, support for AI frameworks, and integration with IoT enable real-time processing and decision-making at the edge, reducing latency and improving the overall performance of HPC applications. Jetson's versatility and scalability make it a valuable tool for researchers, developers, and engineers in the HPC domain. For the latest information, please refer to Nvidia's official resources and news updates.

3 Operating Systems for High-Performance Computing

Operating systems (OS) play a crucial role in HPC environments by managing hardware resources, providing an interface for software applications, and facilitating efficient task scheduling and resource allocation. OS choices significantly impact the performance, scalability, and reliability of HPC systems.

3.1 Linux in HPC: Advantages and Adaptability

Linux has become the de facto standard operating system for HPC due to its open-source nature, flexibility, and extensive support for HPC-related software and tools. Linux distributions such as Debian, Ubuntu, CentOS, and Red Hat Enterprise Linux are widely used in HPC clusters. Linux provides high-performance networking capabilities, efficient memory management, and robust support for parallel processing. Linux is open-source, which means that users have access to the source code and can modify it to suit their specific needs. This feature is crucial in HPC environments where customization and optimization are essential. Researchers and system administrators can customize the Linux kernel and software stack to maximize the performance of HPC clusters. This adaptability allows for fine-tuning the system to meet specific computational requirements. Linux is renowned for its stability and reliability. In HPC, where computations may run for days, weeks, or even months, having a robust operating system is critical to prevent crashes and data loss. Linux-based HPC clusters can easily scale to accommodate increasing computational demands. The open-source nature of Linux allows users to add new hardware seamlessly and adapt the system accordingly. Linux offers low-level control over hardware, making it possible to optimize the operating system for the specific architecture of the HPC cluster. This level of control is essential for squeezing the maximum performance out of the hardware. It also boasts a vast repository of open-source software packages and libraries, many of which are tailored for scientific and computational workloads. Researchers can easily access and integrate these tools into their HPC workflows. Linux benefits from continuous security updates and a large community of developers who actively monitor and patch vulnerabilities. This is crucial in HPC environments, where data integrity and confidentiality are paramount. Linux is cost-effective as it eliminates licensing fees associated with proprietary operating systems. This cost savings can be redirected toward hardware upgrades or research initiatives [6, 7].

As far as the Aadaptability of Linux in HPC is concerned it supports a wide range of hardware architectures, making it adaptable to different types of HPC systems, including ×86, ARM, and GPUs. This flexibility enables researchers to select the hardware that best suits their computational needs. Linux distributions like Debian, CentOS, Ubuntu, and Red Hat offer HPC-specific versions equipped with cluster management tools such as Slurm, Torque, and OpenHPC. These tools simplify the administration and maintenance of HPC clusters. Technologies like Docker and Kubernetes, which are natively supported on Linux, facilitate the deployment of containerized applications in HPC environments. Containers enhance portability and reproducibility of research workloads. Linux provides robust support for parallel processing and message-passing libraries like MPI (Message Passing Interface). This is essential for executing complex parallelized simulations and computations. Linux can seamlessly integrate with cloud computing platforms, allowing HPC workloads to leverage the scalability and resources of the cloud when necessary [7].

3.2 Nvidia Jetson Supported Operating Systems

Nvidia Jetson supports multiple operating systems, including Ubuntu, which is often the preferred choice for HPC applications. Ubuntu offers a rich ecosystem of software packages, libraries, and development tools for HPC. Additionally, Nvidia provides JetPack SDK, a comprehensive software development package, specifically tailored for Jetson platforms. Title: NVIDIA Jetson Supported Operating Systems. The NVIDIA Jetson platform is a series of powerful and energy-efficient embedded computing modules designed for AI and edge computing applications. These modules require specific operating systems that are compatible with their hardware capabilities. This note explores the supported operating systems for NVIDIA Jetson devices and highlights their features and use cases.

NVIDIA JetPack

NVIDIA JetPack is a comprehensive software development package that includes an Ubuntu-based Linux distribution, CUDA Toolkit, cuDNN, TensorRT, and various libraries and tools optimized for AI and machine learning.

- JetPack provides a full-stack software solution that leverages the GPU capabilities of Jetson modules.
- It is well-suited for AI research, deep learning model development, and computer vision applications.
- NVIDIA provides regular updates and support for JetPack, ensuring compatibility with the latest hardware and software developments.

Ubuntu-based Linux Distributions

- Many Jetson users prefer Ubuntu-based Linux distributions due to their familiarity and extensive software support.

- Ubuntu-based distributions like Ubuntu LTS (Long-Term Support) and Ubuntu Core are readily available for Jetson platforms.
- Ubuntu LTS is suitable for general-purpose development and deployment, while Ubuntu Core is a minimal, transactional version designed for IoT and security-focused applications.

NVIDIA DeepStream

- NVIDIA DeepStream is an SDK for building scalable, high-performance AI-based video analytics applications.
- It runs on top of the Ubuntu-based Linux distribution and is optimized for Jetson devices.
- DeepStream simplifies the development of video analytics solutions for smart cities, surveillance, and industrial applications.

NVIDIA Isaac SDK

- The NVIDIA Isaac SDK is a robotics software development kit tailored for autonomous machines and robot.
- It includes a real-time operating system (RTOS) designed to run on Jetson devices.
- The RTOS ensures precise control and low-latency communication, making it suitable for robotics and autonomous navigation tasks.

Yocto Project and Poky

- For users who require custom Linux distributions and high levels of customization, the Yocto Project and Poky provide a framework to build a tailored OS for Jetson devices.
- This approach is ideal for embedded systems where every component of the OS needs to be carefully selected and configured.

Third-Party and Community Distributions

- The Jetson community is active and has developed various third-party and community-supported operating systems.
- These may include alternative Linux distributions, real-time operating systems (RTOS), and custom firmware.
- While not officially supported by NVIDIA, these options can be valuable for specialized use cases and experimentation.

So NVIDIA Jetson devices are versatile platforms for AI and edge computing applications, and they are compatible with a range of operating systems to suit various development and deployment needs. The choice of an operating system depends on factors such as the specific application, hardware capabilities, and user preferences. NVIDIA's official software stack, including JetPack and specialized SDKs like DeepStream and Isaac, provide comprehensive support for AI and robotics development, making Jetson an attractive choice for developers and researchers in these domains. Additionally, the flexibility to use custom distributions and third-party options expands the possibilities for Jetson-based projects [8–12].

3.3 Selection Criteria for Choosing an OS

High-Performance Computing (HPC) clusters are at the forefront of scientific research, engineering simulations, and data-intensive computations. Selecting the right operating system (OS) for an HPC environment is a critical decision that can impact performance, stability, and usability. This note outlines the key criteria to consider when choosing an operating system for HPC. When selecting an operating system for HPC with Nvidia Jetson, several factors need to be considered. These include compatibility with software dependencies, availability of drivers and support for Nvidia GPU technologies, ease of installation and maintenance, performance optimizations, and community support. Title: Selection Criteria for Choosing an Operating System in HPC.

Hardware Compatibility

The chosen OS must be compatible with the hardware architecture of the HPC cluster. Most HPC systems use x86_64 processors, but some may use ARM, PowerPC, or other architectures. Users need to ensure that the OS supports the specific hardware components, such as network interfaces, GPUs, and accelerators, to maximize performance and compatibility.

Performance Optimization

The OS should allow for low-level customization and performance tuning to extract the maximum computational power from the hardware. Look for an OS that supports technologies like CPU affinity, InfiniBand, RDMA, and NUMA (Non-Uniform Memory Access) for efficient parallel processing.

Cluster Management Tools

We must consider OS options that integrate well with cluster management software like Slurm, Torque, or OpenHPC. These tools simplify resource allocation, job scheduling, and monitoring in an HPC cluster.

User and Application Support

The OS should have a well-established user community and a rich ecosystem of software packages and libraries relevant to HPC workloads. The OS should support the programming languages, compilers, and libraries commonly used in scientific computing, such as MPI, OpenMP, CUDA, and Python.

Stability and Reliability

HPC clusters often run long and resource-intensive computations. Therefore, the OS must be highly stable and reliable to minimize the risk of system crashes or data corruption. OS versions with long-term support (LTS) should be chosen to ensure ongoing security updates and stability.

Security

Data security is paramount in HPC, especially in research involving sensitive or confidential information. Select an OS with robust security features and a track record of prompt security patching. Consider OS options that provide mandatory access controls (e.g., SELinux, AppArmor) and support for encryption protocols.

Scalability

HPC clusters often grow over time. The selected OS should be scalable, allowing for the addition of new nodes and hardware components without significant reconfiguration. One should evaluate how well the OS handles large-scale parallelism and distributed computing.

Containerization Support

Container technologies like Docker and Singularity have become popular for packaging and deploying HPC applications. Ensure that the chosen OS supports these containerization solutions.

Licensing Costs

Though most of the sources on HPC can be open source like Ubuntu but some operating systems, particularly proprietary ones, may involve licensing fees. There should be a consideration for the budget constraints and licensing models when selecting an OS for HPC.

Documentation, Support and Future Compatibility

A robust ecosystem of documentation, online forums, and vendor support can be invaluable in troubleshooting issues and optimizing system performance. Evaluate the availability of such resources. It is important to consider the roadmap of the OS and its compatibility with emerging hardware and software technologies. The OS should be adaptable to future HPC requirements.

In general, choosing the right operating system for an HPC cluster is a critical decision that impacts performance, stability, and usability. A well-informed selection process should take into account hardware compatibility, performance optimization, cluster management tools, user and application support, stability, security, scalability, containerization support, licensing costs, documentation, and future compatibility. Ultimately, the choice should align with the specific needs and goals of the HPC project or organization to ensure the successful execution of scientific research and data-intensive computations.

4 Software Dependencies in HPC

4.1 Definition and Significance

Software dependencies in HPC refer to the libraries, frameworks, and other software components that are required for running HPC applications efficiently. These dependencies provide access to optimized algorithms, parallel computing frameworks, GPU-accelerated libraries, and machine learning frameworks, enabling developers to harness the full potential of HPC systems. High-Performance Computing (HPC) systems are used for complex scientific simulations, data analysis, and other computationally intensive tasks. These applications often rely on a multitude of software packages, libraries, and tools to function effectively. Software dependencies refer to the relationships and requirements between different software components within an HPC environment. Understanding these dependencies is critical for ensuring the stability, performance, and reproducibility of HPC workloads [13].

Software dependencies in HPC can be defined as the relationships between various software components, where one component relies on the presence and proper functioning of another to operate as intended. These dependencies can manifest in several ways:

Library Dependencies: Many HPC applications depend on external libraries to perform specific functions. For example, scientific computing applications might rely on libraries like LAPACK or FFTW for linear algebra or fast Fourier transform operations.

Compiler and Toolchain Dependencies: HPC software often requires specific compilers (e.g., GCC, Intel compilers) and toolchains (e.g., MPI implementations) for building and running applications.

Operating System Dependencies: The choice of the operating system (OS) can also introduce dependencies. Different OS versions and distributions may offer varying levels of support for HPC-related features and libraries.

Version Dependencies: The specific version of a software component can impact compatibility and functionality. In some cases, HPC applications may only work with a particular version of a library or tool.

As far as the significance of Software Dependencies in HPC is concerned, understanding software dependencies is crucial for optimizing HPC applications. Utilizing the right versions of libraries and compilers can significantly improve computational performance. Reproducibility is vital in scientific research. By documenting and managing software dependencies, researchers can ensure that their experiments and simulations can be reproduced accurately, facilitating peer review and validation. HPC systems often involve complex software stacks. Ensuring that all software components are compatible with each other and with the hardware is essential to prevent crashes and errors.

Managing dependencies helps in efficient resource allocation. It ensures that only necessary software components are installed, reducing storage and memory usage and streamlining system maintenance. In the event of errors or performance issues, having a clear understanding of software dependencies can aid in debugging. Identifying conflicts or missing dependencies is crucial for resolving problems. Vulnerabilities in software dependencies can pose security risks. Regularly updating and patching software components is essential to address security vulnerabilities and protect sensitive data on HPC systems. Version control systems like Git can help manage and track changes in software dependencies. This aids in managing software updates and ensuring that the correct versions are used consistently. HPC applications may need to run on different systems. Managing dependencies helps ensure that applications remain portable and can be deployed on various HPC clusters without major modifications. Efficiently managing software dependencies can lead to reduced resource utilization and faster job execution on HPC clusters, ultimately improving overall cluster throughput [14].

So, software dependencies in HPC are a fundamental aspect of managing and optimizing computational workflows. By understanding and effectively managing these dependencies, researchers and HPC administrators can ensure that their applications run smoothly, perform efficiently, and remain reproducible and secure. Proper documentation and version control practices are key to handling software dependencies successfully in the context of high-performance computing.

4.2 *Libraries and Frameworks for HPC*

Various libraries and frameworks are available for HPC development, catering to different domains and requirements. Examples include Message Passing Interface (MPI) for distributed computing, OpenMP for shared memory parallelism, and OpenCL for heterogeneous computing. These libraries enable developers to parallelize their code and efficiently utilize the available hardware resources. High-Performance Computing (HPC) relies heavily on a wide range of software libraries and frameworks to optimize and enhance the performance of scientific simulations, data analytics, and other computationally intensive tasks. These libraries and frameworks play a crucial role in accelerating computations, facilitating parallel processing, and providing access to essential mathematical and scientific functions. This note explores some of the key libraries and frameworks commonly used in HPC environments. Though the list is very exhaustive, some have been discussed below for the reader's understanding [15–17].

Few example of libraries for HPC

1. MPI (Message Passing Interface):

- MPI is a fundamental library for distributed memory parallel computing. It enables communication and data exchange between parallel processes in a cluster.
- Popular MPI implementations include Open MPI, MPICH, and Intel MPI.

2. OpenMP:

- OpenMP is an API for shared-memory parallelism that simplifies the development of multithreaded applications. It allows developers to specify parallel regions in code that can be executed concurrently.
- It is commonly used for parallelizing loops and other sections of code.

3. BLAS (Basic Linear Algebra Subprograms):

- BLAS libraries provide optimized routines for common linear algebra operations, such as matrix–vector multiplication and matrix factorizations. These routines are essential for scientific and engineering computations.
- Notable implementations include OpenBLAS and Intel Math Kernel Library (MKL).

4. LAPACK (Linear Algebra Package):

LAPACK builds upon BLAS and provides a higher-level interface for solving systems of linear equations, eigenvalue problems, and singular value decompositions. It is widely used in numerical simulations.

5. FFTW (Fastest Fourier Transform in the West):

- FFTW is a highly optimized library for performing fast Fourier transforms (FFT), a key operation in signal processing, simulations, and data analysis.
- It offers different algorithms optimized for various hardware architectures.

6. PETSc (Portable, Extensible Toolkit for Scientific Computation):

- PETSc is a framework for the scalable solution of partial differential equations (PDEs) and nonlinear equations. It provides a wide range of numerical solvers and data structures.
- PETSc is often used in simulations of complex physical phenomena.

7. Trilinos:

Trilinos is another HPC framework for solving large-scale, complex mathematical problems. It includes a wide range of packages for linear and nonlinear solvers, optimization, and more. [14–17]

Few examples of frameworks for HPC:

1. CUDA (Compute Unified Device Architecture):

Developed by NVIDIA, CUDA is a parallel computing framework designed for GPUs (Graphics Processing Units). It allows developers to harness the power of GPUs for highly parallel tasks, such as deep learning and scientific simulations.

2. OpenACC:

OpenACC is a directive-based framework for parallel programming that targets accelerators like GPUs and multi-core CPUs. It simplifies the process of porting code to heterogeneous architectures.

3. Hadoop and Spark:

Hadoop and Apache Spark are frameworks for distributed data processing and analytics. While not exclusive to HPC, they are used for handling and analyzing large datasets in HPC environments.

4. TensorFlow and PyTorch:

These deep learning frameworks are increasingly being used in HPC for AI and machine learning applications. They offer support for distributed training on HPC clusters.

5. MPI-Based Libraries:

Some HPC frameworks, like ParaView and VisIt, are built on top of MPI to provide visualization and data analysis capabilities for large-scale simulations.

6. Containers (Docker, Singularity):

Containerization frameworks like Docker and Singularity are used to package and deploy HPC applications and their dependencies, simplifying the deployment of complex software stacks on HPC clusters.

So, software libraries and frameworks are the backbone of high-performance computing, enabling researchers and engineers to leverage the full computational power of HPC clusters. The choice of libraries and frameworks should align with the specific needs of the computational tasks at hand, the hardware architecture of the HPC cluster, and the expertise of the development team. Properly utilizing and optimizing these software components can significantly enhance the performance, scalability, and efficiency of HPC applications.

4.3 CUDA and CuDNN: Nvidia's GPU Computing Technologies

Nvidia's CUDA (Compute Unified Device Architecture) is a parallel computing platform and API that enables developers to harness the power of Nvidia GPUs for general-purpose computing. CUDA provides a programming model and libraries that allow developers to write GPU-accelerated code in languages such as C, C++, and Python. Additionally, Nvidia's cuDNN (CUDA Deep Neural Network) library provides optimized primitives for deep neural networks, further enhancing the performance of deep learning applications on Nvidia GPUs as shown in Fig. 2. NVIDIA's CUDA (Compute Unified Device Architecture) and cuDNN (CUDA Deep Neural

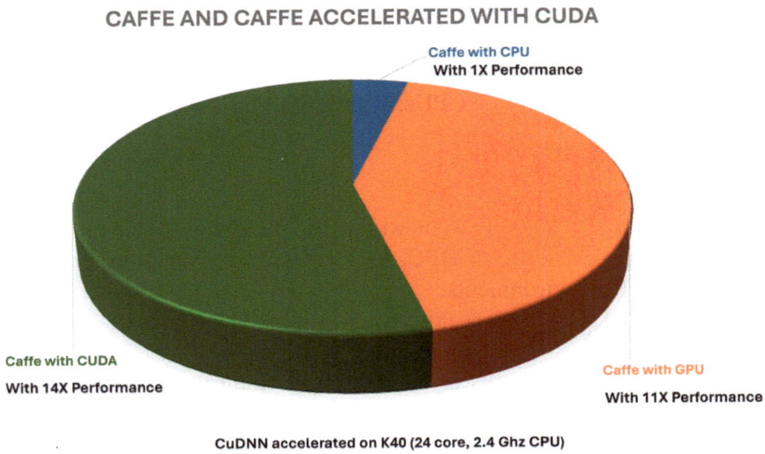

Fig. 2 cuDNN performance comparison in Caffe (*Source* https://developer.nvidia.com/blog/accelerate-machine-learning-cudnn-deep-neural-network-library/)

Network) are pivotal technologies in the realm of GPU computing. These software frameworks have revolutionized the way developers harness the power of NVIDIA GPUs for a wide range of applications, including scientific simulations, machine learning, deep learning, and artificial intelligence. This section provides an in-depth overview of CUDA and cuDNN, their significance, and their impact on GPU-accelerated computing.

Definition and Overview:

CUDA is a parallel computing platform and API developed by NVIDIA. It enables developers to leverage the computational power of NVIDIA GPUs for general-purpose computing tasks. Unlike traditional graphics APIs, CUDA allows programmers to write custom kernels and algorithms that execute on the GPU, significantly accelerating computations.

Key Features and Capabilities:

- Parallelism: CUDA unlocks massive parallelism, with GPUs comprising thousands of cores. This parallelism is ideal for tasks that can be divided into many smaller parallel threads.
- CUDA C/C++: Developers can write CUDA code in C/C++ with specialized extensions, making it accessible to a wide range of programmers.
- Unified Memory: CUDA introduces unified memory management, enabling seamless data sharing between CPU and GPU memory, simplifying data transfer.
- Dynamic Parallelism: CUDA allows kernels to launch other kernels, creating dynamic execution flows and supporting complex algorithms.

- GPU Libraries: NVIDIA provides GPU-accelerated libraries like cuBLAS (linear algebra), cuFFT (Fast Fourier Transform), and cuSPARSE (sparse matrix operations) for use with CUDA.

Applications: CUDA has transformed scientific computing, enabling faster simulations in fields such as physics, chemistry, and engineering. It has also revolutionized deep learning by accelerating neural network training and inference, making it a cornerstone of modern AI research. CUDA has applications in medical imaging, computational finance, weather modeling, and more [18–21].

cuDNN (CUDA Deep Neural Network):

Definition and Overview: cuDNN is a GPU-accelerated library developed by NVIDIA specifically for deep neural network (DNN) computations. It provides highly optimized implementations of essential DNN functions, including convolution, pooling, normalization, and activation functions.

Key Features and Capabilities:

- Performance Optimization: cuDNN is engineered for speed, utilizing low-level GPU optimizations and hardware-specific libraries to deliver the highest performance for DNN workloads as shown in Fig. 2.
- Integration with Deep Learning Frameworks: cuDNN is seamlessly integrated with popular deep learning frameworks like TensorFlow, PyTorch, and Caffe, allowing developers to harness GPU acceleration without low-level programming.

Applications: cuDNN has played a pivotal role in the rapid advancement of deep learning. It underpins the training and inference of deep neural networks in applications ranging from image recognition to natural language processing. Its speed and efficiency have made it a cornerstone technology for large-scale machine learning and AI research.

Significance and Impact of both

1. Accelerated Computing: CUDA and cuDNN have democratized GPU computing, enabling developers to unlock the tremendous computational potential of GPUs for a broad spectrum of applications.
2. Scientific and Research Advancements: CUDA has accelerated scientific simulations, while cuDNN has driven breakthroughs in deep learning and AI. These technologies have empowered researchers to explore complex problems more efficiently.
3. Widespread Adoption: CUDA is widely adopted in academia and industry, while cuDNN has become an integral part of the deep learning ecosystem. This adoption has led to a thriving community and an extensive library of GPU-accelerated software.
4. Innovation and Performance: The continuous development and optimization of CUDA and cuDNN have pushed the boundaries of what's possible in terms of computation and machine learning, driving innovation in various fields.

CUDA and cuDNN are therefore foundational technologies that have reshaped the landscape of GPU computing and deep learning. They have not only enabled remarkable advancements in scientific research and artificial intelligence but have also empowered developers to leverage GPU acceleration in a user-friendly manner. These technologies continue to evolve, driving progress and innovation in the fields of high-performance computing, machine learning, and AI [18–21].

4.4 TensorRT: Deep Learning Inference Optimizer

TensorRT is a high-performance deep learning inference optimizer and runtime library developed by Nvidia. It optimizes and accelerates deep learning models for inference on Nvidia GPUs. TensorRT performs optimizations such as layer fusion, precision calibration, and dynamic tensor memory management, resulting in faster and more efficient inference in HPC applications. TensorRT, short for Tensor Runtime, is a high-performance deep learning inference optimizer and runtime library developed by NVIDIA. It is designed to optimize and accelerate deep learning models for deployment on GPUs in production environments. TensorRT is a key component of NVIDIA's AI platform and plays a crucial role in enhancing the efficiency and speed of deep learning inference. This section below provides a detailed overview of TensorRT, its features, benefits, and applications [22].

Key Features and Capabilities of TensorRT:

1. Model Optimization: TensorRT optimizes deep learning models to improve inference performance while maintaining or even enhancing model accuracy. It applies a range of optimizations, including precision calibration, layer fusion, and kernel auto-tuning.
2. Mixed-Precision Support: TensorRT supports mixed-precision inference, allowing models to use lower-precision data types like FP16 (half-precision) and INT8 (quantized) to accelerate computations while minimizing memory usage.
3. Dynamic Tensor Memory Management: TensorRT efficiently manages GPU memory during inference, minimizing memory overhead and enabling the execution of large models on GPUs with limited memory.
4. Layer Fusion: TensorRT performs layer fusion, which combines multiple layers of a neural network into a single optimized kernel, reducing the number of memory transfers and computations.
5. INT8 Quantization: TensorRT enables quantization to INT8, which reduces the memory footprint and computation requirements while maintaining inference accuracy within acceptable ranges.
6. GPU-Accelerated Libraries: TensorRT integrates with other GPU-accelerated libraries, such as cuBLAS and cuDNN, to further optimize deep learning workloads and improve inference speed.

7. ONNX and TensorFlow Integration: TensorRT can import models from popular deep learning frameworks like TensorFlow and ONNX, making it compatible with a wide range of model architectures and training pipelines.

Benefits of Using TensorRT:

1. Faster Inference: TensorRT's optimizations significantly accelerate deep learning inference, making it suitable for real-time applications like object detection, natural language processing, and autonomous driving.
2. Reduced Memory Footprint: By optimizing memory usage and supporting lower-precision data types, TensorRT allows for the deployment of large models on GPUs with limited memory resources.
3. Energy Efficiency: TensorRT's efficiency improvements translate to reduced power consumption, making it a suitable choice for edge devices and data centers focused on energy efficiency.
4. Production-Ready: TensorRT is designed for production use, ensuring reliability, stability, and support for multi-GPU deployment.
5. Compatibility: Its compatibility with popular deep learning frameworks simplifies the deployment of trained models and allows developers to leverage existing model architectures.

Applications of TensorRT: A few applications of TensorRT includes:

1. Autonomous Vehicles: TensorRT is used for real-time object detection, semantic segmentation, and path planning in autonomous driving systems.
2. Natural Language Processing (NLP): It accelerates NLP tasks like language translation, sentiment analysis, and chatbots.
3. Computer Vision: TensorRT is widely used for image classification, object recognition, and image generation.
4. Medical Imaging: In the medical field, TensorRT aids in tasks such as disease diagnosis, anomaly detection, and radiology image analysis.
5. Recommendation Systems: TensorRT is applied to enhance recommendation algorithms for personalized content delivery.

TensorRT is therefore a powerful deep learning inference optimizer that significantly improves the efficiency and speed of deep learning models on NVIDIA GPUs. Its wide range of optimizations, mixed-precision support, memory management capabilities, and compatibility with popular deep learning frameworks make it an essential tool for deploying deep learning models in production environments. Whether in autonomous vehicles, natural language processing, computer vision, or other AI applications, TensorRT plays a crucial role in delivering high-performance and energy-efficient inference solutions. Figure 3 shows the deep earning performance of TensorRT compared to cuDNN [22].

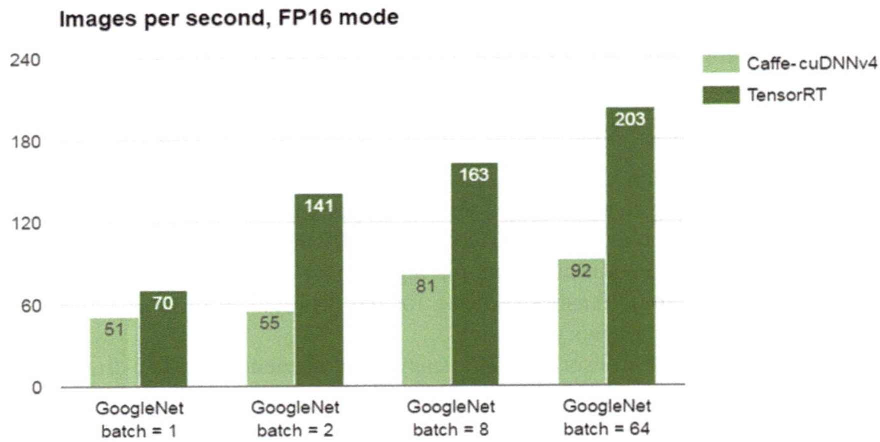

Fig. 3 NVIDIA TensorRT Delivers Twice the Deep Learning Inference for GPUs & Jetson TX1, even cuDNNs (*Source* https://www.linkedin.com/pulse/nvidia-tensorrt-delivers-twice-deep-learning-gpus-jetson-franklin)

4.5 Other Software Dependencies for IoT Integration

For IoT integration in HPC, additional software dependencies may be required. These can include MQTT (Message Queuing Telemetry Transport) for lightweight and efficient communication between edge devices and cloud servers, IoT-specific frameworks like AWS IoT Greengrass or Azure IoT Edge for managing edge deployments, and edge analytics frameworks for real-time data processing and decision-making at the edge. The integration of Internet of Things (IoT) devices into High-Performance Computing (HPC) environments is becoming increasingly important in various domains, including scientific research, industrial automation, and smart cities. However, successfully integrating IoT into HPC necessitates careful consideration of software dependencies. This note explores the software components and dependencies essential for IoT integration in HPC and their significance.

IoT Device Drivers and Middleware: To communicate with IoT devices, HPC systems must have the necessary device drivers installed. These drivers enable the hardware components to interact with the OS and software applications. Middleware software, such as MQTT (Message Queuing Telemetry Transport) brokers or CoAP (Constrained Application Protocol) servers, is essential for managing IoT data communication and ensuring reliable message delivery. Middleware bridges the gap between IoT devices and HPC applications.

Communication Protocols: MQTT is a lightweight and widely used messaging protocol for IoT. It allows IoT devices to publish data to topics and subscribe to topics for data retrieval, facilitating efficient communication between devices and HPC systems. CoAP is another protocol designed for resource-constrained IoT devices and is particularly suitable for low-power and low-bandwidth scenarios. It simplifies data exchange between IoT devices and HPC applications.

Edge Computing Frameworks: Apache NiFi is an open-source data integration tool that enables the flow of data between IoT devices and HPC systems. It can preprocess, filter, and route data from IoT devices to the appropriate destinations within the HPC cluster. EdgeX Foundry is an open framework for building edge computing solutions. It simplifies IoT device management, data processing, and integration with HPC applications.

IoT Data Processing and Analytics: Apache Kafka is a distributed streaming platform that can handle high-throughput, real-time data streams from IoT devices. It serves as a message broker for data ingestion, processing, and analytics within HPC systems.

Stream Processing Engines: Tools like Apache Flink, Apache Spark Streaming, and TensorFlow Streaming enable real-time processing and analytics of IoT data within HPC clusters, allowing for timely insights and decision-making.

Containerization and Orchestration: Docker and Kubernetes: Containerization technologies like Docker and orchestration platforms like Kubernetes are valuable for managing IoT integration software components. They ensure portability, scalability, and easy deployment across HPC environments.

Security Frameworks:

- IoT Security: Implementing security measures for IoT devices is crucial to protect against vulnerabilities and cyber threats. Secure device management and authentication are vital aspects of IoT integration.
- HPC Security: The integration should also consider HPC security practices to safeguard data and infrastructure. This includes network segmentation, firewall rules, and intrusion detection systems.

Data Storage and Databases:

- Time Series Databases: IoT data often involves time-series information. Time-series databases like InfluxDB and TimescaleDB are designed to efficiently store and query time-stamped data from IoT sensors.
- Big Data Storage: HPC clusters may need scalable storage solutions, such as Hadoop Distributed File System (HDFS) or distributed object stores, to accommodate the large volumes of IoT data generated.

HPC Middleware and Libraries:

- MPI and CUDA: Depending on the nature of the HPC applications, libraries like MPI and CUDA may be essential for parallel processing and GPU acceleration.
- HPC Simulation Tools: For scientific simulations involving IoT data, HPC-specific simulation tools may be required to integrate IoT data into computational workflows.

In conclusion, successfully integrating IoT devices into HPC environments requires careful consideration of software dependencies and the orchestration of various components. These dependencies encompass device drivers, middleware, communication protocols, edge computing frameworks, data processing and analytics tools, containerization, security measures, data storage solutions, and HPC-specific libraries and middleware. A well-structured integration strategy that addresses these software dependencies can unlock the potential of IoT data for enhancing the capabilities of HPC systems in research, industry, and smart city applications.

5 Integration of Nvidia Jetson and IoT

5.1 Internet of Things (IoT)

The Internet of Things (IoT) refers to the network of interconnected physical devices embedded with sensors, software, and connectivity capabilities. IoT enables the collection, exchange, and analysis of data from these devices, leading to improved efficiency, automation, and decision-making in various domains. Integrating Nvidia Jetson with IoT allows for real-time data processing, analytics, and decision-making at the edge, reducing the latency associated with cloud-based processing.

5.2 IoT Applications in High-Performance Computing

IoT integration in HPC brings several benefits. For example, in scientific research, IoT devices can gather data from distributed sensors and feed it to Jetson-powered HPC systems for real-time analysis and modeling. In manufacturing, IoT devices can monitor and control industrial processes, with Jetson providing the computational power for predictive maintenance, quality control, and optimization. The combination of Jetson and IoT enables edge AI, which is particularly beneficial for applications requiring low latency and offline operability.

5.3 Nvidia Jetson for IoT Edge Computing

Nvidia Jetson's powerful GPU architecture and low-power consumption make it an ideal platform for edge computing in IoT applications. Jetson's embedded nature allows it to be deployed directly at the edge, enabling real-time processing and inference on the data generated by IoT devices. This reduces the need for transferring large volumes of data to the cloud, improving response times and reducing network bandwidth requirements. Title: NVIDIA Jetson for IoT Edge Computing. As discussed in previous sections, IoT (Internet of Things) edge computing is a paradigm that involves processing data closer to its source, such as IoT devices, rather than sending it to a centralized cloud server. NVIDIA Jetson is a family of embedded computing platforms specifically designed for IoT edge computing applications. One of the primary distinguishing features of Jetson is its powerful GPU architecture. It includes NVIDIA's CUDA cores, making it ideal for high-performance computing tasks, including deep learning and computer vision. Jetson platforms are optimized for AI and deep learning workloads. They support popular deep learning frameworks like TensorFlow, PyTorch, and Caffe, enabling the development of AI-driven IoT applications. Jetson devices are small, energy-efficient, and designed for deployment in space-constrained IoT edge environments. They are suitable for applications like robotics, drones, and smart cameras. Jetson platforms offer multiple camera and sensor interfaces, allowing them to capture and process data from various sources, such as cameras, lidar, and radar sensors. The devices come with a range of I/O options, including USB, Ethernet, GPIO (General Purpose Input/Output), and PCIe (Peripheral Component Interconnect Express) for connecting to peripherals and external devices. Jetson platforms can perform real-time data processing, making them suitable for applications that require low latency and immediate decision-making, such as autonomous vehicles and industrial automation. NVIDIA provides pre-trained AI models and software libraries, such as TensorRT and cuDNN, to accelerate AI inference on Jetson platforms. This simplifies the development of AI-powered IoT applications. benefits from a thriving developer community and ecosystem. NVIDIA offers extensive documentation, tutorials, and tools to support IoT edge computing development. Figure 4 shows the IoT architecture, layers and components with functionalities.

Significance of NVIDIA Jetson in IoT Edge Computing:

1. Low Latency and Real-Time Processing: Jetson's ability to process data locally enables low-latency responses, making it suitable for applications like autonomous vehicles, industrial automation, and robotics, where real-time decision-making is critical.
2. Privacy and Data Security: IoT edge computing with Jetson allows data to be processed and analyzed locally, reducing the need to transmit sensitive data to remote servers, thus enhancing data privacy and security.
3. Bandwidth Efficiency: By processing data at the edge, Jetson reduces the volume of data that needs to be transmitted to the cloud, leading to more efficient bandwidth usage and cost savings.
4. Versatile Applications: Jetson's AI capabilities open the door to a wide range of IoT applications, from image and video analytics to natural language processing and predictive maintenance.
5. Scalability: Jetson offers scalability options, with devices ranging from entry-level platforms to high-performance models like the NVIDIA Jetson AGX Xavier, accommodating a variety of IoT use cases.

Applications of NVIDIA Jetson in IoT Edge Computing:

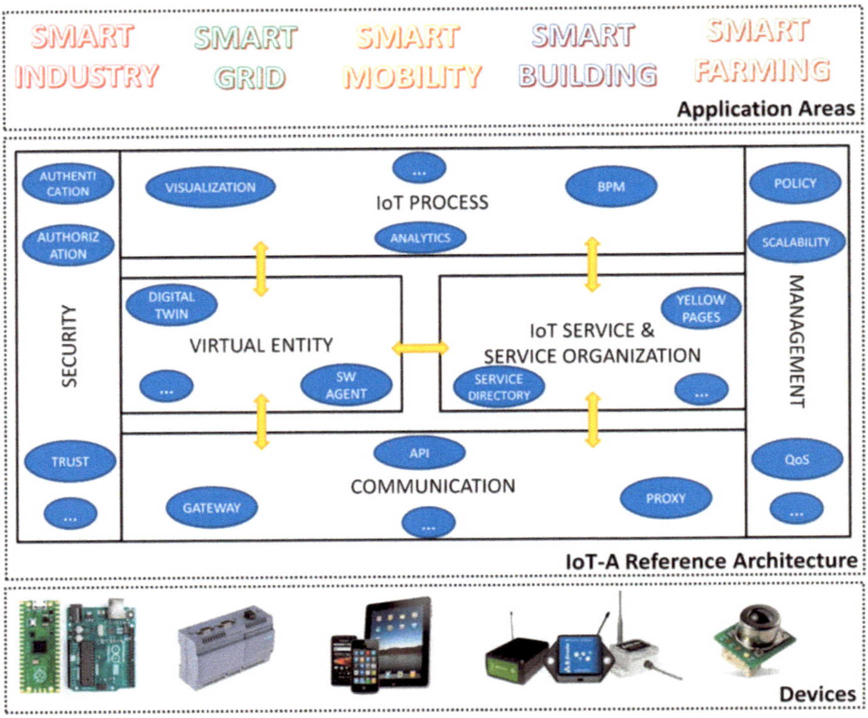

Fig. 4 IOT architecture, layers and components (*Source* https://doi.org/10.3390/s22062196)

1. Smart Cities: Jetson devices enable real-time video analytics for traffic management, public safety, and environmental monitoring in smart city deployments.
2. Industrial IoT (IIoT): In manufacturing and industrial settings, Jetson platforms support predictive maintenance, quality control, and automation tasks. Figure 5 depicts the RSM in Industrial IoT.
3. Agriculture: Jetson can be used for precision agriculture applications, including crop monitoring, pest detection, and automated farming equipment.
4. Healthcare: Jetson powers medical devices and wearable health technology for remote monitoring and diagnostics.
5. Autonomous Vehicles: Jetson's real-time processing capabilities are crucial for autonomous vehicles, enabling object detection, obstacle avoidance, and decision-making on the edge.

So NVIDIA Jetson plays a pivotal role in IoT edge computing by providing a powerful, compact, and AI-capable platform for processing data at the source. Its low latency, real-time processing capabilities, and support for deep learning make it a valuable tool for a wide range of IoT applications, from smart cities to healthcare and autonomous vehicles. As IoT continues to grow and demand for edge computing

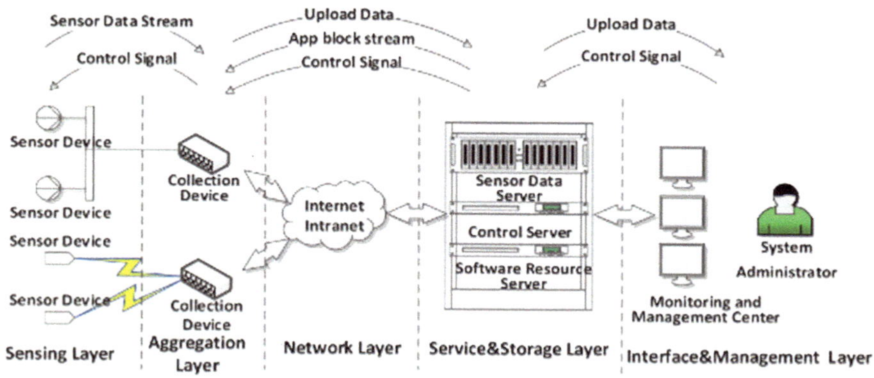

Fig. 5 Resource Service Model in the Industrial IoT System Based on Transparent Computing [23]

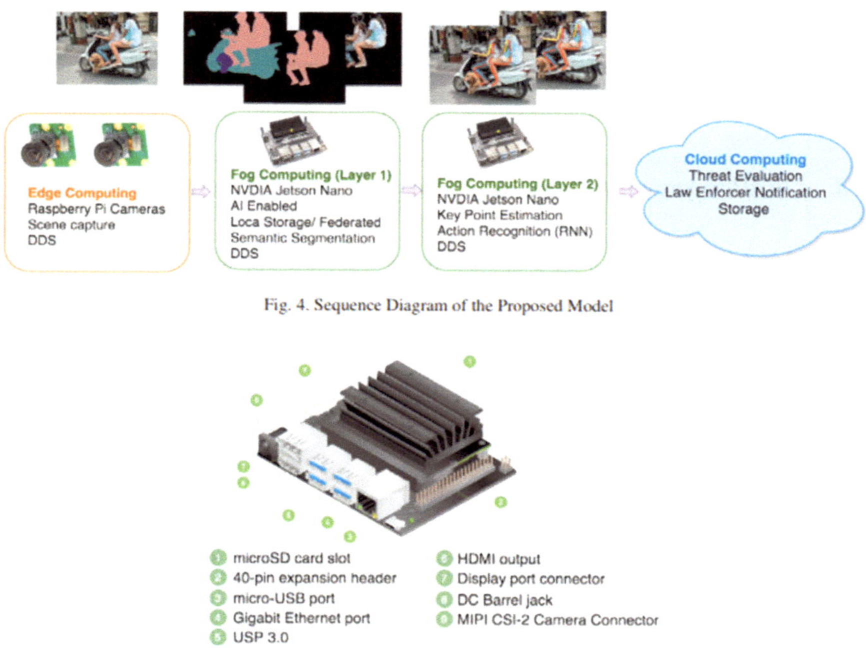

Fig. 6 Sequence diagram of the proposed model in the case study and the Nvidia Jetson Nano Kit [36]

solutions increases, Jetson's role in enabling intelligent and efficient edge devices becomes even more significant.

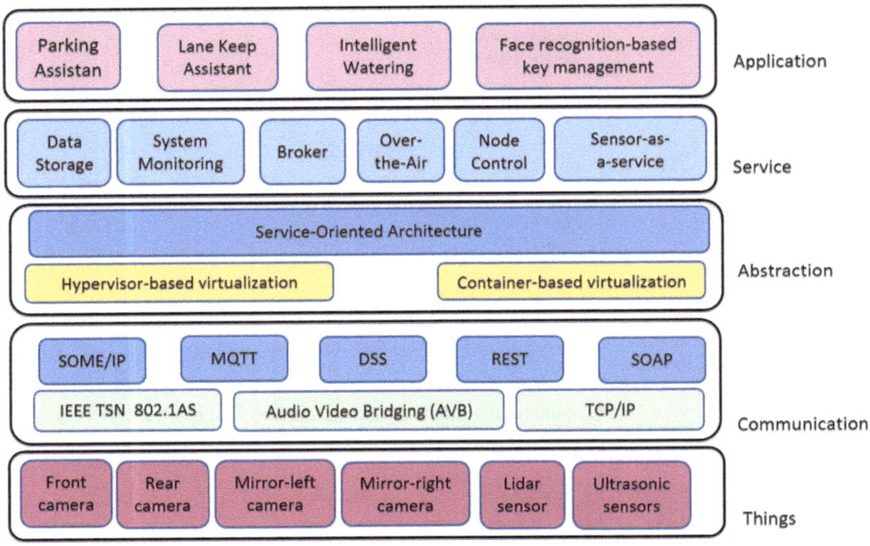

Fig. 7 Hub-OS logical architecture [37]

5.4 Challenges and Considerations for IoT Integration

Integrating Nvidia Jetson with IoT also poses challenges and considerations. Edge devices often have limited resources, including power, memory, and storage. Optimizing software dependencies, implementing efficient data transfer and synchronization mechanisms, and ensuring security and privacy are crucial factors to address when integrating Jetson with IoT. Additionally, managing a heterogeneous edge ecosystem, with different devices and communication protocols, requires careful planning and compatibility testing.

6 Optimizing Software Dependencies for HPC with Nvidia Jetson and IoT

6.1 Performance Optimization Techniques

To achieve optimal performance in HPC with Nvidia Jetson and IoT, several techniques can be employed. These include code parallelization, vectorization, algorithmic optimizations, and workload balancing. Additionally, leveraging GPU-specific optimizations like shared memory usage, memory coalescing, and efficient memory transfers can significantly improve performance in GPU-accelerated applications.

6.2 Memory Management and GPU Utilization

Efficient memory management is crucial in HPC to minimize data transfer overheads and maximize GPU utilization. Techniques such as memory pooling, data compression, and data streaming can help optimize memory usage. Furthermore, employing techniques like overlapping computation and communication or using asynchronous GPU operations can further enhance performance and reduce idle times.

6.3 Power and Thermal Management

Power and thermal management are critical considerations for HPC systems, particularly in energy-constrained edge environments. Optimizing power usage, reducing thermal hotspots, and employing techniques like dynamic voltage and frequency scaling (DVFS) can help mitigate power and thermal challenges. Additionally, techniques like workload consolidation and dynamic resource allocation can optimize resource utilization and reduce overall power consumption.

6.4 Code Profiling and Debugging

Profiling and debugging tools are essential for identifying performance bottlenecks, analyzing resource utilization, and optimizing code in HPC applications. Tools like Nvidia Nsight, GNU Profiler (gprof), and performance analysis frameworks like TAU or Scalasca can help developers profile their code, identify hotspots, and optimize performance in Jetson-based HPC applications.

6.5 Monitoring and Analytics for IoT Integration

Effective monitoring and analytics are essential for ensuring the smooth operation of IoT-integrated HPC systems. Monitoring tools can provide real-time insights into system performance, resource utilization, and network connectivity. Analytics frameworks, such as Apache Kafka or Apache Flink, can enable real-time stream processing and complex event processing, enabling actionable insights and efficient decision-making at the edge [24–35].

7 Some Case Studies: HPC with Nvidia Jetson and IoT Integration

7.1 Case Study 1: Real-Time Image Processing for the Internet of Things

In this case study, Nvidia Jetson is integrated with IoT devices for real-time image processing in a smart surveillance system. Jetson's GPU capabilities enable rapid object detection and tracking, while IoT devices collect data from surveillance cameras and perform local analytics. The combination of Jetson and IoT allows for quick response times, reduced bandwidth requirements, and enhanced security in the surveillance system [36].

This case study, as described in reference [36], presents a deep learning-based framework for intelligent video surveillance capable of processing real-time frames on two consecutive fog layers. One fog layer is dedicated to action recognition, while the other focuses on generating responses to criminal threats. The proposed architecture comprises three key modules. The initial module is responsible for capturing surveillance videos using RaspberryPi cameras deployed within a distributed network. The second module is dedicated to action recognition and employs a deep learning-based model installed on NVIDIA Jetson Nano devices positioned on the two fog layers. Ultimately, the security response is generated and transmitted to the law enforcement agency. To assess the effectiveness of the proposed model, experiments were conducted, specifically in the realm of semantic segmentation-based scene object recognition. The experimental results revealed a well-suited recognition model suitable for deployment within the fog layers of our proposed framework, as discussed in reference [36].

7.2 Case Study 3: Edge AI for Industrial Automation

This case study introduces an innovative architecture for an IoT computing platform, known as Hub-OS. It is designed to efficiently manage devices and applications from various vendors, allowing for seamless integration without requiring modifications to either the Hub-OS software components or the vendor-specific applications and devices. Additionally, Hub-OS can host IoT applications that require real-time processing when necessary. To assess Hub-OS, we implemented a case study for a smart vehicle and conducted a quantitative evaluation using diverse configurations. The quantitative evaluation demonstrates that Hub-OS successfully integrates applications and devices from different vendors, resulting in reduced latency, decreased CPU usage, and improved efficiency in IoT application startup and migration times. [37].

8 Future Trends and Challenges

8.1 Emerging Technologies in HPC and IoT

The field of HPC and IoT integration is rapidly evolving, with several emerging technologies shaping its future. These include advancements in AI and deep learning, edge computing frameworks, 5G and edge networking, and the emergence of new IoT protocols and standards. These technologies offer new opportunities for HPC with Nvidia Jetson and IoT integration, enabling more complex and sophisticated applications.

8.2 Challenges in Scaling HPC with IoT Integration

Despite the numerous benefits of HPC with Nvidia Jetson and IoT integration, several challenges need to be addressed. These include managing heterogeneous edge ecosystems, ensuring data security and privacy, optimizing resource utilization, addressing scalability concerns, and developing efficient algorithms and software architectures for distributed computing. Additionally, the complexity of integrating different hardware and software components necessitates standardized interfaces and interoperability frameworks.

8.3 Potential Solutions and Research Directions

To overcome the challenges in scaling HPC with Nvidia Jetson and IoT integration, researchers and industry practitioners are exploring various solutions and research directions. These include the development of edge computing architectures and frameworks tailored for HPC, advancements in hardware accelerators and GPUs, development of intelligent resource management techniques, and the integration of AI and machine learning algorithms for automated decision-making in dynamic edge environments. Standardization efforts and collaborative research can also drive the adoption and integration of HPC with Nvidia Jetson and IoT.

9 Conclusion

This chapter provided a comprehensive overview of operating systems and software dependencies in high-performance computing with Nvidia Jetson and IoT integration. It discussed the importance of operating systems, software dependencies, and the role of Nvidia Jetson in HPC. The integration of Jetson with IoT for edge computing in HPC applications was explored, along with the challenges and considerations associated with this integration. Operating systems play a crucial role in managing hardware resources and providing an interface for HPC applications. Software dependencies, including libraries, frameworks, and Nvidia-specific technologies, enable developers to optimize and accelerate HPC applications. The integration of Nvidia Jetson and IoT brings real-time data processing and decision-making to the edge, enhancing the overall performance and efficiency of HPC systems. The future of HPC with Nvidia Jetson and IoT integration looks promising, with emerging technologies and advancements in AI, edge computing, and IoT protocols. Addressing challenges such as heterogeneous edge ecosystems, security concerns, and scalability will pave the way for broader adoption and innovative applications in HPC. Collaborative research and industry efforts will further drive the integration of Nvidia Jetson and IoT in HPC, enabling advancements in various domains.

References

1. Shi, W., Cao, J., Zhang, Q., Li, Y., Xu, L.: Edge computing: vision and challenges. IEEE Internet Things J. **3**, 637–646 (2016)
2. Fernando, R.: GPU Gems: Programming Techniques, Tips and Tricks for Real-Time Graphics. Addison Wesley Professional (2004)
3. Plazolles, B., El Baz, D., Spel, M., Rivola, V.: Comparison between GPU and MIC on balloon envelope drift descent analysis. LAAS report (2014)
4. Boyer, V., El Baz, D.: Recent advances on GPU computing in Operations Research. In: Proceedings of the 27th IEEE Symposium IPDPSW 2013, Boston USA, 20–24 May 2013, pp. 1778–1787
5. Boyer, V., El Baz, D., Elkihel, M.: Solving knapsack problems on GPU. Comput. Oper. Res.. Oper. Res. **39**(1), 42–47 (2012)
6. Xavier, M.G., Neves, M.V., Rossi, F.D., Ferreto, T.C., Lange, T., De Rose, C.A.F.: Performance Evaluation of container-based virtualization for high performance computing environments. In: 2013 21st Euromicro International Conference on Parallel, Distributed, and Network-Based Processing, Belfast, UK, pp. 233–240 (2013). https://doi.org/10.1109/PDP.2013.41
7. Beserra, D., Moreno, E.D., Endo, P.T., Barreto, J., Fernandes, S., Sadok, D.: Performance analysis of Linux containers for high performance computing applications. Int. J. Grid Util. Comput. **8**(4), 321–329 (2017)
8. Wolfer, J.: A heterogeneous supercomputer model for high-performance parallel computing pedagogy. In: 2015 IEEE Global Engineering Education Conference (EDUCON), Tallinn, Estonia, pp. 799–805 (2015)https://doi.org/10.1109/EDUCON.2015.7096063
9. Süzen, A.A., Duman, B., Şen, B.: Benchmark analysis of Jetson TX2, Jetson Nano and raspberry PI using Deep-CNN. In: 2020 International Congress on Human-Computer Interaction, Optimization and Robotic Applications (HORA), Ankara, Turkey, pp. 1–5 (2020). https://doi.org/10.1109/HORA49412.2020.9152915
10. Cetre, C., Ferreira, F., Sevin, A., Barrere, R., Gratadour, D.: Real-time high performance computing using a Jetson Xavier AGX. In: 11th European Congress Embedded Real Time System (ERTS2022), Jun 2022, Toulouse, France. ⟨hal-03693764⟩ (2022)
11. Ullah, S., Kim, D.-H.: Benchmarking Jetson platform for 3D point-cloud and hyper-spectral image classification. In: 2020 IEEE International Conference on Big Data and Smart Computing (BigComp), Busan, Korea (South), pp. 477–482 (2020). https://doi.org/10.1109/BigComp48618.2020.00-21

12. Shin, D.-J., Kim, J.-J.: A deep learning framework performance evaluation to use YOLO in Nvidia Jetson platform. Appl. Sci. **12**(8), 3734 (2022). https://doi.org/10.3390/app12083734
13. Zakaria, F., Scogland, T.R.W., Gamblin, T., Maltzahn, C.: Mapping out the HPC dependency chaos. In: SC22: International Conference for High Performance Computing, Networking, Storage and Analysis, Dallas, TX, USA, pp. 1–12 (2022). https://doi.org/10.1109/SC41404.2022.00039
14. Mohr, B., Malony, A.D., Shende, S., et al.: Design and prototype of a performance tool interface for OpenMP. J. Supercomput.Supercomput. **23**, 105–128 (2002). https://doi.org/10.1023/A:1015741304337
15. https://citeseerx.ist.psu.edu/document?repid=rep1&type=pdf&doi=6d33c40811 0f6dff869177c4edf3fcd9845bb05a
16. Merchant, F., Chattopadhyay, A., Raha, S., Nandy, S.K., Narayan, R.: Access accelerating BLAS and LAPACK via efficient floating point architecture design. Parallel Process. Lett. **27**(03n04), 1750006 (2017)
17. Dechow, D.R., Abell, D.T., Stoltz, P., McInnes, L.C., Norris, B., Amundson, J.F.: A beam dynamics application based on the Common Component Architecture. In: 2007 IEEE Particle Accelerator Conference (PAC), Albuquerque, NM, USA, pp. 3552–3554 (2007)https://doi.org/10.1109/PAC.2007.4440489.h
18. Suita, S., et al.: Efficient cuDNN-compatible convolution-pooling on the GPU. In: Wyrzykowski, R., Deelman, E., Dongarra, J., Karczewski, K. (eds.) Parallel Processing and Applied Mathematics. PPAM 2019. Lecture Notes in Computer Science, vol. 12044. Springer, Cham (2020). https://doi.org/10.1007/978-3-030-43222-5_5
19. Awan, A.A., Subramoni, H., Panda, D.K.: An in-depth performance characterization of CPU- and GPU-based DNN training on modern architectures. In: Proceedings of the Machine Learning on HPC Environments (MLHPC'17). Association for Computing Machinery, New York, NY, USA, Article 8, 1–8 (2017). https://doi.org/10.1145/3146347.3146356
20. Kim, H., Nam, H., Jung, W., Lee, J.: Performance analysis of CNN frameworks for GPUs. In: 2017 IEEE International Symposium on Performance Analysis of Systems and Software (ISPASS), Santa Rosa, CA, USA, pp. 55–64 (2017)https://doi.org/10.1109/ISPASS.2017.7975270
21. Koo, Y., You, C., Kim, S.: OpenCL-Darknet: an OpenCL implementation for object detection. In: 2018 IEEE International Conference on Big Data and Smart Computing (BigComp), Shanghai, China, pp. 631–634 (2018). https://doi.org/10.1109/BigComp.2018.00112
22. Liu, Z., Ding, D. (2022) TensorRT acceleration based on deep learning OFDM channel compensation. In: Journal of Physics: Conference Series 2303 012047. https://doi.org/10.1088/1742-6596/2303/1/012047
23. Li, W., Wang, B., Sheng, J., Dong, K., Li, Z., Hu, Y.: A resource service model in the industrial IoT system based on transparent computing. Sensors **18**(4), 981 (2018). https://doi.org/10.3390/s18040981
24. Hammad, M., Iliyasu, A.M., Elgendy, I.A., Abd El-Latif, A.A.: End-to-end data authentication deep learning model for securing IoT configurations. Hum. Cent. Comput. Inf. Sci. **12**(4) (2022)
25. Anusha, A., Guptha, A., Rao, G.S., Tenali, R.K.: A model for smart agriculture using IOT. Int. J. Innov. Technol. Explor. Eng. **8**, 6 (2019)
26. Guillermo, J.C., García-Cedeño, A., Rivas-Lalaleo, D., Huerta, M., Clotet, R.: IoT architecture based on wireless sensor network applied to agricultural monitoring: a case of study of cacao crops in Ecuador. In: International Conference of ICT for Adapting Agriculture to Climate Change, pp. 42–57. Springer, Cham, Switzerland (2018)
27. El Azzaoui, A., Choi, M.Y., Lee, C.H., Park, J.H.: Scalable lightweight blockchain-based authentication mechanism for secure VoIP communication. Hum. Cent. Comput. Inf. Sci. **12**, 8 (2022)
28. Li, G., Yang, K.: Study on data processing of the IOT sensor network based on a Hadoop cloud platform and a TWLGA scheduling algorithm. J. Inf. Processing Syst. **17**, 1035–1043 (2021)
29. La, H.J., An, K.H., Kim, S.D.: Design patterns for mitigating incompatibility of context acquisition schemes for IoT devices. KIPS Trans. Softw. Data Eng. **5**, 351–360 (2016)

30. Shin, S., Eom, S., Choi, M.: Soft core firmware-based board management module for high performance blockchain/fintech servers. Hum. Cent. Comput. Inf. Sci. **12**, 3 (2022)
31. Choi, M., Kiran, S.R., Oh, S.-C., Kwon, O.-Y.: Blockchain-based badge award with existence proof. Appl. Sci. **9**, 2473 (2019)
32. Keswani, B., Mohapatra, A.G., Mohanty, A., Khanna, A., Rodrigues, J.J.P.C., Gupta, D., de Albuquerque, V.H.C.: Adapting weather conditions based IoT enabled smart irrigation technique in precision agriculture mechanisms. Neural Comput. Appl.Comput. Appl. **31**, 277–292 (2018)
33. Heble, S., Kumar, A., Prasad, K.V.D., Samirana, S., Rajalakshmi, P., Desai, U.B.: A low power IoT network for smart agriculture. In: Proceedings of the 2018 IEEE 4th World Forum on Internet of Things (WF-IoT), Singapore, 5 February 2018, pp. 609–614. IEEE, New York, NY, USA (2018)
34. Jawad, H.M., Nordin, R., Gharghan, S.K., Jawad, A.M., Ismail, M., Abu-AlShaeer, M.J.: Power reduction with sleep/wake on redundant data (SWORD) in a wireless sensor network for energy-efficient precision agriculture. Sensors **18**, 3450 (2018)
35. Nam, J., Jun, Y., Choi, M.: High performance iot cloud computing framework using pub/sub techniques. Appl. Sci. **12**(21), 11009 (2022). https://doi.org/10.3390/app122111009
36. Nurnoby, M.F., Helmy, T.: A real-time deep learning-based smart surveillance using fog computing: a complete architecture. Procedia Comput. Sci. **218**, 1102–1111 (2023). https://doi.org/10.1016/j.procs.2023.01.089
37. Abdelqawy, D., El-Korany, A., Kamel, A., Makady, S.: Hub-OS: an interoperable IoT computing platform for resources utilization with real-time support. J. King Saud Univ.-Comput. Inf. Sci. **34**(4), 1498–1510 (2022). https://doi.org/10.1016/j.jksuci.2022.02.011

GPU and ASIC as a Boost for High Performance Computing

Rajkumar Sampathkumar

Abstract This chapter explores the fascinating intersection of biology and computer science, where nature's design principles are harnessed to solve complex computational problems. This chapter provides an overview of bio-inspired computing techniques, including genetic algorithms, neural networks, swarm intelligence, and cellular automata. It delves into the core concepts of each approach, highlighting their biological counterparts and demonstrating their applications across various domains. Furthermore, this chapter discusses the evolution of bio-inspired algorithms, emphasizing their adaptation to contemporary computing paradigms such as machine learning and artificial intelligence. It examines how these algorithms have been employed to address real-world challenges, ranging from optimization problems and pattern recognition to robotics and autonomous systems. In addition to theoretical insights, the chapter offers practical guidance on implementing bio-inspired algorithms, including algorithmic design considerations and the integration of bio-inspired approaches with traditional computing methods. It also discusses the ethical and societal implications of bio-inspired computing, touching upon topics like algorithm bias and data privacy.

Keywords Arithmatic Logic Unit · Graphical Processing Unit · Application Specific Integrated Circuits · OpenCL

1 Introduction

Since the beginning of high-performance computing systems, clusters of CPUs were used for accelerating the performance of computing tasks. Since the beginning of usage of GPUs for general purpose, the overall performances of HPC platforms have increased drastically. When comparing the computing performance, a GPU tend to outperform any CPU. This is due to the large number of processing cores

R. Sampathkumar (✉)
Centre for Computational Engineering Sciences, Cranfield University, College Road, Cranfield MK43 0AL, UK
e-mail: r.sampathkumar.161@cranfield.ac.uk

K. A. Ahmad et al. (eds.), *High Performance Computing in Biomimetics*, Series in BioEngineering, https://doi.org/10.1007/978-981-97-1017-1_9

present in the GPU and the massively parallel processing architecture which allows the GPU to process multiple threads simultaneously or process a single thread with multiple data elements. Newer generation of HPCs incorporate ASICs (Application Specific Integrated Circuits) which are custom built for specific tasks and can provide unparalleled performances, which accelerated the overall performance of the system. In this chapter, we are going to discuss the advantages of GPU and ASIC based HPC platforms and how they impact the performance and efficiency of the system.

2 GPU and ASIC Acceleration

Central Processing Unit (CPU) is the integral component of the modern computer which does all the calculations. The number of cores and the clock speed determines the computing power of it. The CPU houses all the necessary components for computations such as Arithmetic Logical Unit (ALU) where the calculation actually happens using logic gates, Control Unit which controls the flow of tasks, Registers where the data is stored and loaded (similar to a RAM), Bus which through-puts the data to and from the units [1].

All these parts work efficiently with minimum to no latency for faster calculations. CPU employs serial processing where a single task is handled at any given time and when the number of tasks tend to overflow or the computation time for each task increases, the amount of time required for calculation also increases. For achieving higher computation power, the physical cores are split into multiple logical cores with the help of simultaneous multithreading to introduce parallel processing in CPU without increase in dynamic power draw and increasing the efficiency [2] (Figs. 1 and 2).

Graphical Processing Units (GPU) are purposefully built for complex mathematical calculations like video processing and rendering, simulations, data analysis etc., which requires high computing power. When compared to CPUs, GPUs will have tens, hundreds or even thousands of cores which can compute at much faster rates (Figs. 3 and 4). Figure 4a, b also shows the control of CPU analyzing registers and then transmitting data respectively.

When the GPU is employed along with a CPU for intensive operations such as data analytics, simulation etc., is called GPU Acceleration. GPU acceleration provides superior performance for software applications. The compute intensive tasks are handled by the GPU and the rest of the tasks are executed in the CPU through serial processing. GPU on the other hand works in parallel processing which can compute several tasks at the a given time effectively decreasing the overall computing time.

While the sequential calculations are performed in the CPU, compute intensive calculations are processed parallelly in the GPU. Another advantage is that the parallel programming model support, which helps application developers to provide exceptional application performance. GPU acceleration is being used for several purposes like video editing and rendering, CFD simulations, data analysis, medical imaging etc.

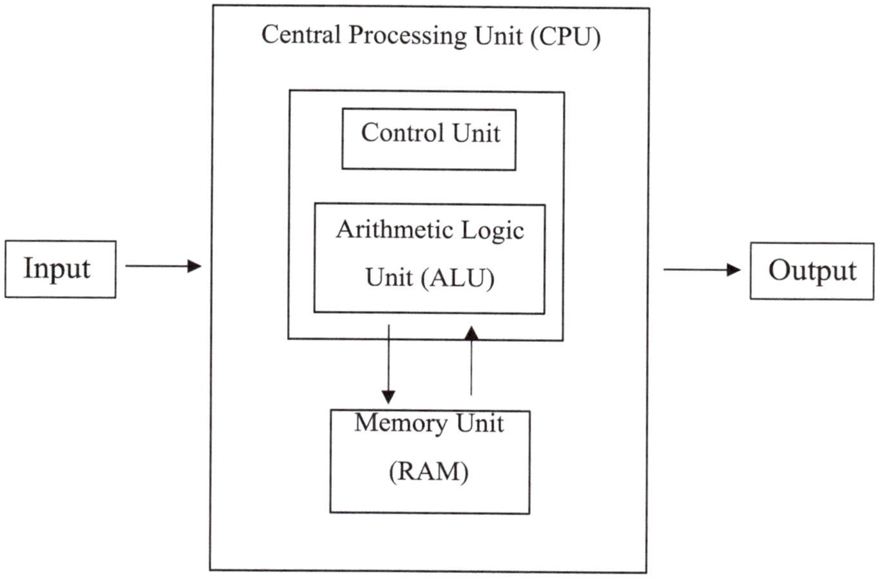

Fig. 1 CPU components

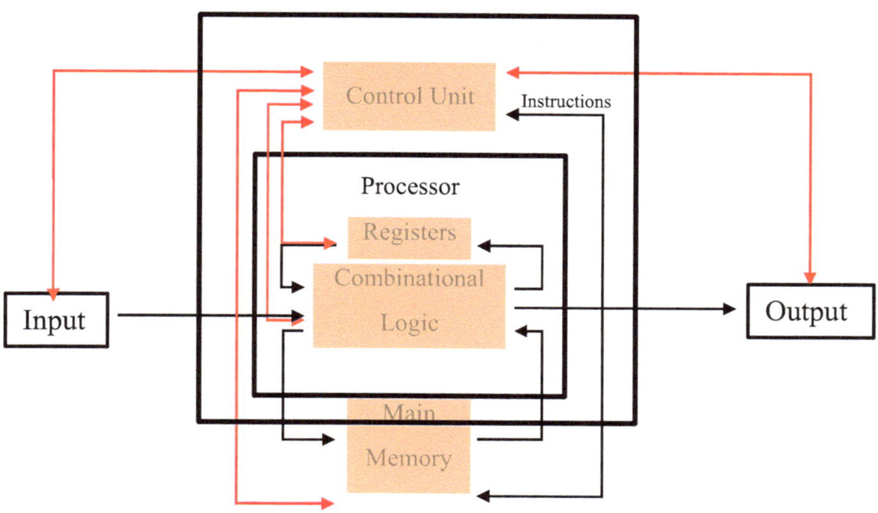

Fig. 2 Anatomy of CPU

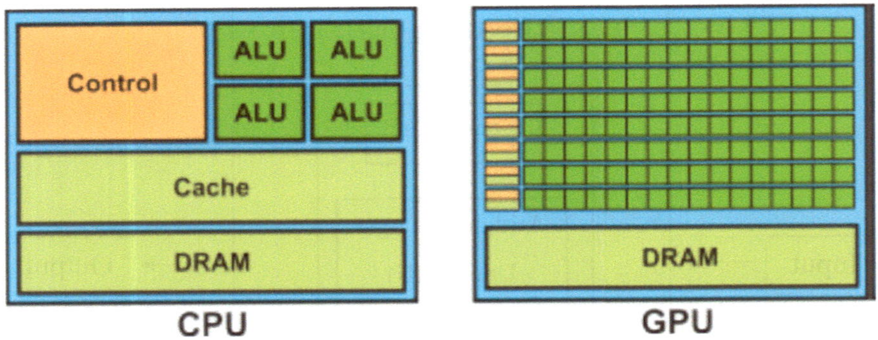

Fig. 3 GPU versus CPU architecture [3] https://doi.org/10.3390/rs10060864

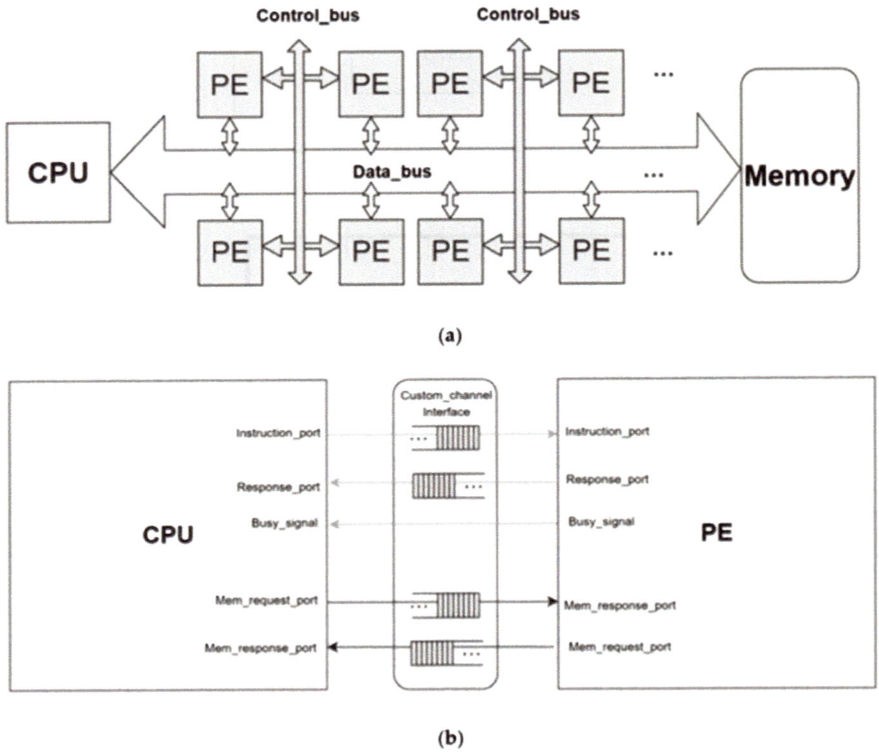

Fig. 4 Typical CPU control architecture [3] https://doi.org/10.3390/electronics11192989

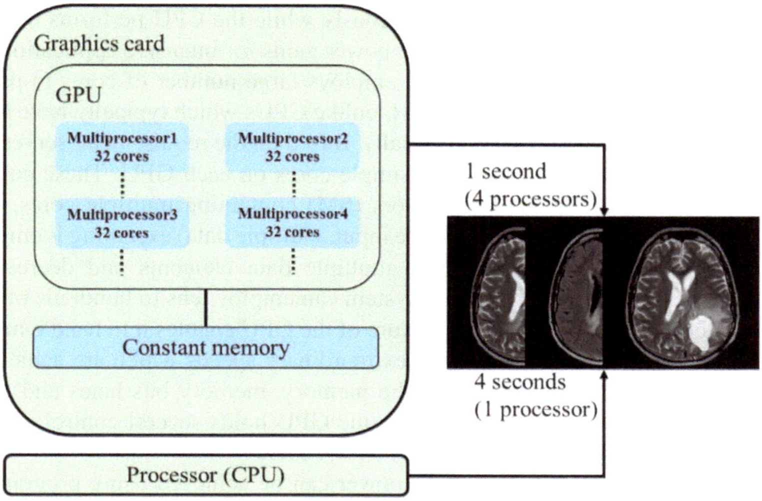

Fig. 5 CPU and GPU combination for a typical MRI processing [3] https://doi.org/10.3390/s24051591

Application Specific Integrated Circuits (ASIC) are processors which are built for one specific application domain, and it eliminates any unwanted and unnecessary components and is optimized for maximum performance and efficiency. They provide unparalleled performance because of their custom design and optimization of the circuit for the domain of work it is developed for. In terms of computing power, ASIC exceeds both CPU and GPU and is the go-to solution for High Performance Computing systems (HPC).

GPU in one way is also an ASIC built for graphical rendering computations. ASICs are preferred for large scale high-cost computing systems which will be used for one specific domain so that the circuit design is optimized for that specific workload. ASIC acceleration works similar to GPU acceleration much higher computing power to the system (Fig. 5). Figure 5 shows the common CPU and GPU combination MRI of human brain.

3 Parallel Processing Capabilities of GPUs

The main features that set GPUs apart from the traditional CPUs is its parallel processing capabilities. CPUs are designed to handle wide range of tasks in multiple different intensity but GPUs on the other hand are designed and optimized for performing complex mathematical calculations and rendering graphics in parallel. This helps the system to run more efficiently, and computing time decreases considerably. Parallel processing in the GPU facilitates the application to run several high compute intensive calculations simultaneously while the CPU performs other less intensive tasks thus resulting in massive power gains for intensive applications.

The parallel processing in the GPU employs large number of cores to perform complex calculations simultaneously [4], unlike CPUs which typically have limited number of cores and processes sequentially. Based on the requirement, server grade GPUs have hundreds to thousands of simple cores on each GPU. These cores are organised into streaming multiprocessors (SM) containing multiple cores. While executing a task or thread, SIMD (Single input, multiple data) executing is employed which can send single instruction to multiple data elements and decrease the processing time. A high-performance system can employ tens to hundreds of GPUs in a single cluster. The parallel architecture of the GPU enables it to handle multiple datasets and complex algorithms with extraordinary speeds which are achieved by high performance and low latency cache memory, memory bus lanes and control units. As we can see from Figs. 1 and 4, the GPU holds several control units and ALUs compared to CPU which only has one of each.

Harnessing this parallel processing power can be achieved using programming models know as Application Programming Interface (API) such as CUDA (Compute Unified Device Architecture) or OpenCL (Open Computing Language) which is written as code to instruct the GPU to divide the workload into numerous threads and process the threads parallelly. These APIs have their own set of libraries and instructions using which a developer can write their own codes and optimise the GPU to work seamlessly for their application.

The parallel processing of GPU extends to a wide range of applications in various fields. Ueda et al. [5], in their work performed optimization of global trajectory framework using GPU and CPU parallel architecture system where the initial trajectory propagation with low fidelity and over a trillion of initial conditions was carried out in the GPU and the solutions around thousands of conditions are sent to the CPU for final trajectory calculation with high fidelity. Figure 6 shows the schematic of the multi-fidelity framework.

Similarly, Quezada et al. [6], in their work showcased the performance of a GPU while using the dynamic parallelism technology by a subdivision model using heterogeneous workloads. They employed CUDA DP (developed by NVIDIA) which employs recursive kernels and compared 4 different GPUs with different computing powers.

Parallel processing of the GPUs also facilitated in computational fluid dynamics domain and can be found on used for various complicated CFD simulations [7] which takes a lot of time and computing resources to operate. Researchers have come up with advanced modified solvers which takes advantage of hardware acceleration [8] and also the advantage of multiple GPU in a single cluster [9]. The hardware acceleration provided by the GPUs shown extraordinary performance gains for lattice Boltzmann simulations [10, 11] which involves huge number of complex calculations per second. Every year, the GPU is improved in its performance by optimizing the new architectures which makes the new GPUs much faster compared to previous generations.

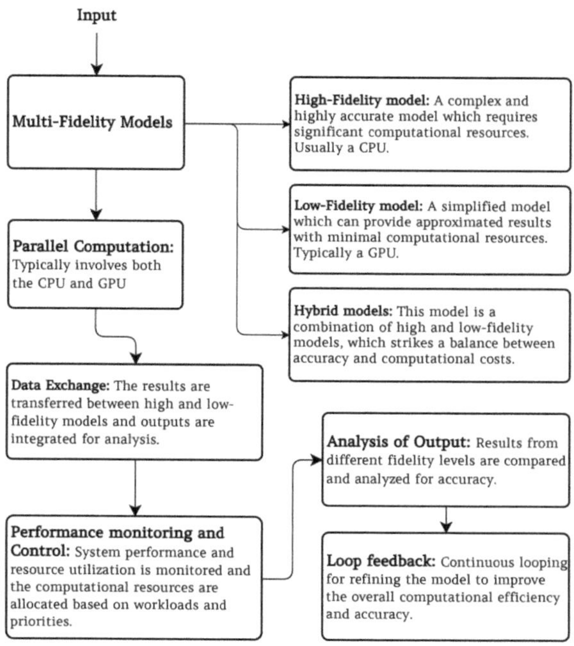

Fig. 6 Schematic flow diagram of the multi-fidelity framework

4 GPU Architecture and HPC Performance

The architecture of a GPU plays a vital role in HPC capabilities. The design of a GPU starts from its base architecture. The architecture design choices impact the overall performance of the GPU, this includes the computational capabilities, memory access and bandwidth, efficient power consumption, data transfer and parallel processing capabilities with multiple GPU. Parallel processing has been the major advantage and explains why HPCs use GPU. The other features of GPU architecture which influence the performance are memory hierarchy, memory bandwidth, power efficiency, specialised units, and scalability.

The memory hierarchy is an import factor of any GPU because of the presence of shared memory and on-chip L1 and L2 cache memory which allows the for faster and low latency data retrievals effectively reducing the memory access time which directly influences the performance. The efficient memory management, cache utilization and minimum memory access also improves the performance. Mawson et al. [12] in 2014 published their research on memory transfer optimisation for lattice Boltzmann solver for fluid flow simulations using NVIDIA's Kepler architecture GPU (Fig. 7).

GPUs incorporate specialized units for accelerating specific operation. The latest NVIDIA GPUs come with Tensor cores which are used for accelerating matrix operations and are effectively used in deep learning. Similarly, the ray tracing cores found in the RTX series NVIDIA GPUs from 2019 are specifically designed for ray

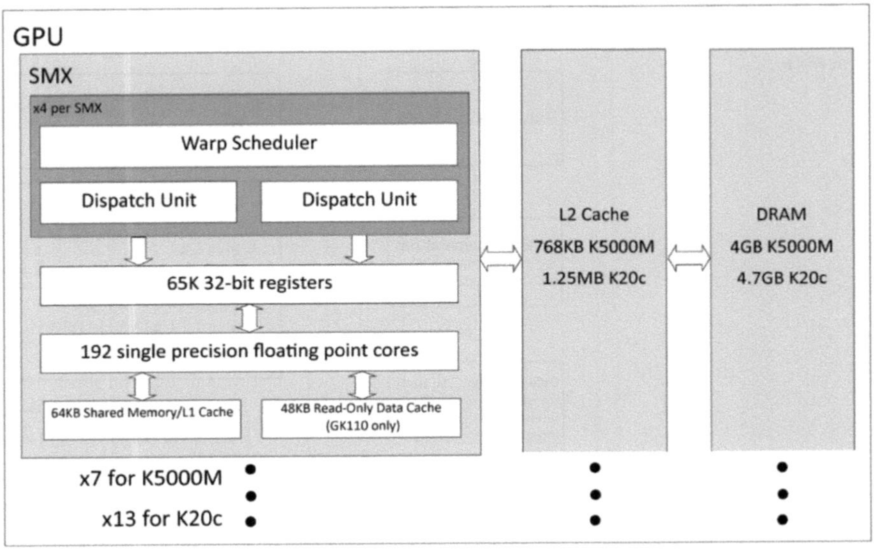

Fig. 7 Kepler GPU Architecture. Reprinted from the Ref. [12]. Copyright (2014), with permission from Elsevier

tracing operations in real-time to enhance the video rendering experiences. While the industry advances, the need for computing power also increases as the applications become more and more power demanding.

Various features found in the mobile and computer applications like speech recognition, automatic caption generation, advertisement recommendation etc. has been efficiently processed with the help of Deep Neural Networks (DNN) which involves lot of matrix operations. Fujita et al. [13] focused on cross-corelation function using NVIDIA's Ampere and Hopper architecture GPUs and found that the tensor cores accelerated the overall performance of the matrix calculations but the theoretical performance of the said GPUs were limited by their shared and global memory bandwidths. Atoofian [14], worked on accelerating the DNNs using sparsity which uses the tensor core to accelerate the matrix calculations.

The ray tracing present in RTX GPUs were particularly used to accelerate video rendering and fluid simulation applications. Wang et al. [15] focused their research on Eulerian–Lagrangian simulations using hardware accelerated ray tracing cores. They developed a fluid simulation model which uses the CUDA and Ray Tracing cores present in the NVIDIA GPU for particle tracking. The CUDA core accelerated the fluid calculations, and the RT cores accelerated the visualization calculations.

5 GPGPU Programming Frameworks: CUDA and OpenCL

General Purpose computing on Graphics Processing Units (GPGPUs) has gained significant popularity in harnessing the parallel processing power of GPUs for non-graphical computing operations [16–18]. Programming the GPUs for such purposes are handled using CUDA and OpenCL programming models which enable developers to utilize the computational power of GPUs for general purpose.

CUDA stands for Compute Unified Device Architecture, developed by NVIDIA is a parallel computing programming model which works for NVIDIA GPUs. It provides the developers various set of APIs, libraries and tools to work with that enable them to write programs for parallel computing using a language similar to C. CUDA gives the developers full control of the thread and data execution on the GPU using its massively parallel computing capabilities. The GPU can be controlled to the maximum level through the CUDA framework, this allows for fine tuning for optimization and performance tuning. Since the introduction of CUDA, it has become the industry standard GPU programming, especially in machine learning, specific computation tasks and other domains. Shi et al. [19] proposed a new framework called vCUDA, which is focused on General purpose GPU acceleration for virtual machines. This framework is based on CUDA and they enabled it to work with virtual machines.

OpenCL stands for Open Computing Language, is an open-source framework and is not specific to any vendor GPU which was developed by Khronos Group. OpenCL allows the developer to write cross platform programming model which works with wide range of GPUs, CPUs and other computing accelerators. The way OpenCL works is that it provides with an abstraction layer where the developers can write in a C-like programming language once and can execute it to multiple devices without any significant modifications. The main strong suite of OpenCL is it supports heterogeneous computing, where it can be executed in multiple different devices such as GPUs and CPUs, which results in maximum performance.

Both of these frameworks provide the user a starting point from which they can develop an efficient model for specific applications and by optimizing the model, the acceleration can be maximum and can yield more performance out the GPUs. Petrovič et al. [20], developed a set of benchmarks for efficiently tuning the CUDA and OpenCL frameworks by a kernel tuning toolkit. A kernel of a system has full control over the application and the components related to that application and by tuning the kernel, the components which are used for the specific application are tuned which increases the performance of the application in that specific hardware. This results in hardware acceleration.

Both of these frameworks, whether it is CUDA's tight integration with NVIDIA or OpenCL's flexible integration with any vendor, they both offer the users with full control over any GPU that they are working with for optimizing and unlocking the full potential computing power for any specific application. The technological improvement of these GPGPUs can be found by improving the architecture and improve the heterogenous computing capabilities[21] for better hardware acceleration.

6 Heterogeneous Computing: CPU-GPU Collaboration

Heterogeneous computing involves the utilization of both CPU and GPU to collaborate together for computing operations. Heterogeneous computing has several advantages in terms of performance gains, efficiency in power usage and the optimization and routing of workloads. When it comes to architectures, GPUs and CPUs are vastly different from each other as their way of processing is different. CPUs are mainly optimized for sequential processing whereas GPUs are built with massively parallel architecture. The sequential processing in the CPU is suited for high single thread performance and complex control flow operations. The parallel processing is well suited for data parallel operations, where a single compute intensive operation is fed to multiple data elements.

Collaborative computing between CPU and GPU involves their respective strengths. The CPU being the main processor managing the I/O operations, overall system and managing the workload execution by the GPU. The GPU works on the highly parallelizable computing operations and offloading the intensive tasks from CPU. This helps the system stability and responsive as the CPU will always have the resources to manage the system. CPU-GPU collaboration can be enabled with the help of CUDA or OpenCL framework through which the developers can optimized the hardware from identifying the intensity of the computing operations and route the tasks to the respective hardware for acceleration if required.

HPCs has a heterogeneous architecture with one or two sever grade CPUs and multiple GPUs for computational tasks. Researchers have been working with collaborative computing in various fields. Borrell et al. [22] has worked on CFD simulations on airplane aerodynamics with a heterogeneous HPC platform. Through their research, they reported an approx. 23% decrease in assembly time when compared to pure GPU execution. They also concluded that the proper utilization of CPU for calculations is similar to having a GPU extra per node in HPC. This increases the power efficiency of the overall system. Skrzypczak et al. [23] focused their work on simulation of crowd movement. They used a heterogeneous CPU-GPU high performance computing and compared the results with both standalone CPU and GPU configurations. They concluded that the heterogeneous configuration outperforms the standalone configurations and its because of the load balance between the CPU and GPU and optimal use of the pinned and shared memory by the host and GPU respectively. Liu et al. [24] focused on developing hybrid models for solving line CFD equations and compared the computing performance of GPU and CPU.

From the graph below, it is clear that the GPU has a significant power advantage over the CPU and the difference increases with the increase in control volume (mesh

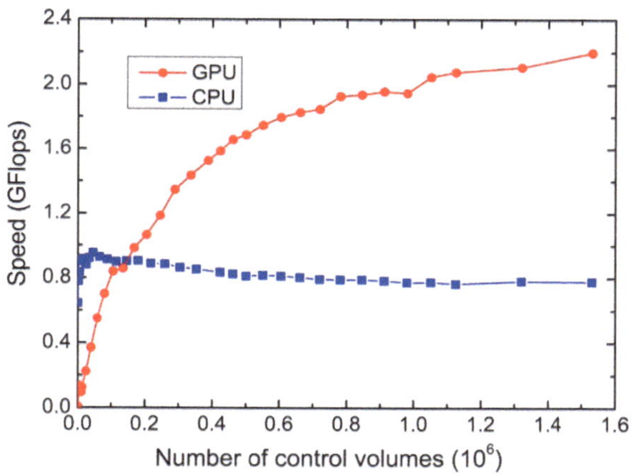

Fig. 8 Speed functions of the CPU cores and GPU processing unit in experiments on the Adonis cluster. Reprinted from Ref. [24]. Copyright (2016), with permission from Elsevier

density). Similarly, the heterogeneous platform has a wide range of applications and primarily used for accelerating the performance for the application [25–27] (Fig. 8).

7 ASICs and Custom Hardware Design

Application Specific Integrated Circuits or ASICs are specialized integrated circuits designed for specific tasks and workloads. ASICs are custom-built processing units for providing optimal performance, power efficiency and functionality for a specific application unlike general purpose GPUs and CPUs which are built for more general applications. The custom hardware design plays an important role in developing the ASICs for more customized and dedicated hardware solutions to applications.

The custom hardware design of ASICs has several steps:

1. To identify the requirements and the specifications of the ASIC. This involves proper understanding of the domain of the application, specific operations to be performed and the desired power efficiency, computing performance, and other system requirements. The architectural design of the ASIC which maps the organisation of the control units, data paths, circuits and interfaces is designed based on the above requirements.
2. RTL (Register Transfer Language) design involves the creation of a detailed digital representation of the ASIC with the help of what is called Hardware Description Language (HDL) like Verilog or VHDL. The behaviour and structure of the ASIC is specified by the RTL at the register transfer level. Through rigorous simulation and formal verification techniques, the design is ensured for its functionality, timing, and accordance to the specifications.
3. After the verification of the design, the RTL is converted into a physical layout which is then fabricated on a silicon wafer. This fabrication process involves floor planning, placement, routing, and optimization. The main aim is to maximum

optimization of the layout for power distribution, integrity of the signals between the components, timing constraints and the utilization of the area while ensuring the lower costs of fabrication and manufacturability.
4. The final stage is the manufacturing and testing. The physical design of ASIC is manufactured using semiconductor fabrication processes. The fabrication includes wafer fabrication, packaging of the chip and finally testing. The testing is the most crucial phase which validates the performance of the ASIC through rigorous functional testing and quality assurance to meet the specified requirements.

8 Advantages of ASICs in HPC Performance

The customized design of ASIC for HPC offers several advantages:

1. The main advantage is power optimization. ASICs as their name suggests application specific, which is tailored for a specific application. Through the customization of the architecture and the circuits, ASICs can showcase exceptional performance when compared to general purpose computing processors.
2. Secondly, through the power optimization, the power consumption is also optimized for the maximum efficiency. This comes in handy in applications which has power or battery life constraints.
3. Thirdly, by customizing the architecture, ASICs can accommodate multiple components and functions in a single chip which results in lesser system complexity, increased reliability of the chipset and more compact form factor or more transistors packed when compared to similar size GPGPUs. They can also be designed for better integration with other system components leading to better system performance.
4. ASICs also provide better protection to intellectual property, which meant the competitors find it difficult to replicate or reverse engineer the design or functionality as the chipset is completely customized.

Though the advantages of ASICs are numerous, yet the limitations have to be looked at. Firstly, it's very complicated to design an entire chipset, huge amount of time is required to finalize the design and finally its very expensive to manufacture which also requires expertise in hardware design, fabrication, and verification. Also, they are very difficult to reprogram and are not flexible when compared to GPGPUs, which makes them undesirable for applications which require frequent changes and adaptations.

9 Comparison of GPUs and ASICs in HPC Applications

When it comes to HPC platforms, GPUs and ASICs are the two most common hardware options. Let's compare them for more insights as shown in Table 1 based on factors like programmability, flexibility, power efficiency etc.

Table 1 Comparison of GPUs and ASICs for HPC based on factors

Factor	GPU	ASIC
Programmability and flexibility	GPUs offer high flexibility and programmability. They are equipped with frameworks like CUDA or OpenCL which enables the developers in writing parallel programs and take advantage of the massively parallel processing architecture of the GPUs. They can also be engaged for a wide range of applications and workloads	ASICs are customized designs which are designed for specific tasks. This results in exceptional performance for the specified task which meant the ASIC lacks flexibility of GPUs. They are also less suitable for a wide range of applications and applications which involve frequent changes
Performance and efficiency	GPUs take advantage of their parallel processing capabilities and with the help of large number of cores, they can execute thousands of threads simultaneously and they can also apply the same operation to multiple data elements which can speed up the computing process. In applications run on HPC which require significant number of parallel computations, GPUs deliver excellent performance	ASICs are built for a specific operation and has the potential of delivering the maximum performance if both GPU and ASIC are involved in the same operation. They are designed from scratch with custom architecture and hardware units that are optimized for both superior power delivery and power efficiency. They are to be chosen when extreme performance is the requirement, but it comes at the cost of little to no flexibility and complexity in designing
Power efficiency	GPUs are built for more generalized purpose and strikes the balance between performance and power consumption. The programming frameworks offers the flexibility to the developers in finding this balance or can unlock the full potential while maintaining some level of power efficiency. This makes GPUs more desirable for HPC platforms	AISCs are application specific hardware which offers the best performance while also minimizing the power consumption. They are tailored for optimal performance with a fine-tuned, custom architecture which can deliver exceptional performance the intended task it was built for. ASICs are chosen for HPCs which are built for more focused computing operations with no deviation from the intended use

10 Integration and Coexistence of GPUs and ASICs in HPC Systems

GPUs and ASICs have shown their strengths in various aspects for high performance computing and their integration and coexistence plays an important role in achieving optimal performance and power efficiency. They bring unique advantages and can put to use in HPC systems to extract more performance. There are several technical aspects for this integration.

1. GPUs and ASICs can complement each other with different processing capabilities, where GPU being general purpose processors and excel in parallel processing, can work on parallel tasks and offload computationally intensive tasks to ASIC for which it is optimized for maximum performance. This results in an efficient workload distribution similar to heterogeneous systems and higher overall system performance.
2. When it comes to data exchange and memory management, GPUs have dedicated memory (VRAM) which usually optimized for large data access with high bandwidth. ASIC being custom built, will have memory which is tailored for its intended computing operation. A proper and efficient communication interface is to be established between GPUs and ASICs for data exchange. This can become tricky as unnecessary latency can evolve, which can be avoided by data prefetching, caching and overlapping computation with data transfers, these techniques can optimize the memory performance and decrease the synchronization overhead.
3. The major aspect is the system level optimization for maximum performance and efficiency. Designing the overall architecture and proper collaboration of the software enables effective integration of GPUs and ASICs. Programming models like CUDA and OpenCL should utilize both the hardware and seamlessly manage the workload transfer.

Moreover, power management for both the hardware has to be implemented for optimizing the power consumption and to have room for more performance. Researchers have been working with software level power capping [28], which improves the efficiency and can be programmed to cap and unleash the power whenever its required. Techniques like Power-aware scheduling and dynamic voltage and frequency scaling (DVFS) can be employed for balancing the performance and power consumption based on the workloads.

By integrating GPUs and ASICs in HPC systems, the overall system efficiency and performance with power efficiency is improved. With this enhanced performance, HPCs can tackle intensive computation tasks in the field of CFD, data analytics, AI, and machine learning, and much more.

11 Conclusion

In this chapter, we discussed about how GPU and ASIC boosts and accelerates the performance of HPC systems. GPUs clearly shows computing performance advantage over conventional CPUs or even high-performance server grade multi-core CPUs. The highly parallelized architecture of the GPU and the large number of cores present in the GPU facilitates for multi-thread processing parallelly and accelerates computationally intensive threads by processing them through multiple data elements offering exceptional performance and power efficiency. This was achieved with the help of parallel processing programming frameworks such as CUDA and OpenCL. When it comes to ASIC, the completely customized architecture which is built from ground up for the specific task, enhanced the processor for optimizing the performance and power efficiency of the system and provides unparalleled performance and system stability. The integration of GPU and ASIC can accelerate the system for exceptional computational performance, while overseeing the technical aspects for a smooth integration of both architectures for enhanced overall system performance. With all the facts considered, the future of HPC systems looks promising and can propel the future discoveries and scientific advancements in the computation domain.

References

1. Ouro, P., Lopez-Novoa, U., Guest, M.F.: On the performance of a highly-scalable computational fluid dynamics code on AMD, ARM and Intel processor-based HPC systems. Comput. Phys. Commun. **269**, 108105 (2021). https://doi.org/10.1016/j.cpc.2021.108105
2. Mantovani, F., et al.: Performance and energy consumption of HPC workloads on a cluster based on Arm ThunderX2 CPU. Futur. Gener. Comput. Syst. **112**, 800–818 (2020). https://doi.org/10.1016/j.future.2020.06.033
3. V./Ml: Exploring the GPU Architecture. https://core.vmware.com/resource/exploring-gpu-architectureforthelatestversion
4. Deluzet, F., Fubiani, G., Garrigues, L., Guillet, C., Narski, J.: Efficient parallelization for 3D–3V sparse grid Particle-In-Cell: single GPU architectures. Comput. Phys. Commun. **289**, 108755 (2023). https://doi.org/10.1016/j.cpc.2023.108755
5. Ueda, S., Ogawa, H.: Multi-fidelity approach for global trajectory optimization using GPU-based highly parallel architecture. Aerosp. Sci. Technol. **116**, 106829 (2021). https://doi.org/10.1016/j.ast.2021.106829
6. Quezada, F.A., Navarro, C.A., Romero, M., Aguilera, C.: Modeling GPU dynamic parallelism for self similar density workloads. Futur. Gener. Comput. Syst. **145**, 239–253 (2023). https://doi.org/10.1016/j.future.2023.03.046
7. Eichstädt, J., Peiró, J., Moxey, D.: Efficient vectorised kernels for unstructured high-order finite element fluid solvers on GPU architectures in two dimensions. Comput. Phys. Commun. **284**, 108624 (2023). https://doi.org/10.1016/j.cpc.2022.108624
8. De Vanna, F., et al.: URANOS: A GPU accelerated Navier-Stokes solver for compressible wall-bounded flows. Comput. Phys. Commun. **287**, 108717 (2023). https://doi.org/10.1016/j.cpc.2023.108717
9. Zhang, X., Guo, X., Weng, Y., Zhang, X., Lu, Y., Zhao, Z.: Hybrid MPI and CUDA paralleled finite volume unstructured CFD simulations on a multi-GPU system. Futur. Gener. Comput. Syst. **139**, 1–16 (2023). https://doi.org/10.1016/j.future.2022.09.005

10. Xu, A., Li, B.-T.: Multi-GPU thermal lattice Boltzmann simulations using OpenACC and MPI. Int. J. Heat Mass Transf. **201**, 123649 (2023). https://doi.org/10.1016/j.ijheatmasstransfer.2022.123649
11. Spinelli, G.G., et al.: HPC performance study of different collision models using the Lattice Boltzmann solver Musubi. Comput. Fluids **255**, 105833 (2023). https://doi.org/10.1016/j.compfluid.2023.105833
12. Mawson, M.J., Revell, A.J.: Memory transfer optimization for a lattice Boltzmann solver on Kepler architecture nVidia GPUs. Comput. Phys. Commun. **185**(10), 2566–2574 (2014). https://doi.org/10.1016/j.cpc.2014.06.003
13. Fujita, K., Yamaguchi, T., Kikuchi, Y., Ichimura, T., Hori, M., Maddegedara, L.: Calculation of cross-correlation function accelerated by TensorFloat-32 Tensor Core operations on NVIDIA's Ampere and Hopper GPUs. J. Comput. Sci. **68**, 101986 (2023). https://doi.org/10.1016/j.jocs.2023.101986
14. Atoofian, E.: PTTS: Power-aware tensor cores using two-sided sparsity. J. Parallel Distrib. Comput. **173**, 70–82 (2023). https://doi.org/10.1016/j.jpdc.2022.11.004
15. Wang, B. et al.: An GPU-accelerated particle tracking method for Eulerian–Lagrangian simulations using hardware ray tracing cores. Comput. Phys. Commun. **271** (2022). https://doi.org/10.1016/j.cpc.2021.108221
16. Chen, X., Ou, W., Fukuda, D., Chan, A.H.C., Liu, H.: Three-dimensional modelling on the impact fracture of glass using a GPGPU-parallelised FDEM. Eng. Fract. Mech. **277** (2023). https://doi.org/10.1016/j.engfracmech.2022.108929
17. Renc, P., Pęcak, T., De Rango, A., Spataro, W., Mendicino, G., Wąs, J.: Towards efficient GPGPU cellular Automata model implementation using persistent active cells. J. Comput. Sci. **59** (2022). https://doi.org/10.1016/j.jocs.2021.101538
18. Liu, H., Ma, H., Liu, Q., Tang, X., Fish, J.: An efficient and robust GPGPU-parallelized contact algorithm for the combined finite-discrete element method. Comput. Methods Appl. Mech. Eng. **395** (2022). https://doi.org/10.1016/j.cma.2022.114981
19. Shi, L., Chen, H., Sun, J., Li, K.: VCUDA: GPU-accelerated high-performance computing in virtual machines. IEEE Trans. Comput. **61**(6), 804–816 (2012). https://doi.org/10.1109/TC.2011.112
20. Petrovič, F., et al.: A benchmark set of highly-efficient CUDA and OpenCL kernels and its dynamic autotuning with Kernel Tuning Toolkit. Futur. Gener. Comput. Syst. **108**, 161–177 (2020). https://doi.org/10.1016/j.future.2020.02.069
21. Khairy, M., Wassal, A.G., Zahran, M.: A survey of architectural approaches for improving GPGPU performance, programmability and heterogeneity. J. Parallel Distrib. Comput. **127**, 65–88 (2019). https://doi.org/10.1016/j.jpdc.2018.11.012
22. Borrell, R., et al.: Heterogeneous CPU/GPU co-execution of CFD simulations on the POWER9 architecture: application to airplane aerodynamics. Futur. Gener. Comput. Syst. **107**, 31–48 (2020). https://doi.org/10.1016/j.future.2020.01.045
23. Skrzypczak, J., Czarnul, P.: Efficient parallel implementation of crowd simulation using a hybrid CPU+GPU high performance computing system. Simul. Model Pract. Theory **123** (2023). https://doi.org/10.1016/j.simpat.2022.102691
24. Liu, X., Zhong, Z., Xu, K.: A hybrid solution method for CFD applications on GPU-accelerated hybrid HPC platforms. Futur. Gener. Comput. Syst. **56**, 759–765 (2016). https://doi.org/10.1016/j.future.2015.08.002
25. Dubois, R., Goncalves da Silva, E., Parnaudeau, P.: High performance computing of stiff bubble collapse on CPU-GPU heterogeneous platform. Comput. Math. Appl. **99**, 246–256 (2021). https://doi.org/10.1016/j.camwa.2021.07.010
26. Acosta-Quiñonez, R.I., Torres-Roman, D., Rodriguez-Avila, R.: HOSVD prototype based on modular SW libraries running on a high-performance CPU+GPU platform. J. Syst. Architect. **113**, 101897 (2021). https://doi.org/10.1016/j.sysarc.2020.101897

27. Huang, Y., Zheng, X., Zhu, Y.: Optimized CPU–GPU collaborative acceleration of zero-knowledge proof for confidential transactions. J. Syst. Archit. **135** (2023). https://doi.org/10.1016/j.sysarc.2022.102807
28. Krzywaniak, A., Czarnul, P., Proficz, J.: Dynamic GPU power capping with online performance tracing for energy efficient GPU computing using DEPO tool. Futur. Gener. Comput. Syst. **145**, 396–414 (2023). https://doi.org/10.1016/j.future.2023.03.041

Biomimetic Modeling and Analysis Using Modern Architecture Frameworks like CUDA

Balbir Singh, Kamarul Arifin Ahmad, and Raghuvir Pai

Abstract Biomimetic modeling, rooted in the emulation of nature's ingenious designs, has emerged as a transformative discipline across various scientific and engineering domains. In this chapter, we explore the convergence of biomimetic modeling and modern architecture frameworks, specifically focusing on CUDA (Compute Unified Device Architecture). CUDA, developed by NVIDIA, has emerged as a powerhouse for parallel computing, significantly enhancing the capabilities of computational modeling and analysis in biomimetics. The chapter begins with an introduction to biomimetic modeling, emphasizing its relevance and growing importance in fields such as robotics, materials science, aerospace, and medicine. Biomimetic modeling involves the creation of computational models that mimic biological systems, offering innovative solutions to complex challenges. However, its widespread adoption has been limited by the intricate nature of biological systems, multiscale complexities, data collection hurdles, and the computational resources needed for simulations. The subsequent section looks into CUDA architecture, elucidating its key features, including parallelism, CUDA cores, and memory hierarchy. CUDA, originally designed for GPU-accelerated graphics rendering, has evolved into a versatile platform for general-purpose computing. Its immense parallel processing capabilities make it an ideal candidate for accelerating the resource-intensive simulations and analyses that biomimetic modeling demands. We then explore the application of CUDA in biomimetic modeling across various domains, including molecular dynamics simulations, neural network training, biomechanics, fluid dynamics, and evolutionary algorithms. CUDA empowers researchers to run complex simulations faster, bridge multiscale gaps, analyze vast datasets, and enable real-time

B. Singh · K. A. Ahmad
Department of Aerospace Engineering, Faculty of Engineering, Universiti Putra Malaysia, Serdang 43400, Selangor, Malaysia

B. Singh
Department of Aeronautical and Automobile Engineering, Manipal Institute of Technology, Manipal Academy of Higher Education, Manipal, Karnataka 576104, India

R. Pai (✉)
Department of Mechanical and Industrial Engineering, Manipal Institute of Technology, Manipal Academy of Higher Education, Manipal, Karnataka 576104, India
e-mail: raghuvir.pai@manipal.edu

K. A. Ahmad et al. (eds.), *High Performance Computing in Biomimetics*, Series in BioEngineering, https://doi.org/10.1007/978-981-97-1017-1_10

interactions. To illustrate the practicality of this integration, two case studies are presented, showcasing the accelerated study of protein folding and the GPU accelerated CFD simulation of insect flight. Challenges and future prospects are also discussed, emphasizing the need for addressing hardware limitations, simplifying software development, and enhancing data integration. Emerging trends like GPU clusters and quantum computing, along with interdisciplinary collaboration, promise to further advance the field.

Keywords Biomimetics · CUDA · Insect flapping flight · HPC platform · NVIDIA · GPU

1 Introduction

1.1 Background

Biomimetics, also known as biomimicry or bio-inspiration, is an interdisciplinary field that draws inspiration from nature's designs, processes, and systems to solve complex engineering, scientific, and technological challenges. By emulating the ingenious solutions found in the natural world, biomimetics has the potential to revolutionize various industries, including robotics, materials science, and computational modeling. In recent years, the field of biomimetics has been greatly enhanced by the advent of modern architecture frameworks like CUDA (Compute Unified Device Architecture). CUDA is a parallel computing platform and application programming interface (API) developed by NVIDIA, primarily for use with their Graphics Processing Units (GPUs). These powerful GPUs are capable of massive parallelism, making them well-suited for accelerating complex simulations and computations, which are fundamental to biomimetic modeling and analysis. This chapter explores the integration of biomimetic modeling and CUDA-based architecture frameworks, providing an in-depth examination of their synergistic potential. We will dig deep into the various aspects of biomimetic modeling, CUDA architecture, and how they intersect to enable advanced simulations and analyses. Furthermore, we will review the latest research and developments in this rapidly evolving field [1].

2 Biomimetic Modeling

2.1 Definition of Biomimetic Modeling

Biomimetic modeling, at its core, involves the creation of computational models that mimic biological systems, processes, or structures. These models can be used to gain a deeper understanding of natural phenomena, develop innovative technologies, and solve complex problems. The term "biomimetic" is derived from "bio" (life) and "mimesis" (imitation), emphasizing the idea of imitating nature's designs and mechanisms.

2.2 The Relevance of Biomimetic Modeling

Biomimetic modeling is an interdisciplinary approach that draws inspiration from nature to solve complex problems. It has gained increasing relevance in various fields due to its potential to offer sustainable and efficient solutions. This innovative approach harnesses the power of evolution, adaptation, and time-tested designs honed by millions of years of natural selection. Let's look at some key areas where biomimetics has made a significant impact:

2.2.1 Robotics

Biomimetic robots are engineered to mimic the locomotion and behaviors of animals, such as birds and insects. By emulating the agility and efficiency of natural systems, these robots have found applications in a wide array of fields. In search and rescue missions, biomimetic robots navigate challenging terrains with grace, providing access to areas otherwise inaccessible to humans. They also excel in environmental monitoring, where they can mimic the movement patterns of specific animals to observe ecosystems with minimal disruption. Moreover, in the realm of medical technology, biomimetic robots have proven invaluable in minimally invasive surgery. By replicating the dexterity and precision of biological systems, these robots enable surgeons to perform intricate procedures with enhanced precision and reduced patient trauma [2, 3].

2.2.2 Materials Science

Nature is a master of materials engineering, offering inspiration for creating stronger, more resilient synthetic materials. Researchers have worked on the intricate structures of biological materials, such as spider silk and seashell composites, to develop novel materials with exceptional properties. For instance, spider silk, known for its remarkable strength and flexibility, has influenced the development of super-strong synthetic fibers that find applications in industries ranging from textiles to military equipment. The study of seashell structures has led to innovative approaches in creating impact-resistant materials, which are invaluable in protective gear and structural engineering.

2.2.3 Aerospace

Biomimetic designs have revolutionized aerospace engineering by mimicking nature's optimal aerodynamics. For instance, bird-inspired wing designs have led to more efficient aircraft, reducing fuel consumption and emissions. By studying the intricate mechanisms that enable birds to soar gracefully, engineers have improved the performance and sustainability of aviation. Furthermore, investigations into bird flocking behavior have paved the way for autonomous drone swarms, enabling enhanced surveillance, search and rescue missions, and even wildlife conservation efforts. The coordination observed in flocks of birds has been replicated to create autonomous drone systems capable of collaborative tasks.

2.2.4 Medicine

Biomimetic modeling is a boon in the field of medicine, where it is employed to simulate biological processes within the human body. This aids in drug discovery, disease modeling, and the design of medical devices. By replicating the complexities of biological systems, researchers can test potential drugs and therapies more efficiently. Disease modeling allows for a deeper understanding of various conditions, leading to improved diagnostics and treatment strategies. Additionally, biomimetics has contributed to the development of medical devices that interact seamlessly with the human body, enhancing patient outcomes and comfort.

Biomimetic modeling harnesses the genius of nature to address real-world challenges in fields as diverse as robotics, materials science, aerospace, and medicine. It exemplifies the power of cross-disciplinary collaboration and the potential for sustainable and efficient solutions inspired by the natural world.

2.3 Challenges in Biomimetic Modeling

Biomimetic modeling is a promising approach with the potential to revolutionize various fields, but it is not without its share of challenges. These challenges are critical to address in order to harness the full potential of biomimetics:

2.3.1 Complex Systems

One of the primary challenges in biomimetic modeling is the inherent complexity of biological systems. These systems are composed of intricate networks of components with numerous interdependencies. Modeling such complexity accurately is a formidable task. Advanced computational techniques, such as agent-based modeling, machine learning, and systems biology, are often required to decipher and replicate these complex interactions. Understanding the nuances of natural systems and translating them into computational models is a constant endeavor, demanding interdisciplinary collaboration and expertise.

2.3.2 Multiscale Modeling

Another significant challenge arises from the multiscale nature of biological processes. Biological phenomena occur across various levels, from molecular and cellular interactions to organ-level functions. Bridging these scales to create a comprehensive model is a formidable challenge. To overcome this, researchers employ a range of techniques, including hierarchical modeling and multi-scale simulation methods. Such approaches allow for the integration of information from different scales to develop cohesive models that accurately represent biological systems, and this is crucial for making biomimetic solutions more effective.

2.3.3 Data Collection

Biomimetic modeling heavily relies on biological data for calibration and validation. Gathering accurate and comprehensive data, however, can be a challenging and time-consuming process. The availability of biological data often varies depending on the system being studied. Data collection methods must be precise, consistent, and ethically sound, and researchers often find themselves investing significant resources to obtain the necessary data. Technological advancements such as high-throughput sequencing and imaging technologies have significantly improved data acquisition, but challenges still persist in ensuring the quality and relevance of the data collected.

2.3.4 Computational Resources

Simulating complex biological systems often demands substantial computational power. This is where specialized architectures like CUDA (Compute Unified Device Architecture) come into play. To accurately replicate biological processes and interactions, researchers require high-performance computing resources. These resources are necessary for running simulations and conducting large-scale modeling. With CUDA architecture, which is optimized for parallel processing on graphics processing units (GPUs), researchers can significantly accelerate their simulations, making it an invaluable tool in biomimetic modeling.

Biomimetic modeling has tremendous potential, it is not without its set of challenges. Addressing the complexity of biological systems, bridging multiscale modeling, acquiring accurate biological data, and having access to adequate computational resources are essential steps in overcoming these challenges. As technology continues to advance and interdisciplinary collaboration grows, the field of biomimetic modeling is poised to unlock innovative and sustainable solutions inspired by nature, including sustainable technologies [4].

Now that we've established the importance of biomimetic modeling, let's explore the CUDA architecture in more detail.

3 CUDA Architecture

3.1 Overview of CUDA

CUDA, short for Compute Unified Device Architecture, is a revolutionary parallel computing platform and Application Programming Interface (API) developed by NVIDIA. Its primary purpose is to enable developers to leverage the immense computational power of Graphics Processing Units (GPUs) for general-purpose computing tasks. Initially designed for rendering graphics, GPUs have evolved to become versatile parallel processors, well-suited for a wide range of applications beyond just gaming and visual computing. CUDA stands out as a pioneer in the world of GPU computing, opening up the door to massive parallelism and significantly accelerating computational tasks. Figure 1 shows the general architecture of NVIDIA CUDA.

The foundation of CUDA's success lies in several key features:

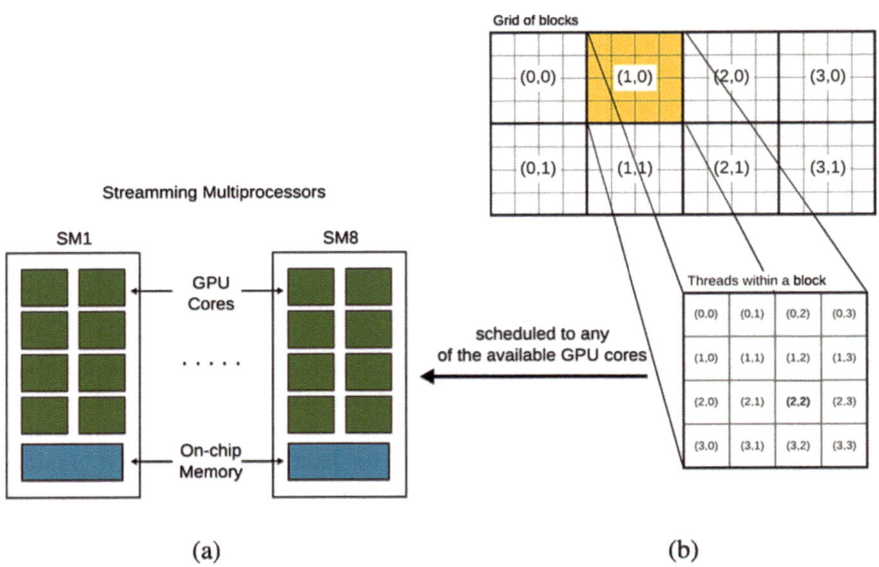

Fig. 1 NVIDIA CUDA architecture. **a** Streaming multiprocessors and **b** organizations of grid, blocks and threads. Adapted from ref [5], copyright (2014), with permission from Springer

- Parallelism: CUDA embraces both data parallelism and task parallelism, making it incredibly versatile. Data parallelism involves performing the same operation on multiple pieces of data simultaneously, while task parallelism allows multiple tasks to be executed concurrently. This makes CUDA suitable for diverse applications, from scientific simulations to deep learning and image processing.
- CUDA Cores: One of the standout features of GPUs is their vast number of CUDA cores. Unlike CPUs, which typically have a few powerful cores, GPUs house thousands of smaller cores. These cores can execute tasks concurrently, making them exceptionally efficient at parallel processing. This parallelism is why GPUs are particularly well-suited for tasks like matrix calculations and complex simulations.
- Memory Hierarchy: CUDA GPUs incorporate a sophisticated memory hierarchy, offering various memory types, each optimized for specific use cases. This hierarchy includes global memory, shared memory, and registers. Global memory serves as the primary storage for data, while shared memory enables fast data sharing among threads within a single thread block. Registers are used to store thread-specific data. This memory hierarchy design enhances data access efficiency, a crucial aspect of high-performance computing.
- CUDA Toolkit: NVIDIA provides a comprehensive toolkit for CUDA development. This toolkit encompasses everything a developer needs to harness the power of GPUs efficiently. It includes libraries tailored for GPU computation, like cuBLAS for linear algebra and cuDNN for deep learning, as well as tools for performance optimization, debugging, and profiling.

CUDA architecture has redefined the landscape of general-purpose computing by making GPUs accessible for a wide range of applications. Its inherent support for parallelism, vast CUDA cores, memory hierarchy, and extensive toolkit contribute to its status as a groundbreaking platform in the world of high-performance computing, enabling developers to unlock the full potential of GPU technology.

3.2 CUDA in Scientific Computing

In the realm of scientific computing, where complex simulations and data analysis are paramount, CUDA has emerged as a game-changer, offering the ability to dramatically accelerate these tasks. Fields such as computational biology, physics, climate modeling, and more have harnessed CUDA's power to achieve significant speedups in their research endeavors. The advantages of incorporating CUDA into scientific computing are manifold and profound:

Massive Parallelism: One of the cornerstones of CUDA's success in scientific computing is its capacity for massive parallelism. GPUs are equipped with thousands of cores, which can work together to process data in parallel. This parallelism is particularly advantageous for scientific simulations, where intricate calculations and data processing can be executed concurrently. As a result, scientists can significantly reduce simulation times, enabling them to run more experiments and refine their models more quickly.

Scalability: CUDA offers a scalable solution for scientific computing. Researchers have the option to utilize multiple GPUs within a single workstation or create GPU clusters, which can aggregate the computational power of numerous GPUs. This scalability is invaluable for handling larger datasets and tackling more computationally intensive problems. In fields like astrophysics or climate modeling, where the amount of data and the complexity of simulations are immense, this ability to scale computational resources is transformative.

Flexibility: CUDA's versatility extends to algorithm development. Scientists can choose to program in a low-level fashion, allowing them fine-grained control over the GPU, or they can take advantage of high-level libraries that simplify the development process. This flexibility enables researchers to adapt CUDA to their specific needs, whether that entails optimizing existing codes or designing novel algorithms. It's an essential characteristic, especially in scientific computing, where tailor-made solutions often lead to the best outcomes.

Performance Optimization: CUDA provides a suite of tools and techniques for performance optimization. Profiling tools enable developers to identify bottlenecks and areas for improvement in their CUDA applications. By fine-tuning their code using optimization techniques like kernel fusion, memory management, and pipelining, scientists can achieve maximum performance. These optimization measures are crucial in scientific computing, where even small gains in efficiency can translate into significant time and resource savings.

CUDA has become an indispensable tool in scientific computing, revolutionizing the way researchers approach complex simulations and data analysis. The advantages it offers, including massive parallelism, scalability, flexibility in algorithm development, and performance optimization, have led to impressive speedups in various scientific domains. CUDA has not only accelerated research but has also opened the door to exploring more intricate and data-intensive problems that were previously computationally prohibitive. As scientific datasets continue to grow and research questions become more intricate, CUDA is likely to remain at the forefront of scientific computing, enabling breakthrough discoveries and innovations [6–9].

3.3 CUDA in Biomimetic Modeling

The integration of CUDA (Compute Unified Device Architecture) into biomimetic modeling has ushered in a new era of innovation, offering several advantages that significantly impact the development of biomimetic models. This powerful parallel computing platform, originally designed for GPU acceleration, has proven to be a game-changer in this interdisciplinary field. Here are some of the key advantages and applications of CUDA in biomimetic modeling:

Speedup: Biomimetic modeling often involves the simulation of intricate biological systems and behaviors. These simulations can be computationally demanding, requiring significant processing power. CUDA provides the means to accelerate these simulations by harnessing the immense parallel processing capabilities of GPUs. This acceleration leads to reduced simulation times, saving valuable research resources and enabling scientists to iterate and refine their models more efficiently.

Multiscale Modeling: Biomimetic models often encompass multiple scales, ranging from molecular interactions to organ-level behaviors. Bridging these scales in a single model is a formidable challenge. CUDA's parallelism can be leveraged to tackle this issue effectively. By distributing the computational load across thousands of GPU cores, researchers can create comprehensive models that simulate intricate biological systems across multiple scales. This enables a more holistic and accurate representation of biomimetic phenomena, essential for the success of many applications.

Data-Driven Analysis: CUDA's computational prowess is particularly advantageous in the context of biomimetic modeling, where the analysis of large datasets is often required. Many biomimetic models rely on biological data for calibration and validation. With CUDA, scientists can efficiently process and analyze vast datasets generated from biological experiments. This accelerates the data-driven aspects of biomimetic modeling, leading to faster insights and improved model accuracy.

Real-time Interaction: In certain biomimetic applications, such as robotics, real-time interaction and control are essential. CUDA enables the development of real-time biomimetic models that can be employed in robotics and other applications. For example, biomimetic robots can use CUDA-based models to mimic the locomotion and behaviors of animals with remarkable precision. This real-time interaction is invaluable in fields like search and rescue, environmental monitoring, and medical applications, where split-second decisions can make a significant difference.

The integration of CUDA into biomimetic modeling is a significant advancement that empowers researchers to address complex challenges with enhanced computational capabilities. The advantages of speedup, multiscale modeling, data-driven analysis, and real-time interaction contribute to the success of biomimetic models in various applications. This integration not only accelerates research but also opens up new possibilities for sustainable and efficient solutions inspired by nature. As technology continues to advance and interdisciplinary collaboration in biomimetics grows, the role of CUDA in biomimetic modeling is poised to expand further, driving innovation and transformative discoveries.

4 Application of CUDA in Biomimetic Modeling

Biomimetic modeling, which draws inspiration from nature to design innovative solutions, has found a powerful ally in CUDA (Compute Unified Device Architecture). CUDA's capacity for parallel computing has led to remarkable advances in various biomimetic applications. In this section, we will explore some of the key applications where CUDA plays a pivotal role in biomimetic modeling:

4.1 Molecular Dynamics Simulation

Molecular dynamics simulation is a cornerstone of biomimetic modeling. It involves the study of the behavior of molecules and atoms in biological systems, offering insights into complex processes like protein folding, drug interactions, and molecular reactions. CUDA has revolutionized this field by significantly accelerating these simulations. For example, Fig. 2 shows the TRP-Cage which was modeled and simulated via Molecular Dynamics (MD) using the AMBER 16 suite of programs with CUDA implementation for GPUs [10].

CUDA-optimized software packages, such as AMBER and GROMACS, are commonly employed in molecular dynamics simulations. These packages harness the power of GPU parallelism to achieve substantial speedups over traditional CPU-based simulations. By distributing the computational workload across the numerous GPU cores, these simulations can be performed at a fraction of the time it would take on a CPU. This acceleration is instrumental in advancing research in fields like structural biology, pharmaceuticals, and material science.

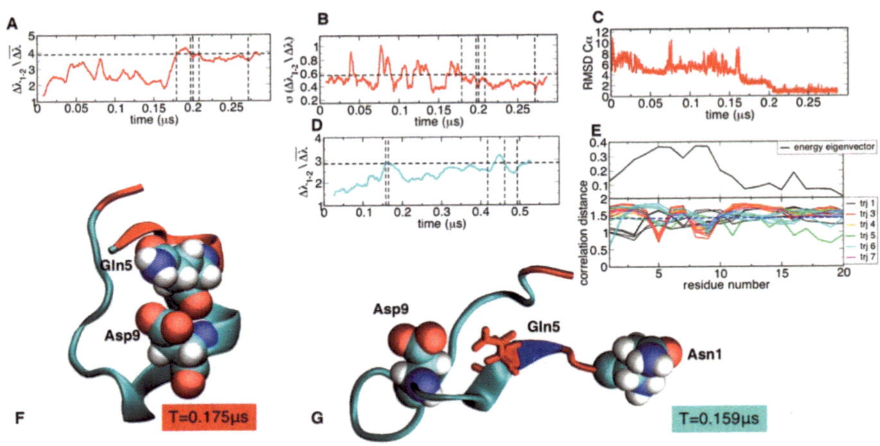

Fig. 2 Trp cage folding analysis: **a** energy gap trajectory 3, **b** running standard deviation of energy gap trajectory 3, **c** RMSD to native structure trajectory 3, and **d** energy gap trajectory 6. Dashed lines indicate the identified transitions. **e** Profiles of the folded energy eigenvector and of the lambda–energy eigenvector correlation for each transition. **f** Folded state identified in trajectory 3. **g** Misfolded state identified in trajectory 6. Contact between Gln 5 and Asp9 is lacking the native-like backbone interaction [10]

4.2 Neural Network Training

Neural networks, inspired by the structure of the human brain, are fundamental in biomimetic modeling for machine learning and artificial intelligence applications. Training deep neural networks is a computationally intensive task that involves complex matrix operations. CUDA offers a significant advantage in this domain by accelerating these operations.

Frameworks like TensorFlow and PyTorch provide GPU support through CUDA, allowing researchers to train large neural networks more efficiently. This becomes especially relevant when modeling neural systems for tasks such as image recognition, natural language processing, and autonomous decision-making. The parallel processing capabilities of GPUs enhance training speed and enable the exploration of more complex neural network architectures, contributing to advances in biomimetic AI.

4.3 Biomechanics and Fluid Dynamics

Biomechanics and fluid dynamics involve the simulation of biological structures and fluids, such as blood flow through arteries or the behavior of muscles during locomotion. These simulations are essential in understanding the mechanics of biological systems and have practical applications in fields like medicine and sports science.

CUDA accelerates these simulations, enabling researchers to run more detailed and accurate models. The parallel processing power of GPUs allows for real-time or near-real-time simulations, which can be critical in medical diagnostics, the design of prosthetics, and sports equipment optimization. By accurately mimicking the behavior of biological structures and fluids, CUDA-enhanced biomimetic models have the potential to drive innovations in medical devices and improve our understanding of human biomechanics.

4.4 Evolutionary Algorithms

Evolutionary algorithms, inspired by the principles of natural selection and genetic processes, are utilized in biomimetic modeling to optimize designs and solutions. These algorithms often involve fitness evaluation steps that require running simulations or complex mathematical computations. CUDA is employed to accelerate these fitness evaluations, leading to more efficient and faster optimization processes.

CUDA-enabled evolutionary algorithms find applications in diverse fields, from optimizing robotic locomotion to designing efficient antenna arrays. By leveraging GPU parallelism, researchers can explore a broader search space and conduct more iterations, resulting in improved designs inspired by natural selection. This application of CUDA in biomimetic modeling allows for rapid evolution and adaptation, mirroring the dynamic processes observed in the natural world.

CUDA into biomimetic modeling has revolutionized the field by offering speedup, efficiency, and scalability in diverse applications, from molecular dynamics simulations to neural network training, biomechanics and fluid dynamics, and evolutionary algorithms. This computational powerhouse has opened the doors to exploring complex and intricate biomimetic systems, ultimately driving innovations and sustainable solutions across multiple domains [11–14].

5 Recent Case Study

To illustrate the practical use of CUDA in biomimetic modeling, let's explore another case study from recent research. The Case Study is from a paper titled "GPU Acceleration in CFD Simulation for Insect Flight".

This case study is based on the research done by Ref. [15] that presents a computational study on the aerodynamics and kinematics of a free-flying model fruit-fly. An existing integrative computational fluid dynamics (CFD) framework was further developed using CUDA technology and adapted for the free flight simulation on heterogeneous clusters. The application of general-purpose computing on graphics processing units (GPGPU) significantly accelerated the insect flight simulation and made it less computational expensive to find out the steady state of the flight using CFD approach.

The authors used NVIDIA CUDA technology to accelerate the numerical simulations for hovering fruitfly as shon in Fig. 3. The FSI problem was solved using the predictor–corrector method because CFD consumes lots of time and storage. In the intermediate Evaluator Step CPU parallelization code (OpenMP) was modified with CUDA technology heterogeneous cluster of the National Supercomputing Centre of Singapore with the NVIDIA Tesla K40 accelerator installed. It is basically a SVD-GFD approach.

Figure 4a–c illustrate that the utilization of parallelization and GPU acceleration significantly expedites the nodal search process and fractional step iteration within the aforementioned CFD scheme. The degree of GPU acceleration increases proportionally with the size of the Cartesian mesh, as the time spent on data transfer between the CPU and GPU becomes less substantial with the expansion of computational tasks. The decline in the speedup during the nodal search operation may be attributed to the increasing overhead related to memory operations within larger arrays.

In terms of the projection method calculation, the use of GPU acceleration results in approximately half the computation time compared to utilizing just 12 threads for CPU parallelization. When considering the wall-time expended during a full FSI iteration, the advantages of GPU acceleration become even more striking, particularly

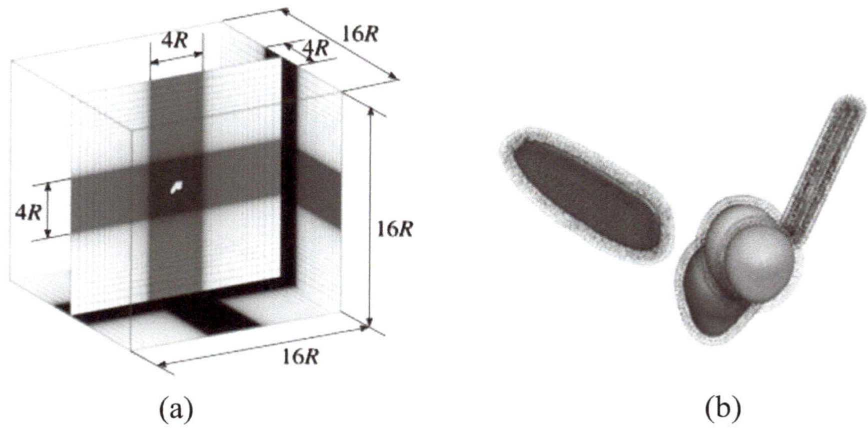

Fig. 3 Mesh used in the study **a** control volume and model, **b** body and wing mesh [15]

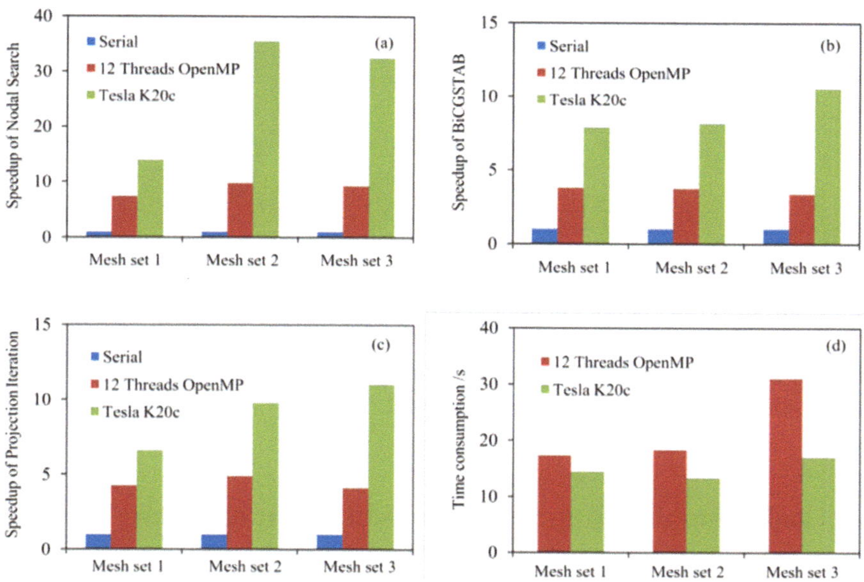

Fig. 4 Code performance in parallel mode [15]

when the grid size is substantial (as indicated in Fig. 4d). This enhanced performance is primarily due to the alleviation of computational bottlenecks, notably those stemming from the SVD procedure executed on CPUs.

Authors here expedited the computationally intensive simulations by the application of CUDA-based GPUs. This greatly accelerated the speed of the CFD 6DoF

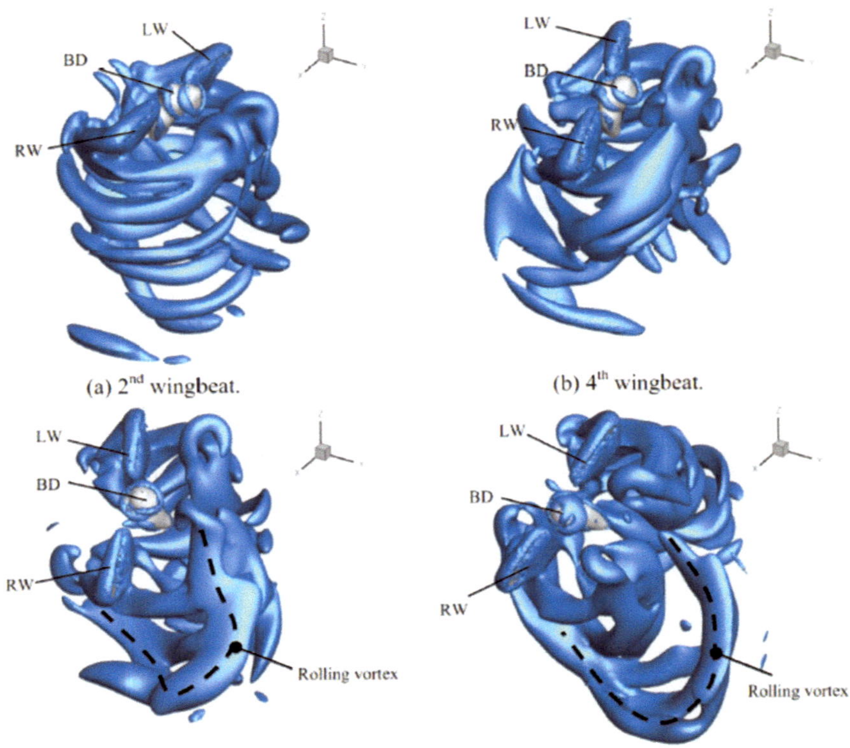

Fig. 5 Iso surfaces of vorticity in banking flight [15]

(Navier–Stokes) solver over that of the original CPU parallelized code. Figure 5 shows the development of vortex wake in banking flight (banking from steady hovering), showingby$\lambda * 2 = -1.8$ iso-surfaces obtained at the mid-downstroke.

6 Challenges and Future Prospects in the Integration of CUDA in Biomimetic Modeling

The integration of CUDA in biomimetic modeling has demonstrated its potential to revolutionize various scientific and engineering disciplines. However, along with its promising advancements, several challenges must be addressed, and there are exciting prospects for the future of this field.

6.1 Challenges

Hardware Limitations: The availability of powerful GPUs is not uniform across all research environments. In resource-constrained settings, the cost and availability of high-performance GPUs can be a limiting factor. To mitigate this challenge, researchers and institutions may need to invest in GPU infrastructure or explore cloud-based solutions to ensure broader access to GPU resources.

- Software Development: Writing CUDA code requires specialized skills in parallel programming and GPU optimization. Researchers from diverse backgrounds may find it challenging to develop CUDA applications. Future prospects involve the development of user-friendly tools and libraries that abstract the complexities of CUDA programming. These tools will enable researchers to leverage GPU acceleration without the need for in-depth GPU programming knowledge.
- Data Integration: Integrating experimental data into CUDA-accelerated biomimetic models is a complex task, especially in the context of personalized medicine and patient-specific simulations. The challenge lies in the seamless fusion of experimental data, such as patient imaging, genetic information, and clinical data, with computational models. Overcoming this challenge requires the development of robust data integration pipelines and interoperable software frameworks that can accommodate various data formats and sources.

6.2 Future Prospects

- GPU Clusters: The use of GPU clusters in biomimetic modeling is a promising avenue for the future. GPU clusters can offer substantial computational power for larger-scale simulations, enabling researchers to tackle even more complex biomimetic systems and simulations. This approach extends the applicability of CUDA in biomimetic modeling to address grand challenges in fields like drug discovery, personalized medicine, and climate modeling.
- Quantum Computing: The future of biomimetic modeling may intersect with emerging quantum computing technologies. Quantum computers have the potential to revolutionize biomimetic modeling by addressing complex problems that are currently computationally intractable. Quantum computing can simulate quantum mechanical processes more efficiently, enabling precise modeling of biochemical interactions and materials at the quantum level. As quantum computing technology matures, it may offer new avenues for biomimetic research.
- Interdisciplinary Collaboration: Collaborations between biologists, computer scientists, and engineers will continue to drive innovations in biomimetic modeling and CUDA applications. The future prospects of this field hinge on interdisciplinary partnerships that bring together experts from diverse domains. Biologists provide valuable insights into natural systems, computer scientists

develop the computational tools, and engineers implement the solutions in real-world applications. Interdisciplinary collaboration will lead to groundbreaking biomimetic models and their practical implementation in various industries.

Overcoming hardware limitations, simplifying software development, and enhancing data integration are crucial for the field's growth. However, the future prospects for CUDA in biomimetic modeling are promising. GPU clusters, quantum computing, and interdisciplinary collaboration hold the potential to drive this field to new heights, leading to groundbreaking discoveries, sustainable solutions, and the continued emulation of nature's wisdom.

7 Conclusion

Biomimetic modeling, with its ability to draw inspiration from nature and apply it to various scientific and engineering domains, holds immense potential. The integration of modern architecture frameworks like CUDA has unlocked new possibilities in simulating and analyzing complex biological systems. As demonstrated by case studies and ongoing research, CUDA-accelerated biomimetic modeling has the power to drive innovation in fields as diverse as molecular biology, robotics, and fluid dynamics. To fully realize this potential, researchers must address the challenges of hardware limitations, software development, and data integration while continuing to push the boundaries of what is possible with CUDA. As technology advances, biomimetic modeling and CUDA will undoubtedly remain at the forefront of cutting-edge research and development, paving the way for a future where nature's solutions inspire the solutions of tomorrow.

References

1. Didari, A., Mengüç, M.P.: A biomimicry design for nanoscale radiative cooling applications inspired by Morpho didius butterfly. Sci. Rep. **8**, 16891 (2018). https://doi.org/10.1038/s41598-018-35082-3
2. Didari, A., Pinar Mengüç, M.: Biomimicry designs for passive optical solutions for nanoscale radiative cooling applications. In: Proceedings of the SPIE 10731, Nanostructured Thin Films XI, 107310C (7 September 2018). https://doi.org/10.1117/12.2320504
3. Zhang, C., Yang, Z., Xue, B., Zhuo, H., Liao, L., Yang, X., Zhu, Z.: Perceiving like a bat: hierarchical 3D geometric–semantic scene understanding inspired by a biomimetic mechanism. Biomimetics **8**(5), 436 (2023). https://doi.org/10.3390/biomimetics8050436
4. Sanderasagran, A.N., Aziz, A.B.A., Oumer, A.N., Mat Sahat, I.: Alternative method of nature inspired geometrical design strategy for drag induced wind turbine blade morphology. Int. J. Automot. Mech. Eng. **19**(2), 9759–9772 (2022)
5. Quesada-Barriuso, P., Argüello, F., Heras, D.B.: Computing efficiently spectral-spatial classification of hyperspectral images on commodity GPUs. In: Tweedale, J., Jain, L. (eds) Recent Advances in Knowledge-based Paradigms and Applications. Advances in Intelligent Systems

and Computing, vol 234. Springer, Cham (2014). https://doi.org/10.1007/978-3-319-01649-8_2
6. Luebke, D.: CUDA: scalable parallel programming for high-performance scientific computing. In: 2008 5th IEEE International Symposium on Biomedical Imaging: From Nano to Macro, Paris, France, pp. 836–838 (2008). https://doi.org/10.1109/ISBI.2008.4541126
7. Yang, Z., Zhu, Y., Pu, Y.: Parallel image processing based on CUDA. In: 2008 International Conference on Computer Science and Software Engineering, Wuhan, China, pp. 198–201 (2008). https://doi.org/10.1109/CSSE.2008.1448
8. Ghorpade, J., Parande, J., Kulkarni, M., Bawaskar, A.: GPGPU processing in cuda architecture. Adv. Comput. Int. J. **3**(1) (2012). https://doi.org/10.5121/acij.2012.3109
9. Che, S., Boyer, M., Meng, J., Tarjan, D., Sheaffer, J.W., Skadron, K.: A performance study of general-purpose applications on graphics processors using CUDA. J. Parallel Distrib. Comput. **68**(10), 1370–1380 (2008). https://doi.org/10.1016/j.jpdc.2008.05.014
10. Meli, M., Morra, G.: Giorgio Colombo simple model of protein energetics to identify ab initio folding transitions from all-atom MD simulations of proteins. J. Chem. Theory Comput. **16**(9), 5960–5971 (2020). https://doi.org/10.1021/acs.jctc.0c00524
11. Harish, P., Narayanan, P.J.: Accelerating large graph algorithms on the GPU using CUDA. In: Aluru, S., Parashar, M., Badrinath, R., Prasanna, V.K. (eds.) High Performance Computing – HiPC 2007. HiPC 2007. Lecture Notes in Computer Science, vol. 4873. Springer, Berlin, Heidelberg (2007). https://doi.org/10.1007/978-3-540-77220-0_21
12. Dafeng, G., Xiaojun, W.: Real-time visual hull computation based on GPU. In: 2015 IEEE International Conference on Robotics and Biomimetics (ROBIO), Zhuhai, China, pp. 1792–1797 (2015). https://doi.org/10.1109/ROBIO.2015.7419032
13. Austin, J., Corrales-Fatou, R., Wyetzner, S., Lipson, H.: Titan: a parallel asynchronous library for multi-agent and soft-body robotics using NVIDIA CUDA. In: 2020 IEEE International Conference on Robotics and Automation (ICRA), Paris, France, pp. 7754–7760 (2020). https://doi.org/10.1109/ICRA40945.2020.9196808
14. Zhang, Z.: Soft-body simulation with CUDA based on mass-spring model and Verlet integration scheme. In: Proceedings of the ASME 2020 International Mechanical Engineering Congress and Exposition. Volume 7A: Dynamics, Vibration, and Control. Virtual, Online. November 16–19, 2020. V07AT07A025. ASME. https://doi.org/10.1115/IMECE2020-23221
15. Yao, Y.: An application of GPU acceleration in CFD simulation for insect flight. Supercomput. Front. Innov. Int. J. **4**(2), 13–26 (2017). https://doi.org/10.14529/jsfi170202

Unsteady Flow Topology Around an Insect-Inspired Flapping Wing Pico Aerial Vehicle

Balbir Singh, Adi Azriff basri, Noorfaizal Yidris, Raghuvir Pai, and Kamarul Arifin Ahmad

Abstract In this study, we employ Computational Fluid Dynamics (CFD) to conduct a topology analysis of unsteady flow instabilities around a mosquito-inspired flapping wing Pico Aerial Vehicle (PAV) called RoboMos. The objective is to gain an understanding of the aerodynamic phenomena that underlie the flight mechanics of PAVs and to elucidate the potential for optimization. The study looks into the intricacies of designed PAVs and their emulation of insect flight, shedding light on the relevance of nature's designs for small-scale aerial vehicles. Through CFD simulations using HPC, we examine the geometry and mesh generation, governing equations, boundary conditions, and simulation parameters. The results highlight the emergence of vorticity shedding, unsteady flow separation, and the interaction between wings, unveiling critical insights into the aerodynamic behavior of insect-inspired PAVs. These insights offer opportunities for optimizing PAV designs, flapping frequencies, and other operational parameters. By understanding the complexities of unsteady flow instabilities, we aim to advance the efficiency, maneuverability, and applicability of PAVs in surveillance, environmental monitoring, and search and rescue missions.

Keywords Unsteady aerodynamics · Flow topology · Flapping wing · PAV · CFD · Vorticity · Vortex shedding · Wing-wing interaction

---

B. Singh · A. A. basri · N. Yidris · K. A. Ahmad (✉)
Department of Aerospace Engineering, Faculty of Engineering, Universiti Putra Malaysia, 43400, Serdang, Selangor, Malaysia
e-mail: aekamarul@upm.edu.my

B. Singh
Department of Aeronautical and Automobile Engineering, Manipal Institute of Technology, Manipal Academy of Higher Education, Manipal, Karnataka 576104, India

A. A. basri · N. Yidris · K. A. Ahmad
Aerospace Malaysia Research Centre, Faculty of Engineering, Universiti Putra Malaysia, 43400, Serdang, Selangor, Malaysia

R. Pai
Department of Mechanical and Industrial Engineering, Manipal Institute of Technology, Manipal Academy of Higher Education, Manipal, Karnataka 576104, India

K. A. Ahmad et al. (eds.), *High Performance Computing in Biomimetics*, Series in BioEngineering, https://doi.org/10.1007/978-981-97-1017-1_11

1 Introduction

In the quest for enhancing the efficiency and maneuverability of small-scale aerial vehicles, researchers have turned to nature for inspiration. Insects, with their remarkable flight capabilities, have served as a rich source of inspiration for developing small-scale aerial vehicles. Among these, the flapping-wing pico aerial vehicle (PAV) has garnered significant attention due to its potential for applications in surveillance, environmental monitoring, and search and rescue missions. Understanding the aerodynamics of insect-inspired flapping wings is crucial for optimizing the design and performance of PAVs. This chapter looks into the topology analysis of unsteady flow instabilities around an insect-inspired flapping wing PAV using Computational Fluid Dynamics (CFD) techniques on HPC platform [1–3].

The small-scale aerial vehicles of today, particularly Pico Aerial Vehicles (PAVs), are designed with a keen focus on efficiency, agility, and versatility. These miniature wonders are envisioned for a myriad of applications, including surveillance, environmental monitoring, and search and rescue missions. One of the most promising designs in this realm is the flapping-wing PAV, which takes inspiration from the agile and versatile flight of insects [4–6].

To truly understand the significance of insect-inspired PAVs, it's essential to appreciate the evolution of these miniature marvels. Over the past few decades, there has been a concerted effort to develop small-scale aerial vehicles that can outperform conventional fixed-wing and rotary-wing drones in terms of maneuverability and energy efficiency. The emergence of flapping-wing PAVs marked a turning point in this journey. Flapping-wing PAVs replicate the way insects fly, with wings that move in an up-and-down flapping motion, as opposed to the conventional fixed wings or rotors used in traditional drones. This biomimetic approach allows PAVs to harness the advantages of unsteady aerodynamics, a domain that was relatively unexplored in conventional aviation. Unsteady aerodynamics takes into account the dynamic and unsteady flow patterns created by flapping wings, which significantly influence the flight performance of these aerial vehicles [7–10].

The advantages of flapping-wing PAVs are abundant. They offer enhanced maneuverability, the ability to hover, and efficient flight at low speeds. This makes them ideal for tasks such as surveillance in cluttered environments, pollination, and search and rescue operations. However, achieving these capabilities requires a deep understanding of the unsteady flow topology around the flapping wings. High-Performance Computing (HPC) platforms play a pivotal role in advancing the capabilities of CFD simulations. This chapter shows the significance of HPC in the context of PAV research and how it enables the exploration of unsteady flow topology. Flapping-wing PAVs introduce complex, unsteady flow dynamics that require substantial computational resources to model accurately. HPC platforms offer the computational horsepower required to run high-fidelity simulations, which can capture the intricate flow patterns around the flapping wings. With the aid of HPC, researchers can perform simulations at high resolutions and over extended durations, allowing for a more comprehensive analysis of the unsteady flow topology. This leads to a deeper understanding of the aerodynamic challenges and opportunities associated with flapping-wing PAVs [11–16]. This chapter looks into the intricate world of insect-inspired flapping wing PAVs and their aerodynamics. Through the lens of Computational Fluid Dynamics (CFD) techniques running on a High-Performance Computing (HPC) platform, we explore the unsteady flow topology around these remarkable creations. Our journey will uncover the fundamental principles that govern the flight of these insect-inspired PAVs, shedding light on the unique unsteady flow phenomena they exhibit.

2 Background and Methodology

Insects possess intricate and highly efficient wing structures that enable them to maneuver in complex environments with astonishing agility. These characteristics have motivated researchers to replicate nature's design in man-made PAVs. To achieve this, it is imperative to thoroughly comprehend the unsteady flow patterns and instabilities generated during flapping wing flight. CFD provides a powerful tool for simulating and analyzing these complex aerodynamic phenomena, enabling the optimization of PAV designs. In this study, we applied a widely-recognized wavelet-based fully adaptive method for solving the incompressible Navier–Stokes equations in intricate, time-varying geometries using high-performance parallel computing [17–19]. This is an openly available code and the method operates based on artificial compressibility and volume penalization. References [17–19] give more details about this solver and its applications in high fidelity computational analysis of flapping insects and inspired models. The synergy of artificial compressibility and volume penalization has been implemented in the RoboMos model. Figure 1, presented below, illustrates the creators of this approach conducting three vortex simulations, showcasing the method's effectiveness and its significance [17–20]

Parallelizing adaptive code presents a considerably greater challenge compared to code employing static grids. The dynamic nature of adaptive simulations often leads to significant variations in the number of blocks required, making it inefficient to maintain a constant number of processes throughout the simulation. The model, as shown, is the 3D time-dependent geometry as in Fig. 3. This code is already validated and benchmarked before in the prior research by its developers [17–19]. The mosquito PAV test model revolves around a single, rigid, rectangular flapping wing with finite thickness and a length of 13.01 mm. There are no imposed inflow conditions, emulating a hovering flight scenario. The wing executes a horizontal stroke in a predefined plane, with its wingbeat motion visualized in Fig. 3. The two half-cycles, commonly referred to as upstroke and downstroke, exhibit symmetry. To approximate an unbounded flow, we have chosen a computational domain size of 4R x 4R x 4R, where R is the wing length. Homogeneous conditions are applied at the outer boundary. In Fig. 2a, you can observe the fundamental aerodynamic

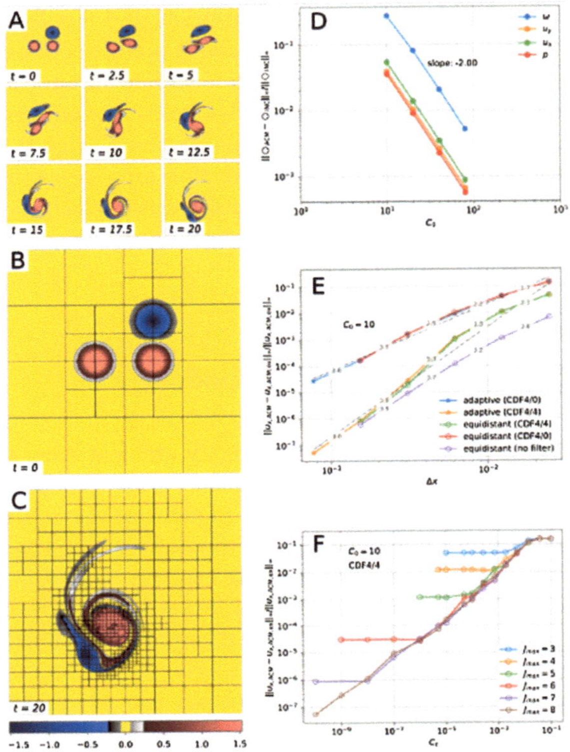

Fig. 1 Three vortices simulations using the CDF 4/ 4 wavelet [17–19]

representation of a mosquito, while Fig. 2b showcases the actual RoboMos version 1 model fabricated at UPM, Malaysia.

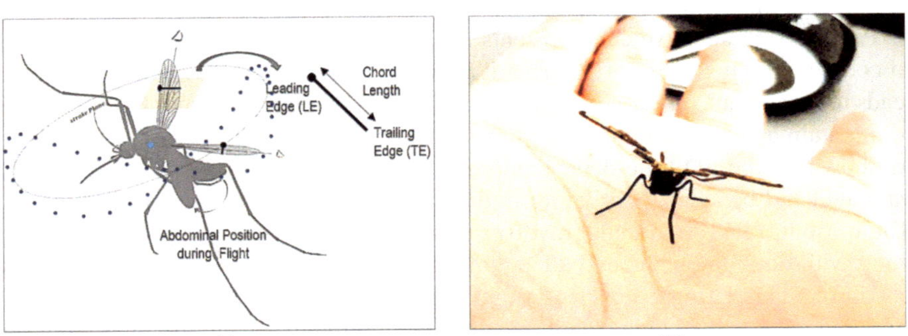

(a) (b)

Fig. 2 **a** Aerodynamic representation of mosquito during flight, mosquito-related flapping cycle phases, formation of leading and trailing edge vortices during phases of the wingbeat cycle of mosquito flight. **b** RoboMos, a mosquito-inspired FWPAV developed at UPM, Malaysia

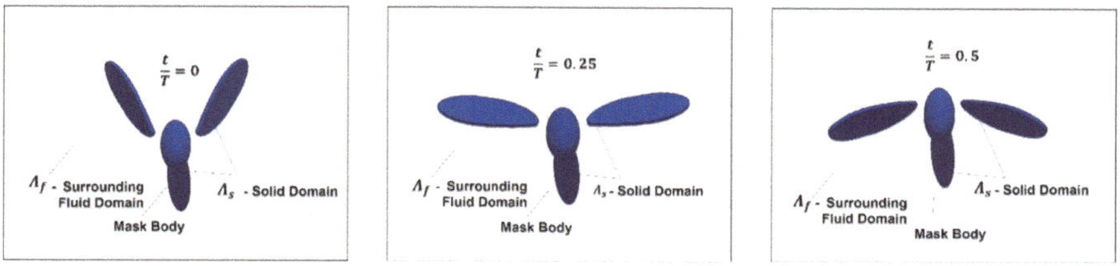

Fig. 3 Representation of fluid and solid domains in flapping till mid-stroke

Wabbit [17–19] is a Computational Fluid Dynamics (CFD) code that stands out for its unique approach to simulating fluid flow phenomena, particularly in cases involving flapping motion. This note highlights the key features of Wabbit, specifically focusing on its utilization of artificial compressibility and volume penalization techniques in the context of flapping simulations. It employs artificial compressibility as a numerical method to solve the modified incompressible Navier–Stokes Eqs. (1–3). In flapping simulations, where the fluid flow around moving objects like wings or fins can be highly dynamic, maintaining incompressibility is crucial. Artificial compressibility allows Wabbit to effectively handle changes in density and pressure while preserving the essential incompressible nature of the flow. This approach simplifies the numerical solution and enhances the stability of the simulations.

$$\frac{\partial \boldsymbol{u}}{\partial \tau} + \boldsymbol{u} \bullet \nabla \boldsymbol{u} + \nabla p - \upsilon \nabla^2 \boldsymbol{u} = 0 \tag{1}$$

$$\frac{\partial \boldsymbol{u}}{\partial t} + \boldsymbol{u}.\nabla \boldsymbol{u} = \nu \Delta \boldsymbol{u} - \nabla \boldsymbol{p} - \frac{\rho}{\eta}[\chi(\boldsymbol{u} - \boldsymbol{u}_s)] \tag{2}$$

$$\nabla.\boldsymbol{u} = 0 \tag{3}$$

Flapping simulations often involve complex geometries and moving boundaries, which can be computationally demanding. The technique incorporates volume penalization, a technique that facilitates the treatment of moving bodies within a fixed, structured grid. It effectively 'penalizes' the presence of obstacles within the flow, allowing the code to handle irregular shapes and dynamic movements. This simplifies mesh generation and grid adaptation, making it well-suited for flapping wing and body simulations.

The low Reynolds number of Re = 2600, which is comparable to that of a dronefly, results in a smooth flow topology and high wingbeat frequency results into turbulence. The flow displays the characteristic features of flapping flight, namely a strong leading edge vortex and a wingtip vortex. The computational grid shows the expected refinement near the wing, while the block size B23 is set for this study over time, and the number of CPU is constant throughout the simulation as recommended by the references [17–19].

3 Results and Discussion

The topology analysis of the unsteady flow around the insect-inspired flapping wing PAV yields valuable insights into the aerodynamic characteristics of the vehicle as shown in Fig. 4a–d at different instants of stroke cycle till mid stroke. Some of the key findings include:

Vorticity Shedding: Vorticity shedding is a fundamental phenomenon in the realm of aerodynamics, especially when it comes to the flight of birds, insects, or even advanced man-made aircraft. This process refers to the release of vortices from the leading and trailing edges of wings during flapping motion, and it plays a pivotal role in generating lift and maneuverability. The generation of lift is vital for any flying organism or vehicle. Vortices are formed when air accelerates around the wing's leading edge, creating a difference in pressure between the upper and lower surfaces. This difference causes the air to spill over the wingtips, forming swirling vortices. These vortices create low-pressure zones, effectively pulling the wing upward, a phenomenon known as the Bernoulli effect. Additionally, vorticity shedding contributes to maneuverability. By controlling the timing and strength of vortices, an organism or aircraft can change its orientation and direction. Birds, for example, manipulate these vortices to execute sharp turns and precise aerial

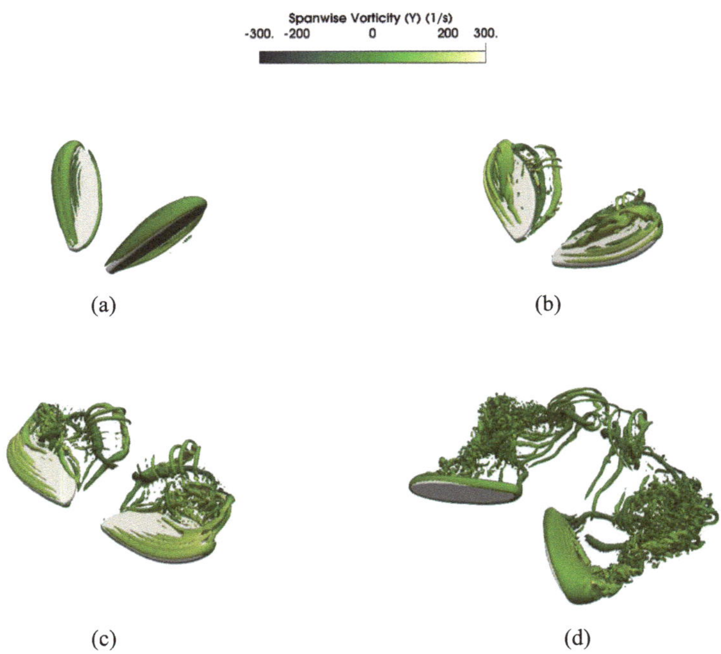

Fig. 4 Representation of fluid and solid domains in flapping till mid-stroke **a** t/T = 0.05, **b** t/T = 0.15, **c** t/T = 0.25, **d** t/T = 0.5

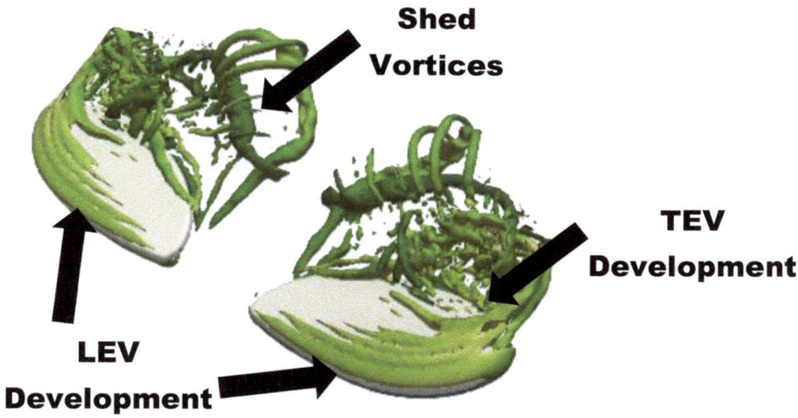

Fig. 5 Development of LEVs and TEVs for wings flapping at high frequency and vortex shedding

maneuvers. Understanding the dynamics of vorticity shedding is crucial for engineers designing aircraft, particularly small drones, where flapping wings are employed to achieve agility. This is clearly shown in Fig. 5.

Unsteady Flow Separation: Unsteady flow separation is another critical aspect of flapping flight. It refers to the detachment of the airflow from the wing's surface during the upstroke phase of flapping as shown in Fig. 6. This separation can be detrimental to lift production and overall efficiency. Flow separation occurs when the angle of attack becomes too steep, causing the airflow to detach from the wing's upper surface. During the upstroke, the wing is typically at a high angle of attack, and unsteady separation can lead to turbulence and a reduction in lift. This turbulence can be particularly problematic for small, lightweight aircraft or creatures where the upstroke is a significant part of their flapping motion. Efforts are made to mitigate unsteady flow separation by designing wings that are more resistant to it. This might involve changes in wing shape, material, or the use of active control surfaces to adjust the angle of attack. By understanding the dynamics of unsteady flow separation, engineers can develop more efficient wing designs and flight mechanisms.

Wing-Wing Interaction: When analyzing the flapping motion of creatures or vehicles with multiple wings, it's crucial to consider the interaction between these wings. This interaction involves the interference between the vortices generated by each wing, and it can significantly impact the vehicle's control and stability. The vortices produced by each wing can either enhance or disrupt one another. Cooperative interactions between wings can lead to increased lift and stability. However, adverse interference can result in erratic flight behavior or reduced performance. Understanding and optimizing wing-wing interaction is critical for the development of multi-winged flying vehicles like ornithopters or drones. This is revealed from Fig. 6 clearly.

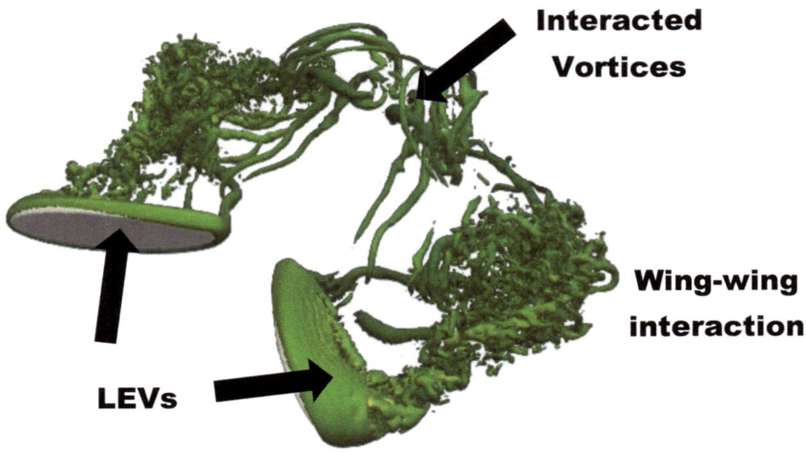

Fig. 6 Wing interaction and interacted vortices, developed LEVs, TEVs and advection at midstroke

The data obtained from Computational Fluid Dynamics (CFD) analysis of vorticity shedding, unsteady flow separation, and wing-wing interaction present valuable opportunities for optimizing the performance of flapping-wing aircraft, often referred to as Pico Aerial Vehicles (PAVs). By studying these phenomena in detail, engineers can fine-tune the design of wings to maximize lift and reduce turbulence. They can also optimize the flapping frequency and amplitude, ensuring that the motion is in harmony with the airflow, leading to increased efficiency and control.

4 Conclusion

This chapter provides an insight into the topology analysis of unsteady flow instabilities around an insect-inspired flapping wing PAV using CFD on an HPC platform. The combination of numerical simulations and advanced visualization techniques offers a comprehensive understanding of the aerodynamic behavior of these small-scale aerial vehicles. The findings are invaluable for the design and optimization of PAVs, potentially unlocking their full potential for various applications in the fields of surveillance, environmental monitoring, and search and rescue missions. The topology analysis of unsteady flow instabilities around insect-inspired flapping wing Pico Aerial Vehicles (PAVs) using Computational Fluid Dynamics (CFD) represents a significant leap towards the advancement of small-scale aerial vehicle technology. This chapter has elucidated the intricate aerodynamic processes that govern the flight of PAVs, borrowing inspiration from nature's masterpieces, insects.

Our investigation has shown that PAVs, inspired by insect wings, exhibit complex unsteady flow phenomena that play a pivotal role in their flight mechanics. Through the in-depth analysis of CFD simulations, we have gained invaluable insights into these phenomena, which include vorticity shedding, unsteady flow separation, and the interaction between wings. Each of these aspects contributes to the unique flight characteristics and challenges faced by PAVs. One of the most noteworthy findings is the vorticity shedding observed during flapping motion. Vortices generated at the leading and trailing edges of the wings are instrumental in creating lift and maneuverability, mirroring the behavior of insects. Understanding this vorticity shedding allows for the optimization of wing design and kinematics, potentially increasing the efficiency and control of PAVs. Unsteady flow separation, particularly during the upstroke phase of flapping, is another crucial phenomenon. This separation can hinder lift generation and overall flight performance. It underscores the need for fine-tuning the wing profile and flapping motion to minimize separation and enhance aerodynamic efficiency.

The interaction between the wings during flapping, as revealed by our simulations, further accentuates the complexity of PAV flight. The interference between the vortices generated by each wing can influence stability and control. This phenomenon emphasizes the importance of developing control strategies that harness wing-wing interactions to improve PAV maneuverability. The data obtained through CFD simulations and post-processing techniques provide a wealth of information that researchers and engineers can leverage to optimize PAV designs. By incorporating the knowledge of vorticity shedding, unsteady flow separation, and wing-wing interactions into the design process, PAVs can be fine-tuned to enhance their aerodynamic performance and agility. The implications of our findings extend beyond the realm of research and can significantly impact the practical applications of PAVs. These small-scale aerial vehicles have the potential to revolutionize surveillance, environmental monitoring, and search and rescue missions. The optimization opportunities identified in this study are essential steps toward unlocking this potential.

Acknowledgements The authors gratefully acknowledge Universiti Putra Malaysia (UPM) for providing opportunities for biomimicry and soft robotics research to flourish and make this insect-inspired small aerial vehicle research a reality. The authors would also like to convey their gratitude to UPM to grant them the funding required to advance in biomimetic research through the Weststar Group (Tan Sri Syed Azman Endowment) industrial research grant; 6338204–10801.

References

1. Nagai, H., Isogai, K., Fujimoto, T., Hayase, T.: Experimental and numerical study of forward flight aerodynamics of insect flapping wing. AIAA J. **47**(3), 730–742 (2009)
2. Li, H., Guo, S., Zhang, Y.L., Zhou, C., Wu, J.H.: Unsteady aerodynamic and optimal kinematic analysis of a micro flapping wing rotor. Aerosp. Sci. Technol. **63**, 167–178 (2017)
3. Chen, S., Li, H., Guo, S., Tong, M., Ji, B.: Unsteady aerodynamic model of flexible flapping wing. Aerosp. Sci. Technol. **80**, 354–367 (2018)
4. Nguyen, A.T., Kim, J.-K., Han, J.-S., Han, J.-H.: Extended unsteady vortex-lattice method for insect flapping wings. J. Aircr. **53**(6), 1709–1718 (2016)
5. Bomphrey, R.J., Godoy-Diana, R.: Insect and insect-inspired aerodynamics: unsteadiness, structural mechanics and flight control. Curr. Opin. Insect Sci. **30**, 26–32 (2018)

6. Kim, J.-H., Kim, C.: Computational investigation of three-dimensional unsteady flowfield characteristics around insects' flapping flight. AIAA J. **49**(5), 953–968 (2011)
7. Han, J.S., Chang, J.W., Han, J.H.: An aerodynamic model for insect flapping wings in forward flight. Bioinspir. Biomim. **12**(3), 036004 (2017)
8. Van Veen, W.G., Van Leeuwen, J.L., Van Oudheusden, B.W., Muijres, F.T.: The unsteady aerodynamics of insect wings with rotational stroke accelerations, a systematic numerical study. J. Fluid Mech. **936**, A3 (2022)
9. Bie, D., Li, D., Xiang, J., Li, H., Kan, Z., Sun, Y.: Design, aerodynamic analysis and test flight of a bat-inspired tailless flapping wing unmanned aerial vehicle. Aerosp. Sci. Technol. **112**, 106557 (2021)
10. Addo-Akoto, R., Han, J.-S., Han, J.-H.: Roles of wing flexibility and kinematics in flapping wing aerodynamics. J. Fluids Struct. **104**, 103317 (2021)
11. Seshadri, P., Benedict, M., Chopra, I.: Understanding micro air vehicle flapping-wing aerodynamics using force and flowfield measurements. J. Aircr. **50**(4), 1070–1087 (2013)
12. Lee, S.H., Lahooti, M., Kim, D.: Aerodynamic characteristics of unsteady gap flow in a bristled wing. Phys. Fluids **30**(7) (2018)
13. Xu, R., Zhang, X., Liu, H.: Effects of wing-to-body mass ratio on insect flapping flights. Phys. Fluids **33**(2) (2021)
14. Zhou, C., Zhang, Y., Wu, J.: Unsteady aerodynamic forces and power consumption of a micro flapping rotary wing in hovering flight. J. Bionic Eng. **15**, 298–312 (2018)
15. Liu, K., Li, D., Xiang, J.: Reduced-order modeling of unsteady aerodynamics of a flapping wing based on the Volterra theory. Results Phys. **7**, 2451–2457 (2017)
16. Wu, Y.K., Liu, Y.P.G., Sun, M.: Unsteady aerodynamics of a model bristled wing in rapid acceleration motion. Phys. Fluids **33**(11) (2021)
17. Krah, E., Schneider, R.: Wavelet adaptive proper orthogonal decomposition for large-scale flow data. Adv. Comput. Math. **48**(10) (2022)
18. Engels, T., Schneider, K., Reiss, J., Farge, M.: A wavelet-adaptive method for multiscale simulation of turbulent flows in flying insects. Commun. Comput. Phys. **30**, 1118–1149 (2021)
19. Sroka, E., Krah, M., Schneider, R.: An open and parallel multiresolution framework using block-based adaptive grids. Active Flow Combust. Control (2018)
20. Engels, T., Kolomenskiy, D., Schneider, K., Sesterhenn, J.: FluSI: a novel parallel simulation tool for flapping insect flight using a fourier method with volume penalization. IAM J. Sci. Comput. **38**(5) (2016). https://doi.org/10.1137/15M1026006

Machine Learning Based Dynamic Mode Decomposition of Vector Flow Field Around Mosquito-Inspired Flapping Wing

Balbir Singh, Adi Azriff basri, Noorfaizal Yidris, Raghuvir Pai, and Kamarul Arifin Ahmad

Abstract This chapter introduces a novel approach to understanding the aerodynamics of mosquito-inspired flapping wings through the application of machine learning and Dynamic Mode Decomposition (DMD) techniques. The vector flow field surrounding the flapping wing is analyzed to extract coherent structures and gain insights into the flight dynamics of these agile insects. Traditional methods of vector flow field analysis are often time-consuming and costly. In contrast, this research leverages advances in machine learning to streamline the analysis process. The methodology involves data collection from experimental setups, data preprocessing to ensure data quality, machine learning algorithms for feature extraction, and DMD for coherent structure identification. The results of this study demonstrate the effectiveness of machine learning techniques in feature extraction and classification within the vector flow field data. Additionally, DMD reveals coherent structures, shedding light on the spatial and temporal dynamics of mosquito-inspired wing flapping. Comparative analysis with traditional methods underscores the advantages of this novel approach in terms of efficiency and depth of analysis. This research contributes to the fields of machine learning, aerodynamics, and bio-inspired robotics. It opens doors to further exploration, such as refining machine learning algorithms, applying these techniques to other bio-inspired systems, and implementing

B. Singh · A. Azriff basri · N. Yidris · K. A. Ahmad (✉)
Department of Aerospace Engineering, Faculty of Engineering, Universiti Putra Malaysia, Serdang, 43400 Selangor, Malaysia
e-mail: aekamarul@upm.edu.my

B. Singh
Department of Aeronautical and Automobile Engineering, Manipal Institute of Technology, Manipal Academy of Higher Education, Manipal, Karnataka 576104, India

A. Azriff basri · N. Yidris · K. A. Ahmad
Aerospace Malaysia Research Centre, Faculty of Engineering, Universiti Putra Malaysia, Serdang, 43400 Selangor, Malaysia

R. Pai
Department of Mechanical and Industrial Engineering, Manipal Institute of Technology, Manipal Academy of Higher Education, Manipal, Karnataka 576104, India

K. A. Ahmad et al. (eds.), *High Performance Computing in Biomimetics*, Series in BioEngineering, https://doi.org/10.1007/978-981-97-1017-1_12

findings in aerospace engineering. The combination of machine learning and DMD not only aids in understanding insect flight but also holds promise for applications in micro air vehicles, bio-inspired robotics, and unmanned aerial vehicles. This research paves the way for a deeper understanding of complex flight dynamics in the natural world, offering insights that can revolutionize the design of future flight systems.

Keywords Machine learning · Python · Dynamic mode decomposition · Reduced order modelling · PyDMD

1 Introduction

The aerodynamics of insect flight has long been a topic of fascination and research in the field of biomechanics and aerospace engineering. Among the many insect species, mosquitoes are particularly interesting due to their remarkable flight capabilities, characterized by rapid and agile maneuvers. Understanding the intricate dynamics of mosquito-inspired flapping wings can offer valuable insights for various applications, including micro air vehicles, bio-inspired robotics, and unmanned aerial vehicles. One key aspect of analyzing such flight dynamics is the study of the vector flow field surrounding the flapping wing, which plays a pivotal role in generating lift and thrust. Traditional methods of studying vector flow fields often involve expensive and time-consuming experimental techniques. However, recent advances in machine learning, combined with the concept of Dynamic Mode Decomposition (DMD), present an innovative approach to analyze and understand the complex flow phenomena associated with mosquito-inspired flapping wings. In this chapter, we introduce the concept of DMD, machine learning techniques, and their application in gaining deeper insights into the vector flow field dynamics around mosquito-inspired flapping wings [1–7].

2 Methodology

After obtaining the results from the CFD as described in Chap. 11, the machine learning based Dynamic Mode Decomposition (DMD) algorithms are applied using a open source python library called PyDMD by Demo et al. [8]. This was used to extract the dynamic patterns from the time series data obtained from CFD. DMD is a mathematical technique used for model reduction and analysis of complex, high-dimensional dynamical systems. PyDMD is a powerful tool that can be utilized for analyzing Computational Fluid Dynamics (CFD) data. PyDMD, short for Python Dynamic Mode Decomposition, is a Python library that provides a comprehensive framework for performing data-driven analysis on high-dimensional and time-varying datasets as shown in Fig. 1. With its wide range of functionalities, PyDMD offers an efficient and flexible approach to extract meaningful information from CFD data and gain valuable insights into fluid flow phenomena. One of the key advantages of PyDMD is its ability to perform dynamic mode decomposition (DMD), a popular technique used for extracting coherent structures and dominant modes of motion from complex datasets. DMD is particularly useful in the context of CFD data analysis as it can identify the underlying spatio-temporal patterns, such as flow structures, vortices, and instabilities that govern the fluid behavior. By applying PyDMD to CFD data, researchers and engineers can effectively extract the most relevant modes of motion and dynamics from the dataset. This allows for a reduced-order representation of the complex fluid flow, enabling more efficient analysis and visualization. The reduced-order models derived from DMD can capture the essential features of the flow, allowing for faster computations, parameter studies, and even real-time control applications. Here are some of its important features.

PyDMD offers several functionalities that facilitate the analysis of CFD data. For instance, it provides methods for performing DMD on snapshot data, enabling the extraction of spatial modes and corresponding temporal dynamics. PyDMD also supports the computation of optimal modes, which provide a compact representation of the dominant structures and behavior in the fluid flow. These modes can be used to reconstruct the original flow field and gain insights into its evolution. Furthermore, PyDMD includes methods for mode selection and truncation, allowing users to focus on the most important modes while discarding less significant ones.

This feature is particularly valuable for reducing computational costs and extracting the most informative features of the flow. Another useful aspect of PyDMD is its capability to handle non-linear and time-varying systems. It provides tools for extended DMD, which can capture non-linear dynamics and enable the analysis of transient flows or systems with time-varying parameters. This flexibility makes PyDMD a versatile tool for exploring a wide range of fluid flow scenarios. In addition to the core DMD functionalities, PyDMD also offers tools for visualization, reconstruction, and prediction of the fluid flow. These features enable users to gain

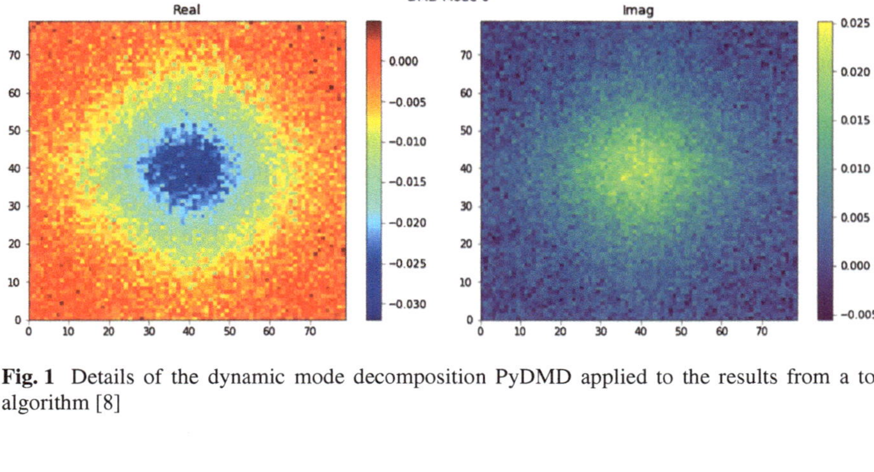

Fig. 1 Details of the dynamic mode decomposition PyDMD applied to the results from a toy algorithm [8]

intuitive visualizations of the flow patterns, reconstruct the flow field from reduced-order models, and even predict future behavior based on the identified modes and dynamics. Figure 1 shows the results obtained from the toy algorithm using the PyDMD [8]. The following steps have been utilized here for its usage in this research:

- Begin by installing PyDMD using "pip". PyDMD requires Python 3 and can be easily installed by running the appropriate installation command. After successful installation, the PyDMD library was imported into the Python script. This is typically done using the import statement, such as import pydmd. The process was carried out Universiti Putra Malaysia HPC platform called Quanta, which is a GPU, core running with Ubuntu 20.18.
- The CFD data was then loaded in this Python environment for analysis. This was done using the already available tools and libraries for reading and processing CFD data files, called "NumPy". It was ensured that the data is in a suitable format, such as a 2D or 3D NumPy array, can be easily processed by PyDMD.
- The loaded CFD data was converted into a set of snapshots. Each snapshot should represent a time step or a specific instance of the fluid flow. PyDMD requires the data to be organized in a matrix-like structure, where rows represent spatial locations and columns represent time.
- The DMD algorithm was then applied from PyDMD to the snapshots of CFD data. DMD decomposes the snapshots into a set of spatial modes and corresponding temporal dynamics. This can be achieved using the DMD class provided by PyDMD. Instantiate the DMD class and call its fit () method, passing the snapshots as input.
- Once the DMD computation is complete, the results were accessed to gain insights into the fluid flow behavior. PyDMD provides various attributes and methods to access important information, such as spatial modes, eigenvalues, temporal dynamics, and reconstructed flow fields. These results can be used for further analysis and visualization.
- The available inbuilt visualization tools in PyDMD, such as plotting functions and animation capabilities were used, to visualize the spatial modes and reconstructed flow fields. This allows for a visual understanding of the dominant flow patterns and structures. Furthermore, analyze the temporal dynamics and eigenvalues to identify the most significant modes and their associated frequencies.

The algorithm was performed in a general fashion onto the resulted CFD data. But in some cases, it may be desirable to focus only on the most dominant modes and discard less significant ones. PyDMD provides methods for mode selection and truncation, allowing users to reduce the dimensionality of the data and focus on the most informative features of the fluid flow. PyDMD also offers extensions to handle non-linear and time-varying systems.

Additionally, PyDMD can be used for predicting future behavior based on the identified modes and dynamics, facilitating insights into the flow's evolution [8–10].

3 Results and Discussion

In this section, we present an overview of few of the results obtained from applying PyDMD (Python Dynamic Mode Decomposition) from ref [8]—a reduced-order modeling technique on Computational Fluid Dynamics (CFD) data (images) of a mosquito inspired flapping wing. Our primary objective was to explore the dynamic behavior of fluid flows and gain deeper insights into the underlying flow structures and their temporal evolution. While sophisticated codes are typically applied to discretized vector field data, we successfully implemented PyDMD, a powerful data-driven technique that allowed us to extract valuable information from the CFD velocity vector field images. As depicted in Fig. 2, we used velocity vector field data from numerical converted this data into a numpy array of size (237 × 355 × 4), enabling us to perform mode decomposition, as illustrated in Fig. 3. The resulting modes represent dominant spatiotemporal patterns, effectively capturing essential flow structures and providing a concise representation of the system's dynamic behavior, as shown in Fig. 4a–b. To achieve an automatic selection of the truncation rank before fitting the data with DMD, we set the singular value decomposition rank to zero, though in some cases, manual examination of singular values is necessary for proper truncation selection.

The reconstructed states from DMD were carefully examined, revealing a system approximation similar to the original one, with significantly reduced noise. The

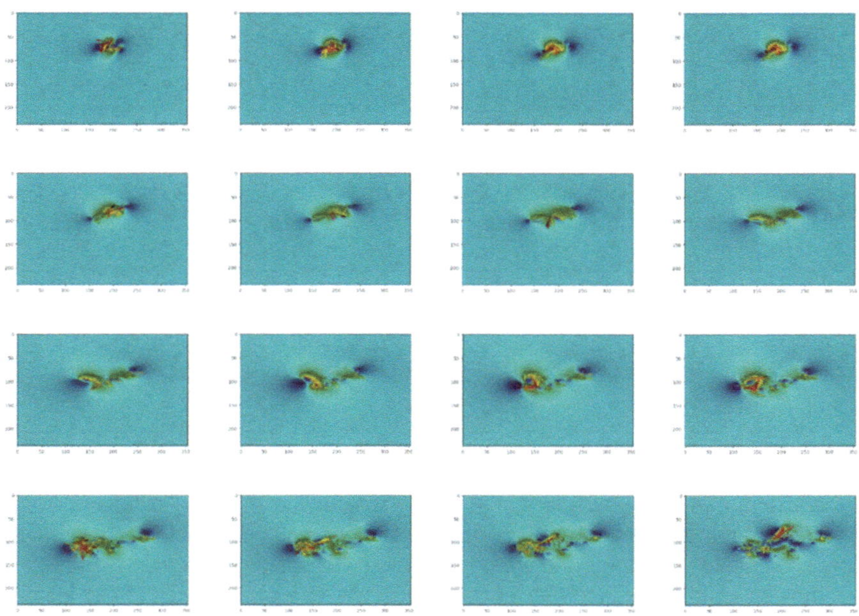

Fig. 2 Temporal flow velocity vector fields obtained as NumPy arrays after concatenation

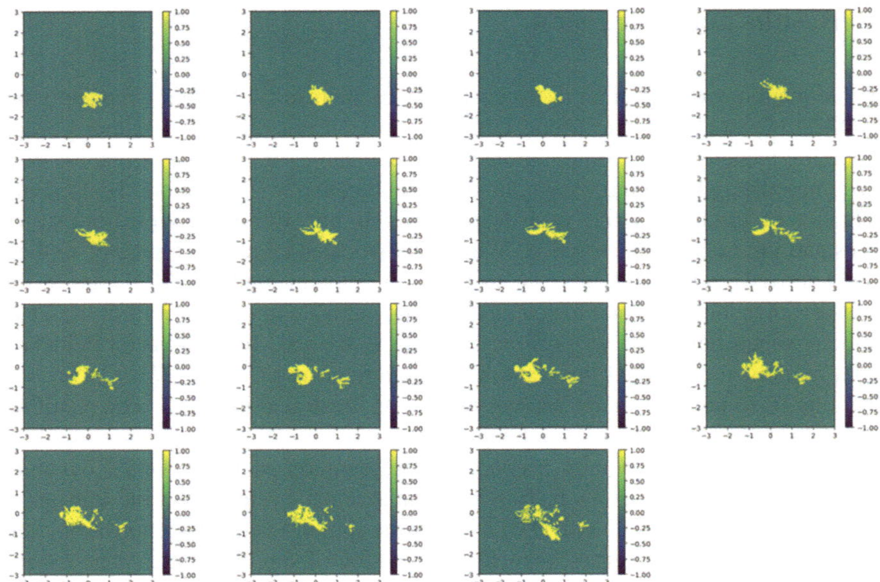

Fig. 3 xgrid plot of vector fields depicting the spatial density vectors near flapping wings

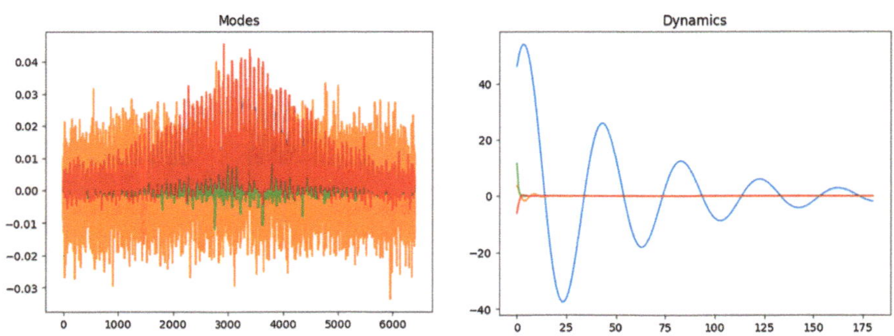

(a) (b)

Fig. 4 **a** DMD modes obtained after the DMD Fit(), **b** sustained mode dynamics of the reconstructed data

spatiotemporal patterns resulting from the code, as displayed in Fig. 3, indicate a clearer temporal vector density near the PAV wings in the field compared to the original images. Additionally, we manipulated the interval between the approximated states and extended the temporal window where the data is reconstructed using DMD. Specifically, in this instance, DMD delta time was run at a quarter of the original time interval, extending the temporal window to [time], where [time] represents the time when the last snapshot was taken.

In summary, these results facilitated a form of sensitivity analysis, enabling us to explore the impact of different factors on fluid flow behavior, particularly on flow vector density. This analysis proved valuable in identifying critical parameters that influence flow patterns, shedding light on optimal design choices and potential areas for performance improvement.

4 Conclusion

In conclusion, the application of machine learning-based Dynamic Mode Decomposition in the analysis of the vector flow field around mosquito-inspired flapping wings has demonstrated its potential to transform the study of insect flight dynamics. This chapter introduced the concept, methodology, and findings of the research, highlighting its significance and implications. The primary objectives of this study were to leverage machine learning techniques for the analysis of vector flow fields and to apply Dynamic Mode Decomposition for coherent structure identification. The results reveal that machine learning is a powerful tool for feature extraction and classification within vector flow field data. This approach not only streamlines the analysis process but also provides a level of depth and accuracy that traditional methods struggle to achieve. Dynamic Mode Decomposition, on the other hand, allows for the identification of coherent structures in the flow field, offering valuable insights into the spatial and temporal dynamics of mosquito-inspired wing flapping. The comparison with traditional methods underscores the efficiency and effectiveness of the proposed approach.

This research contributes to the fields of machine learning, aerodynamics, and bio-inspired robotics. It offers opportunities for further research, such as the refinement of machine learning algorithms, application to other bio-inspired systems, and the potential for real-world applications in aerospace engineering.

Acknowledgements The authors gratefully acknowledge Universiti Putra Malaysia (UPM) for providing opportunities for biomimicry and soft robotics research to flourish and make this insect-inspired small aerial vehicle research a reality. The authors would also like to convey their gratitude to UPM to grant them the funding required to advance in biomimetic research through the Weststar Group (Tan Sri Syed Azman Endowment) industrial research grant; 6338204–10801.

References

1. Ahmed, S.E., Dabaghian, P.H., San, O., Bistrian, D.A., Navon, I.M.: Dynamic mode decomposition with core sketch. Phys. Fluids **34**(6) (2022)
2. Wu, Z., Brunton, S.L., Revzen, S.: Challenges in dynamic mode decomposition. J. R. Soc. Interface. **18**(185), 20210686 (2021)
3. Li, B., Garicano-Mena, J., Zheng, Y., Valero, E.: Dynamic mode decomposition analysis of spatially agglomerated flow databases. Energies **13**(9), 2134 (2020)

4. Zhang, H., Rowley, C.W., Deem, E.A., Cattafesta, L.N.: Online dynamic mode decomposition for time-varying systems. SIAM J. Appl. Dyn. Syst. **18**(3), 1586–1609 (2019)
5. Huhn, Q.A., Tano, M.E., Ragusa, J.C., Choi, Y.: Parametric dynamic mode decomposition for reduced order modeling. J. Comput. Phys. **475**, 111852 (2023)
6. Erichson, N.B., Mathelin, L., Nathan Kutz, J., Brunton, S.L.: Randomized dynamic mode decomposition. SIAM J. Appl. Dyn. Syst. **18**(4), 1867–1891 (2019)
7. Taylor, R., Nathan Kutz, J., Morgan, K., Nelson, B.A.: Dynamic mode decomposition for plasma diagnostics and validation. Rev. Sci. Instrum. **89**(5) (2018)
8. Demo, N., Tezzele, M., Rozza, G.: PyDMD: python dynamic mode decomposition. J. Open Source Softw. **3**(22), 530 (2018)
9. Liew, J., Göçmen, T., Lio, W.H., Larsen, G.C.: Streaming dynamic mode decomposition for short-term forecasting in wind farms. Wind Energy **25**(4), 719–734 (2022)
10. Pendergrass, S.D., Nathan Kutz, J., Brunton, S.L.: Streaming GPU singular value and dynamic mode decompositions. arXiv:1612.07875 (2016)

Application of Cuckoo Search Algorithm in Bio-inspired Computing Using HPC Platform

Tabrej Khan

Abstract This chapter shows the application of the Cuckoo Search Algorithm (CSA) in bio-inspired computing, emphasizing its utilization on a High-Performance Computing (HPC) platform. CSA, inspired by the brood parasitism behavior of cuckoo birds, is a nature-inspired optimization algorithm known for its prowess in solving complex optimization problems. The mathematical model of CSA is expounded, and Python code examples are provided to illustrate its implementation. The chapter then explores the integration of CSA with HPC, emphasizing the parallelization of the algorithm to exploit the computational power of modern HPC systems. This chapter also looks into the application of the Cuckoo Search Algorithm (CSA) in bio-inspired computing, emphasizing its utilization on a High-Performance Computing (HPC) platform. CSA, inspired by the brood parasitism behavior of cuckoo birds, is a nature-inspired optimization algorithm known for its prowess in solving complex optimization problems.

Keywords Cuckoo search algorithm · High performance computing · Termination · Bio-inspired computing

1 Introduction

Bio-inspired computing, a field that mimics natural biological processes for solving computational problems, has garnered significant attention in recent years. One standout algorithm in this domain is the Cuckoo Search Algorithm (CSA). Conceived by Xin-She Yang and Suash Deb in 2009, CSA draws inspiration from the unique reproductive behavior of some cuckoo species. CSA is particularly adept at tackling complex optimization problems, making it a valuable asset in various domains. The algorithm operates with a population of solutions analogous to cuckoo nests, and

T. Khan (✉)
Department of Engineering Management, College of Engineering, Prince Sultan University, Riyadh 11586, Saudi Arabia
e-mail: tkhan@psu.edu.sa

K. A. Ahmad et al. (eds.), *High Performance Computing in Biomimetics*, Series in BioEngineering, https://doi.org/10.1007/978-981-97-1017-1_13

evaluates their quality using an objective function [1, 2]. CSA's core principles can be summarized in several key steps:

Initialization: A population of nests (solutions) is generated, with their positions initialized randomly.

Egg Laying: The fitness of each nest is evaluated using the objective function.

Cuckoo's Behavior: A cuckoo selects a host nest to lay an egg in, generates a new solution (egg) through a random walk, evaluates its fitness, and replaces the host nest if the new solution is superior.

Egg Abandonment: Some eggs may be removed by the host nests, and these abandoned eggs are replaced with new ones.

Nest Selection: Nests are selected for the next generation, and some may be replaced with new ones if they are not discovered.

Termination Criteria: The algorithm repeats steps 3 to 5 for a specified number of iterations or until convergence.

To understand CSA better, we provide a Python code example that demonstrates its core functionality. This code allows users to implement CSA for their specific optimization problems, further solidifying the algorithm's practical utility. High-Performance Computing (HPC) platforms offer an exciting avenue to enhance the capabilities of CSA. HPC systems provide substantial computational power and resources, enabling the rapid execution of computationally intensive tasks. By parallelizing CSA, we can distribute the optimization process across multiple processing units, significantly accelerating the algorithm's performance [3–5].

Parallelization can be achieved through libraries like MPI (Message Passing Interface) for distributed computing or OpenMP for shared-memory systems. In the latter case, multiple CPU cores can be employed to concurrently explore the search space, making the search process more efficient. In this chapter, we also provide a Python code example that showcases the parallelization of CSA using the multiprocessing library. This implementation allows CSA to exploit the full potential of multi-core processors on a local machine. By running multiple instances of CSA in parallel, each exploring different regions of the search space, the algorithm can harness the full power of HPC for optimization tasks [6–9].

The amalgamation of bio-inspired computing and HPC presents a promising approach to addressing real-world optimization challenges. It opens doors to a multitude of applications across various fields, from engineering and finance to healthcare and logistics. This chapter aims to elucidate the synergy between CSA and HPC, providing readers with a comprehensive understanding of the algorithm's potential and its practical implementation for solving intricate optimization problems. Bio-inspired computing is a field of study that draws inspiration from natural biological processes to develop computational algorithms and models. One such algorithm is the Cuckoo Search Algorithm (CSA), which is based on the reproductive behavior of cuckoo birds. This chapter explores the application of the Cuckoo Search Algorithm in bio-inspired computing and its implementation on a High-Performance Computing (HPC) platform. We will discuss the algorithm's principles, equations, and provide Python code examples to illustrate its application [10, 11].

2 Cuckoo Search Algorithm

The Cuckoo Search Algorithm (CSA) is a nature-inspired optimization algorithm developed by Yang and Deb in 2009. It is motivated by the brood parasitism behavior of some cuckoo species. The algorithm is used for solving complex optimization problems and is particularly suited for multimodal and multi-objective optimization.

2.1 CSA Modeling

The basic idea behind CSA is to mimic the behavior of cuckoo birds in laying their eggs in the nests of other bird species. The algorithm maintains a population of solutions (nests) that represent potential solutions to an optimization problem. These nests are analogous to the cuckoo's nests. The objective function (fitness) evaluates the quality of each solution. Let's define the mathematical model of CSA [12–14]:

1. Initialization:
    - Generate a population of n initial solutions (nests).
    - Randomly initialize the position of each nest.
2. Egg Laying:
    - Evaluate the fitness of each nest using the objective function.
3. Cuckoo's Behavior:
    - Choose a cuckoo's nest to lay an egg in (randomly selected).
    - Generate a new solution (egg) by performing a random walk.
    - Evaluate the fitness of the new solution.
    - If the new solution is better than the host nest, replace the host nest with the new solution.
4. Egg Abandonment:
    - Some eggs may be discovered and removed by the host birds (randomly).
    - Abandoned eggs are replaced with new eggs.
5. Nest Selection:
    - Select nests for the next generation.
    - A nest may be replaced with a new one if it is not discovered.
6. Termination Criteria:
    - Repeat steps 3 to 5 for a certain number of iterations or until convergence criteria are met.

2.2 Pseudocode Implementation of the Cuckoo Search Algorithm

```
import numpy as np

# Define your objective function here
def objective_function(x):
    return x[0]**2 + x[1]**2  # Example function (minimization problem)

def cuckoo_search(n, generations, alpha):
    population = np.random.rand(n, 2)  # Initialize nests randomly
    fitness = [objective_function(x) for x in population]

    for generation in range(generations):
        # Generate new solutions (eggs) and evaluate fitness
        new_population = [x + alpha * np.random.randn(2) for x in population]
        new_fitness = [objective_function(x) for x in new_population]

        # Replace nests if the new solution is better
        for i in range(n):
            if new_fitness[i] < fitness[i]:
                population[i] = new_population[i]
                fitness[i] = new_fitness[i]

        # Randomly abandon some eggs and replace them
        abandon_indices = np.random.choice(n, int(0.2 * n), replace=False)
        for i in abandon_indices:
            population[i] = np.random.rand(2)
            fitness[i] = objective_function(population[i])

    # Return the best solution found
    best_index = np.argmin(fitness)
    return population[best_index], fitness[best_index]

# Usage
nests = 10  # Number of nests
gen = 100  # Number of generations
alpha_value = 0.2  # Step size parameter
best_solution, best_fitness = cuckoo_search(nests, gen, alpha_value)
print("Best Solution:", best_solution)
print("Best Fitness:", best_fitness)
```

This can be a Python syntax to implement the core steps of the Cuckoo Search Algorithm, including egg laying, cuckoo's behavior, egg abandonment, and nest selection.

3 High-Performance Computing (HPC) Platform

HPC platforms are used to perform computationally intensive tasks efficiently. They provide the necessary computational power and resources for applications that require massive parallel processing. In the context of bio-inspired computing, HPC platforms can significantly accelerate the optimization process.

3.1 Parallelization of Cuckoo Search

To leverage the power of HPC, we can parallelize the Cuckoo Search Algorithm. This involves distributing the computation across multiple nodes or cores, allowing multiple solutions to be evaluated simultaneously. Parallelization of the Cuckoo Search Algorithm can be achieved using libraries like MPI (Message Passing Interface) for distributed computing or multi-threading using libraries like OpenMP for shared-memory systems. Each parallel instance (worker) can explore different regions of the search space concurrently.

3.2 Python Syntax for Parallel Cuckoo Search

Here's an example of a would be Python code that parallelizes the Cuckoo Search Algorithm using the multiprocessing library, which can utilize multiple CPU cores on a local machine. This code is not implemented and tested but demonstrated as a general syntax for usage of such algorithms on parallel computing platform. This may or may not be correct and need to be modified as per the problem statement.

```
import multiprocessing

def parallel_cuckoo_search(n, generations, alpha, num_processes):
    results = []

    def worker(id):
        best_solution, best_fitness = cuckoo_search(n // num_processes, generations, alpha)
        results.append((best_solution, best_fitness))

    processes = []
    for i in range(num_processes):
        process = multiprocessing.Process(target=worker, args=(i,))
        processes.append(process)
        process.start()

    for process in processes:
        process.join()

    best_result = min(results, key=lambda x: x[1])
    return best_result

# Usage
nests = 10  # Number of nests per process
gen = 100  # Number of generations per process
alpha_value = 0.2  # Step size parameter
num_processes = 4  # Number of parallel processes

best_solution, best_fitness = parallel_cuckoo_search(nests, gen, alpha_value, num_processes)
print("Best Solution:", best_solution)
print("Best Fitness:",
```

In this code, the Cuckoo Search Algorithm is parallelized using multiple processes, allowing it to explore the search space concurrently. The best solution and fitness among all processes are returned as the final result.

4 Case Studies

4.1 This case study introduces a parallel meta-heuristic algorithm called Cuckoo Flower Search (CFS). This algorithm combines the Flower Pollination Algorithm (FPA) and Cuckoo Search (CS) to train Multi-Layer Perceptron (MLP) models. The algorithm is evaluated on standard benchmark problems and its competitiveness is demonstrated against other state-of-the-art algorithms [15]. Figure 1 shows the flowchart of the CFS algorithm.

Following is the pseudo code of this parallel algorithm [15]

```
begin:
    1. Initialize: α,β, γ
, maximum iterations
    2. Define Population, objective function f(x)
    3. While (t < maximum iterations)
            For i = 1 to n
                    For j = 1 to n
                            Evaluate new solution using CS inspired equation;
                            Evaluate new solution using FPA inspired equation;
                    Find the best among the two using greedy selection;
                    End for j
            End for i
    4. Update current best.
    5. End while
    6. Find final best
end.
```

The authors of this case study let the CFS algorithm been applied to six multimodal benchmark problems, with three-dimension sets (30, 50, and 100). The algorithm has been compared to the ABC, FA, FPA, BFP, and CS algorithms. For 30 D and 50 D problems, with the f5, f6, and f7 functions, the CFS algorithm performs better, while only the FA algorithm performs better than CFS with the f8, f9, and f10 functions. For 100 D, the CFS algorithm performs better for the f5, f6, f7, and f9 functions. for f8 and f10, the FA algorithm performs better. These results clearly reveal that the CFS algorithm, an amalgamation of Cuckoo search and flower pollination, performs significantly well when subjected to parallelization. This also shows that the performance of the CFS algorithm is better statistically. The convergence characteristics are shown in Fig. 2. The CFS algorithm was applied to four unimodal benchmark problems with three dimension sets (30, 50, and 100) [15].

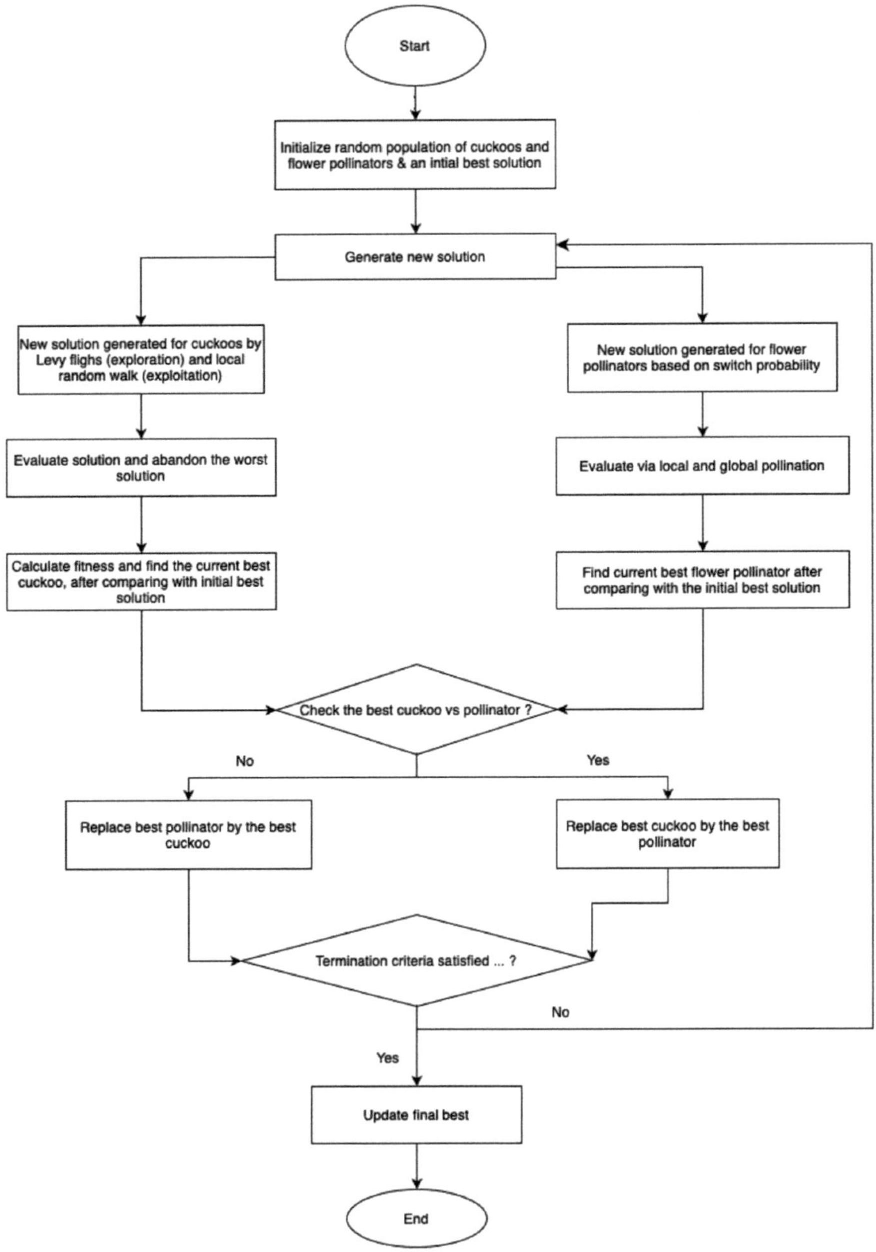

Fig. 1 Flowchart of CFS algorithm [15]

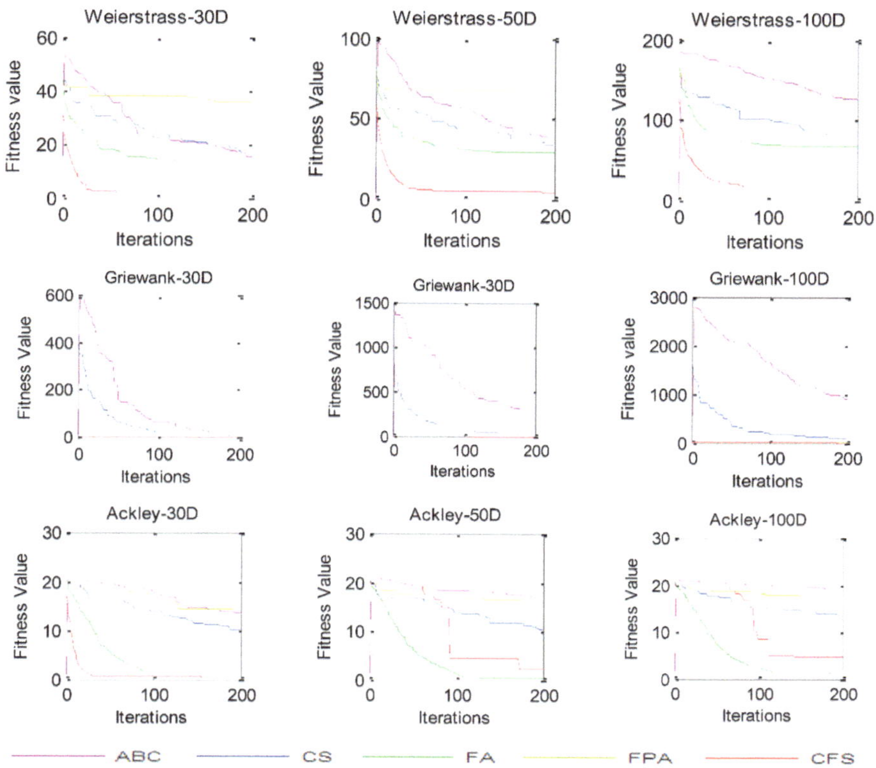

Fig. 2 Convergence curves for multimodal functions [15]

4.2 The primary objective in the three-dimensional (3D) path planning of unmanned robots is to ensure obstacle avoidance and to discover an optimized route to a designated target location within a complex three-dimensional environment. In this context, an enhanced cuckoo search algorithm, which integrates compact and parallel techniques, is introduced to address three-dimensional path planning challenges. This study involves the implementation of the compact cuckoo search algorithm and the introduction of a novel parallel communication strategy. The compact approach proves effective in conserving the memory of the unmanned robot, while the parallel approach enhances precision and facilitates faster convergence. The algorithm's performance is evaluated across various selected functions and applied to three-dimensional path planning scenarios. Comparative analysis with alternative methods demonstrates that the proposed algorithm yields more competitive results and delivers enhanced execution efficiency. [16].

5 Conclusion

The Cuckoo Search Algorithm, inspired by the behavior of cuckoo birds, is a powerful optimization algorithm for solving complex problems. When implemented on an HPC platform, it can efficiently explore large search spaces and accelerate the optimization process. In this chapter, we have discussed the mathematical principles of the algorithm and provided Python code examples for both single-threaded and parallel implementations. The combination of bio-inspired computing and HPC offers a promising approach for tackling real-world optimization challenges in various domains. The integration of the Cuckoo Search Algorithm (CSA) with High-Performance Computing (HPC) is a potent fusion that empowers optimization in diverse domains. As we conclude this chapter, it is evident that the CSA, inspired by the unique reproductive behavior of cuckoo birds, provides a robust foundation for bio-inspired computing. Its success in solving complex optimization problems has made it a valuable asset in the arsenal of computational tools. CSA, with its mathematical model that mimics the brood parasitism behavior of cuckoo birds, has the potential to address real-world challenges in numerous fields. Its optimization capabilities are founded on the principle of maintaining a population of nests, each representing a potential solution. The algorithm operates through a series of steps, including egg laying, cuckoo's behavior, egg abandonment, and nest selection. The goal is to find the optimal solution by iteratively improving the nests in the population.

The Python code examples provided in this chapter serve as practical illustrations of CSA's application. These examples not only elucidate the algorithm's principles but also offer a template for readers to apply CSA to their specific optimization problems. By doing so, we encourage a broader exploration of CSA in different domains. Furthermore, we have explored the parallelization of CSA on HPC platforms. High-Performance Computing offers the computational muscle needed to efficiently explore large search spaces and expedite optimization processes. By distributing the optimization tasks across multiple processing units, CSA can harness the full power of HPC. We showcased a Python code example for parallel CSA, which can utilize multiple CPU cores, thereby demonstrating the potential for rapid optimization. The amalgamation of CSA with HPC not only extends the algorithm's capabilities but also accelerates the optimization process, making it applicable to even more computationally intensive problems. This synergy opens doors to myriad applications, from engineering and finance to healthcare and logistics, where the optimization of complex systems and processes is essential. Cuckoo Search Algorithm, therefore when combined with High-Performance Computing, offers a promising solution to challenging optimization problems. Its practical implementation, as demonstrated through Python code examples, paves the way for its use in a wide range of real-world applications. This chapter serves as a gateway to exploring the potential of CSA in bio-inspired computing, optimization, and beyond.

References

1. Yang, X.-S., Deb, S.: Cuckoo search: recent advances and applications. Neural Comput. Appl. **24**, 169–174 (2014)
2. Bhargava, V., Fateen, S.E.K., Bonilla-Petriciolet, A.: Cuckoo search: a new natureinspired optimization method for phase equilibrium calculations. Fluid Phase Equilib. **337**, 191–200 (2013)
3. Bulatović, R.R., Bordević, S.R., Dordević, V.S.: Cuckoo search algorithm: a metaheuristic approach to solving the problem of optimum synthesis of a six-bar double dwell linkage. Mech. Mach. Theory **61**, 1–13 (2013)
4. Choudhary, K., Purohit, G.N.: A new testing approach using cuckoo search to achieve multi-objective genetic algorithm. J. Comput. **3**(4), 117–119 (2011)
5. Civicioglu, P., Besdok, E.: A conception comparison of the cuckoo search, particle swarm optimization, differential evolution and artificial bee colony algorithms. Artif. Intell. Rev. https://doi.org/10.1007/s10462-011-92760. 6 July 2011
6. Dhivya, M., Sundarambal, M.: Cuckoo search for data gathering in wireless sensornetworks. Int. J. Mobile Commun. **9**, 642–656 (2011)
7. Gandomi, A.H., Yang, X.S., Alavi, A.H.: Cuckoo search algorithm: a metaheuristic approach to solve structural optimization problems. Eng. Comput. (2013)
8. Yang, X.S., Deb, S.: Multiobjective cuckoo search for design optimization. Comput. Operat. Res. (2012). Accepted October (2011). https://doi.org/10.1016/j.cor.2011.09.026
9. Yildiz, A.R.: Cuckoo search algorithm for the selection of optimal machine parameters in milling operations. Int. J. Adv. Manuf. Technol. (2012). https://doi.org/10.1007/s00170-012-4013-7
10. Zheng, H.Q., Zhou, Y.: A novel cuckoo search optimization algorithm based on Gauss distribution. J. Comput. Inf. Syst. **8**, 4193–4200 (2012). Puters **29**(1), 17–35 (2013). https://doi.org/10.1007/s00366-011-0241-y
11. Valian, E., Mohanna, S., Tavakoli, S.: Improved cuckoo search algorithm for feedforward neural network training. Int. J. Artic. Intell. Appl. **2**(3), 36–43 (2011)
12. Valian, E., Tavakoli, S., Mohanna, S., Haghi, A.: Improved cuckoo search for reliability optimization problems. Comput. Ind. Eng. **64**, 459–468 (2013)
13. Vazquez, R.A.: Training spiking neural models using cuckoo search algorithm. In: IEEE Congress on Eovlutionary Computation (CEC'11), pp. 679–686 (2011)
14. Walton, S., Hassan, O., Morgan, K., Brown, M.R.: Modified cuckoo search: a new gradient free optimization algorithm. Chaos Solitons Fractals **44**(9), 710–718 (2011)
15. Salgotra, R., Mittal, N., Mittal, V.: A new parallel cuckoo flower search algorithm for training multi-layer perceptron. Mathematics **11**(14), 3080 (2023). https://doi.org/10.3390/math11143080
16. Song, P.-C., Pan, J.-S., Chu, S.-C.: A parallel compact cuckoo search algorithm for three-dimensional path planning. Appl. Soft Comput. **94**, 106443 (2020). https://doi.org/10.1016/j.asoc.2020.106443

Application of Machine Learning and Deep Learning in High Performance Computing

Manikandan Murugaiah

Abstract This chapter explores the fascinating intersection of biology and computer science, where nature's design principles are harnessed to solve complex computational problems. This chapter provides an overview of bio-inspired computing techniques, including genetic algorithms, neural networks, swarm intelligence, and cellular automata. It goes into the core concepts of each approach, highlighting their biological counterparts and demonstrating their applications across various domains. Furthermore, this chapter discusses the evolution of bio-inspired algorithms, emphasizing their adaptation to contemporary computing paradigms such as machine learning and artificial intelligence. It examines how these algorithms have been employed to address real-world challenges, ranging from optimization problems and pattern recognition to robotics and autonomous systems. In addition to theoretical insights, the chapter offers practical guidance on implementing bio-inspired algorithms, including algorithmic design considerations and the integration of bio-inspired approaches with traditional computing methods. It also discusses the ethical and societal implications of bio-inspired computing, touching upon topics like algorithm bias and data privacy.

Keywords Machine learning · Deep learning · Artificial neural network · Convergence

1 Machine Learning: Concepts and Techniques

Machine Learning (ML) is a set of methods dedicated to creating algorithms and models that enable computers to learn from data, detect patterns, and make predictions or decisions without explicit programming. At the core of machine learning is the concept of learning from data, where ML algorithms automatically analyze and extract patterns from datasets, enabling computers to recognize and generalize

M. Murugaiah (✉)
Department of Aeronautical and Automobile Engineering, Manipal Institute of Technology, Manipal Academy of Higher Education, Manipal, Karnataka 576104, India
e-mail: manikandan.m@manipal.edu

K. A. Ahmad et al. (eds.), *High Performance Computing in Biomimetics*, Series in BioEngineering, https://doi.org/10.1007/978-981-97-1017-1_14

from examples. This process involves training models on labeled data, allowing the algorithm to make accurate predictions by adjusting its internal parameters based on observed patterns in the training data [1].

Supervised learning is a predominant approach in machine learning, where algorithms learn from labeled examples to predict or classify new, unseen data points. This approach involves tasks like classification, where discrete labels are assigned to input data based on learned patterns, and regression, which predicts continuous values [2, 3].

Unsupervised learning, on the other hand, deals with unlabeled data and aims to discover inherent patterns, structures, or relationships in the dataset. Popular techniques in the unsupervised learning approach include clustering, which groups similar data points together based on their characteristics, dimensionality reduction, which reduces the number of input variables for better visualization or processing efficiency, and anomaly detection, which identifies rare or abnormal instances in a dataset [4].

Another vital aspect of machine learning is model evaluation and selection. Various evaluation metrics such as precision, accuracy, recall, and F1 score are used to assess the performance of ML models. Cross-validation techniques, including k-fold cross-validation, help estimate the model's generalization performance and prevent overfitting, where the model becomes too specialized to the training data and performs poorly on new data [5].

Machine learning has evolved over time, with deep learning (DL) emerging as a powerful subset of ML-based Artificial Neural Network (ANN). DL utilizes ANNs with multiple layers to learn complex representations from data. Convolutional Neural Networks (CNNs) [6–8] excel in image and video processing tasks, while Recurrent Neural Networks (RNNs) [9–11] are effective in sequence-based data analysis, such as natural language processing.

ML and DL techniques find applications in various domains, including computer vision [12, 13], natural language processing [14, 15], healthcare [16, 17], finance [18, 19], and more. Integrating ML with other technologies like high-performance computing (HPC) further enhances the capabilities and performance of machine learning models. HPC provides the necessary computational resources for large-scale data analysis, training complex models, and accelerating the execution of ML and DL algorithms.

Machine learning incorporates a broad range of concepts and techniques that enable computers to learn from data and make predictions or decisions. The field is rapidly advancing due to ongoing research, technological advancements, and an increasing demand for intelligent systems. By understanding and utilizing the concepts and techniques of ML, researchers and practitioners can unlock valuable insights from data and develop innovative solutions across various domains [20, 21].

2 Deep Learning: Neural Networks and Architectures

Deep learning (DL) is a subset of machine learning that focuses on training Artificial Neural Networks (ANN) with multiple layers to learn complex representations from data [22]. At the core of deep learning are artificial neural networks, inspired by the structure and functionality of the human brain. Perceptron is a first mathematical model of a biological neuron and the building blocks of ANN, introduced in 1957 by Frank Rosenblatt. DL has become a method of choice, and Deep Neural Networks (DNNs) have better accuracy than traditional methods. Neural networks consist of interconnected nodes called neurons, organized in layers. Each neuron receives inputs, applies an activation function, and produces an output that is passed on to the next layer. Neural networks learn by adjusting the weights and biases of these connections, optimizing them to minimize a loss or error function during the training process.

Feedforward neural networks [23] are the simplest type of neural network, where information flows in one direction, from the input layer to the output layer. They are often used for tasks such as regression and classification. Convolutional Neural Networks (CNNs) [24] excel in processing grid-like data, such as images and videos. CNNs utilize convolutional layers to extract spatial features from input data, enabling effective object recognition and image analysis.

Recurrent Neural Networks (RNNs) [10] are designed to handle sequential data, where the order of the input is crucial. RNNs use recurrent connections that allow information to flow in loops, enabling them to capture temporal dependencies in the data. Long Short-Term Memory (LSTM) networks and Gated Recurrent Units (GRUs) are popular RNN variants that address the vanishing gradient problem and improve the modeling of long-term dependencies.

Generative Adversarial Networks (GANs) [25] are a class of deep learning architectures that involve a generator network and discriminator networks working in tandem. GANs are capable of generating realistic synthetic data by learning from a training dataset. The generator network generates samples, while the discriminator network tries to distinguish between real and generated samples. Through iterative training, GANs achieve improved generation quality.

Transfer learning [26, 27] is another important aspect of deep learning, where pre-trained models are utilized as a starting point for new tasks or domains. By leveraging knowledge learned from large-scale datasets, transfer learning reduces the need for extensive training on limited data and accelerates the development of new models.

The improvement in the domain of deep learning can be attributed to the availability of large-scale datasets, advances in computing power, and improvements in optimization algorithms. Graphics Processing Units (GPUs) have significantly accelerated the training and inference processes of deep learning models, allowing for faster and more efficient computations.

Deep learning has demonstrated remarkable performance in various domains, including computer vision [28], speech recognition [29], natural language processing

[30], and reinforcement learning [31]. It has achieved state-of-the-art results in object detection, image classification, machine translation, and game-playing.

Deep learning has revolutionized the field of machine learning by enabling the training of neural networks with multiple layers, leading to the development of powerful models capable of learning complex representations from input data. With advancements in neural network architectures and optimization techniques, deep learning continues to push the boundaries of AI capabilities and finds widespread applications across diverse domains.

3 Parallelism in DL and Distributed Computing

Parallelism and distributed computing play a crucial role in harnessing the computational power of high-performance computing (HPC) systems for machine learning (ML) and deep learning (DL) tasks.

Data Parallelism: Data parallelism strategies basically divide data into parts and then processing using different processors. The objective is to spread tasks across these processors for scalability and speed. For example, determination of entropy using surrogate input signals as shown in Fig. 1. Data parallelism is a common approach in which the dataset is partitioned across multiple compute nodes, and each node processes a subset of the data. The model parameters are shared among the nodes, and the gradients and the loss of data are computed and synchronized during training. This approach enables efficient utilization of compute resources and scales well with the size of the dataset. This approach has also several challenges, such as accuracy degradation with increasing batch size. Techniques such as gradient aggregation and model averaging are used to synchronize the parameters across nodes and ensure consistency.

Model Parallelism: In model parallelism, the ML/DL model is divided among multiple nodes, where each node is responsible for computing specific operations of the model. This method is especially valuable for exceptionally large models that cannot fit into the memory of a single node or when different parts of the model can be processed independently. The division of the model can be done in two ways: horizontal splitting or vertical splitting. Figure 2a, b depicts the differences between intra-operator data and model parallelism respectively. However, this partitioning process is complex and can lead to load imbalance issues, restricting the scalability of the approach [33, 34]. To successfully implement model parallelism, careful design is essential to optimize communication and minimize overhead. To achieve this, various techniques have been proposed, such as layer-wise parallelism and pipeline parallelism, which effectively distribute the computational load across nodes. One such example is the reduction in error rate of image classification using pipeline parallelism shown in Fig. 3.

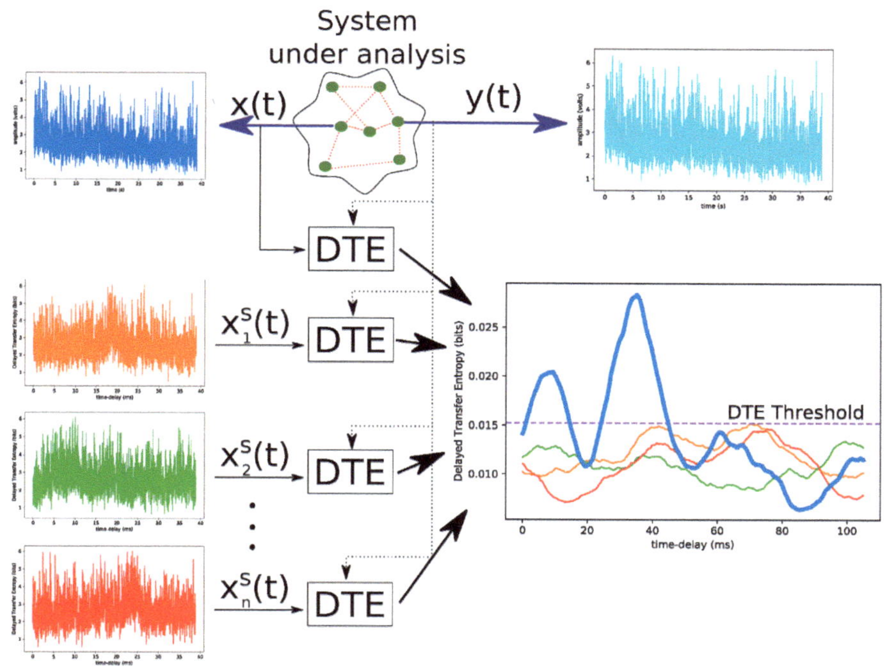

Fig. 1 Data analysis using surrogate input signals and determination of time delayed transfer entropy [32] https://doi.org/10.3390/a12090190

Distributed Training: Distributed training refers to the use of multiple compute nodes to accelerate the training process of ML/DL models. It involves distributing the data and model across nodes, performing parallel computations, and exchanging gradients or model updates. Efficient synchronization and communication protocols are critical to ensure convergence and scalability. Techniques such as parameter servers, bulk synchronous parallel (BSP) models, and asynchronous stochastic gradient descent (ASGD) have been explored to enable distributed training in HPC environments.

Efficient Communication: Communication is a significant factor affecting the performance and scalability of distributed ML/DL in HPC. The frequent exchange of gradients or model updates between nodes can introduce communication overhead. Various techniques have been developed to optimize communication, including quantization, compression, and topology-aware communication strategies. Specialized communication libraries, such as Message Passing Interface (MPI), provide low-level primitives for efficient communication and synchronization.

Task Scheduling and Load Balancing: Efficient task scheduling and load balancing are essential for achieving high performance in distributed ML/DL on HPC systems. Load imbalance, where some nodes are overloaded while others are underutilized,

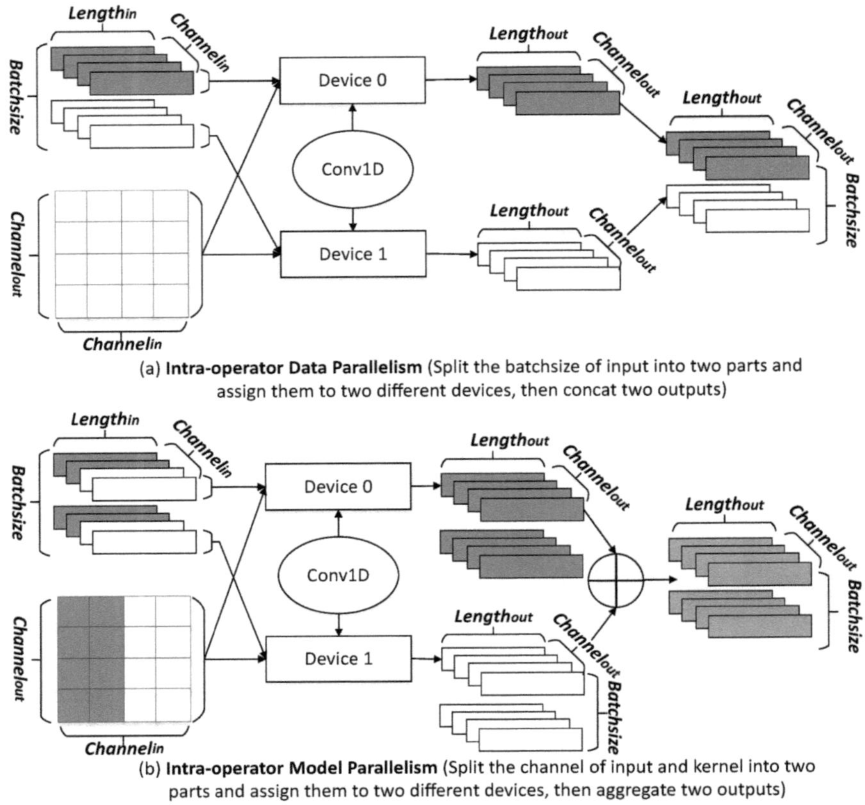

Fig. 2 Data and model parallelism schematics [32]https://doi.org/10.3390/math10244788

can lead to resource wastage and slower overall execution. Techniques such as dynamic load balancing, data-aware scheduling, and task migration aim to distribute the workload evenly across nodes and optimize resource utilization.

Heterogeneous Architectures: HPC systems often comprise heterogeneous architectures, including multi-core CPUs, GPUs, and specialized accelerators. Leveraging these diverse hardware resources efficiently is crucial for achieving high performance in ML/DL. Frameworks and libraries, such as TensorFlow and PyTorch, provide abstractions and optimizations for different hardware platforms. Techniques like hybrid parallelism, where different parts of the ML/DL workflow are executed on suitable devices, enable efficient utilization of heterogeneous architectures.

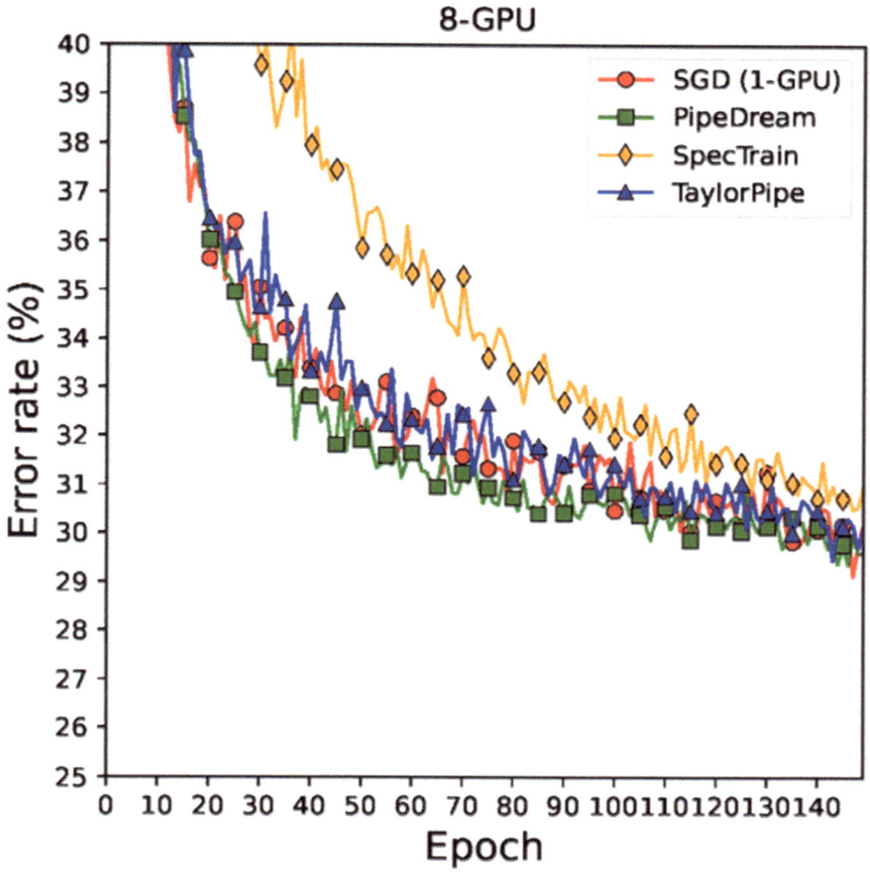

Fig. 3 Image classification error rate using pileline parallelism based algorithm [32] https://doi.org/10.3390/app132111730

4 Training and Inference in ML and DL

Training and inference are two crucial stages in the lifecycle of machine learning (ML) and deep learning (DL) models. Training is the first step where ML and DL models learn from labeled or unlabeled data to capture patterns and make accurate predictions. During training, the model iteratively adjusts its internal parameters, known as weights and biases, based on the observed patterns in the training data. The objective of the correction of internal parameters is to minimize a defined loss or error function, which measures the discrepancy between predicted outputs and ground truth labels.

The optimization of model parameters is typically achieved through optimization algorithms such as gradient-based solvers and their variants. Gradient-based solver computes the gradient of the error function with respect to the model parameters and updates them in a direction that minimizes the loss. Stochastic Gradient Descent (SGD), RMSprop, and Adam are popular optimization algorithms used in training ML and DL models.

Batching and mini-batching techniques [35] are employed to improve training efficiency. Instead of updating the model's internal parameters after each individual training sample, batching involves processing a subset of samples simultaneously. Mini-batching further divides the data into smaller subsets, enabling more frequent parameter updates and reducing memory requirements.

Regularization techniques are employed to prevent overfitting, where the model becomes too specialized to the training data and performs poorly on new, unseen data. Regularization methods such as L1 and L2 regularization, dropout, and early stopping help in controlling model complexity and improving generalization performance [36].

In DL, training deep neural networks with many layers poses challenges such as vanishing and exploding gradients. Techniques like weight initialization, batch normalization, and skip connections mitigate these issues and facilitate effective training of deep networks [37].

Once the model is trained, it enters the inference phase, where it applies the learned parameters to make predictions on new, unseen data. Inference involves feeding the input data through the trained model and obtaining the corresponding output or prediction. The inference process aims to provide accurate and efficient predictions by leveraging the learned representations and patterns.

In DL, inference can be computationally intensive due to the complex architectures and the large number of parameters. To address this, techniques like model compression, quantization, and hardware acceleration (e.g., specialized chips like GPUs and TPUs) are employed to optimize the inference process and improve its efficiency [38].

Both training and inference processes benefit from advancements in hardware and software. High-performance computing (HPC) platforms and distributed computing frameworks enable parallelization and accelerate the training and inference stages. Specialized hardware architectures, such as Graphics Processing Units (GPUs) and Tensor Processing Units (TPUs), provide significant speedup for training and inference in DL.

5 Convergence of ML/DL and HPC

The convergence of machine learning (ML) and deep learning (DL) with high-performance computing (HPC) has emerged as a powerful combination in the field of computational science and data analysis. ML and DL techniques require substantial computational resources for training complex models and processing large-scale datasets. HPC provides the necessary infrastructure, including high-performance computing clusters, distributed computing frameworks, and specialized hardware accelerators, to meet the computational demands of ML and DL workloads.

HPC platforms offer parallel computing capabilities that enable the efficient training and inference of ML and DL models. By utilizing distributed computing frameworks like Apache Spark or TensorFlow distributed, ML/DL algorithms can be scaled to utilize thousands of processors, reducing the training time significantly.

Parallelization techniques, such as data parallelism and model parallelism, play a crucial role in leveraging HPC architectures for ML and DL. Data parallelism involves distributing the training data across multiple compute nodes and updating model parameters in parallel. Model parallelism, on the other hand, divides the model across different compute nodes, allowing each node to process a specific portion of the model.

The integration of ML/DL with HPC has also led to the development of specialized hardware architectures. Graphics Processing Units (GPUs) and Tensor Processing Units (TPUs) are widely used for accelerating ML and DL computations. GPUs excel in parallel processing and can significantly speed up the training and inference processes, while TPUs offer specialized hardware designed specifically for ML workloads.

HPC also facilitates large-scale data processing and analytics required in ML/DL workflows. The ability to efficiently store, retrieve, and pre-process massive datasets is critical for training accurate and robust ML/DL models. HPC storage solutions, such as parallel file systems or distributed object storage systems, provide high-speed data access and enable efficient data processing.

The convergence of ML/DL and HPC has found applications in diverse domains, including scientific simulations, image and signal processing, genomics, natural language processing, and more. ML/DL techniques are employed to analyze and extract insights from complex datasets generated by HPC simulations or experiments.

Challenges related to the convergence of ML/DL and HPC include algorithm design, performance optimization, and resource management. Developing ML/DL algorithms that efficiently utilize distributed computing resources and harness the potential of specialized hardware remains an active area of research.

6 Motivation for Integrating ML/DL with HPC

The motivation for integrating ML/DL with HPC arises from the need for efficient, scalable, and accelerated computation for training complex models, processing large-scale datasets, and extracting valuable insights. The convergence of ML/DL with HPC offers researchers and practitioners a powerful combination to tackle challenging problems, accelerate scientific discoveries, and advance various domains such as healthcare, finance, engineering, and more.

Scalability: ML and DL algorithms often require extensive computational resources to process large-scale datasets and train complex models. HPC provides the scalability necessary to distribute computations across multiple compute nodes, enabling efficient parallel processing and faster model training. By harnessing the computational power of HPC clusters, researchers and practitioners can scale ML/DL workloads to handle massive datasets and accelerate training times.

Accelerated Computation: The convergence of ML/DL with HPC leverages specialized hardware accelerators such as Graphics Processing Units (GPUs) and Tensor Processing Units (TPUs). These accelerators are designed to perform parallel computations efficiently, significantly speeding up the training and inference processes. By offloading computations to GPUs or TPUs, ML/DL models can benefit from their massive parallelism, resulting in faster time-to-solution and improved performance.

Complex Model Training: DL models, with their deep architectures and numerous parameters, often require extended training times on traditional computing systems. HPC platforms offer the computational power needed to train deep neural networks efficiently. Furthermore, HPC provides the ability to perform hyperparameter tuning and model selection in parallel, reducing the time required to find optimal model configurations.

Large-Scale Data Analytics: HPC infrastructure enables efficient storage, retrieval, and processing of massive datasets, which is crucial for ML/DL applications that require analyzing extensive amounts of data. By integrating ML/DL with HPC, researchers can leverage distributed file systems, parallel data processing frameworks, and efficient data transfer mechanisms to handle the storage and processing demands of big data analytics.

Simulation and Experimentation: HPC simulations and experiments generate vast amounts of data that can be effectively analyzed using ML/DL techniques. By combining the power of HPC simulations with ML/DL algorithms, researchers can extract insights from complex data, perform predictive modeling, and optimize simulations by training ML/DL models on simulation outputs.

Insight Extraction: ML/DL can enhance HPC simulations and experiments by uncovering hidden patterns, relationships, or anomalies in the data. ML/DL algorithms can analyze vast datasets generated by HPC simulations, providing valuable insights and helping researchers understand complex phenomena. The integration of ML/DL with HPC enables researchers to extract more meaningful information from simulations and experiments, leading to improved scientific understanding and decision-making.

7 Benefits and Challenges of ML/DL in HPC

Integrating ML/DL with HPC offers numerous benefits in terms of scientific insights, accelerated data analysis, improved modeling, and scalability. However, challenges related to computational resource requirements, algorithm design, data management, and interdisciplinary collaboration need to be addressed to fully leverage the synergies between ML/DL and HPC. Overcoming these challenges will pave the way for transformative advancements and discoveries across various domains.

Benefits:

Enhanced Scientific Insights: ML/DL techniques applied within HPC enable researchers to extract valuable insights from vast amounts of complex data. By leveraging ML/DL algorithms, scientists can uncover hidden patterns, correlations, and anomalies that may be difficult to identify using traditional analysis methods. This leads to a deeper understanding of scientific phenomena, accelerated discoveries, and more informed decision-making.

Accelerated Data Analysis: ML/DL algorithms, when combined with the computational power of HPC, facilitate faster data analysis. HPC platforms can handle large-scale datasets, enabling efficient parallel processing and reducing the time required for complex computations. This acceleration in data analysis enables researchers to analyze and process data more quickly, improving productivity and time to insight.

Improved Prediction and Modeling: ML/DL techniques have demonstrated remarkable capabilities in predictive modeling. By leveraging HPC resources, researchers can train and fine-tune ML/DL models on massive datasets, resulting in improved accuracy and performance. This enables more accurate predictions, simulations, and modeling, which are crucial in fields such as climate modeling, drug discovery, and financial forecasting.

Scalability and Parallelism: HPC systems offer the scalability and parallel computing capabilities required for training and deploying ML/DL models at scale. By leveraging distributed computing frameworks and parallel processing techniques, ML/DL workloads can be efficiently distributed across multiple compute nodes, accelerating training times and enabling real-time inferencing on large datasets.

Challenges:

Computational Resource Requirements: ML/DL models, especially deep neural networks, demand significant computational resources, including memory and processing power. HPC systems need to be equipped with high-performance processors, specialized accelerators (e.g., GPUs or TPUs), and sufficient memory capacity to handle the resource-intensive nature of ML/DL workloads.

Algorithm Design and Optimization: Designing ML/DL algorithms that effectively utilize HPC architectures and distributed computing frameworks remains a challenge. Optimizing algorithms to exploit parallelism, minimizing communication overhead, and reducing resource contention are critical areas of research. Efficient algorithms and optimization techniques are essential to fully harness the power of HPC in ML/DL applications.

Data Management and I/O: Managing and processing large-scale datasets within HPC environments can be complex. Efficient data storage, retrieval, and input/output (I/O) mechanisms are necessary to handle the data-intensive nature of ML/DL workloads. Designing scalable data management systems and optimizing I/O operations are crucial for reducing latency and enabling efficient ML/DL training and inference.

Interdisciplinary Expertise: The convergence of ML/DL and HPC requires interdisciplinary collaboration between domain experts, data scientists, and HPC specialists. Bridging the gap between ML/DL expertise and HPC infrastructure knowledge is essential for effectively integrating ML/DL algorithms into HPC workflows. Collaborative efforts and knowledge exchange are vital to address the challenges and realize the full potential of ML/DL in HPC.

8 Advances in ML/DL for HPC

Machine learning (ML) and deep learning (DL) techniques have seen significant advances in the context of high-performance computing (HPC).

Scalable Distributed Training: Researchers have made substantial progress in developing scalable distributed training techniques for ML/DL in HPC environments. This includes advancements in data parallelism, model parallelism, and hybrid parallelism, allowing efficient training of large-scale models across distributed systems. Techniques such as decentralized training algorithms, elastic scaling, and communication optimizations have been explored to improve scalability and reduce training time.

Optimized Communication Protocols: Communication is a critical aspect of distributed ML/DL for HPC, and recent advances have focused on optimizing communication protocols. Researchers have proposed techniques such as hierarchical and topology-aware communication strategies, reducing the communication overhead and enabling efficient synchronization among distributed nodes. Additionally, advancements in low-latency interconnect technologies, such as InfiniBand and RDMA, have further improved the efficiency of communication in distributed ML/DL systems.

Hardware Acceleration: ML/DL workloads in HPC can benefit from hardware accelerators such as graphics processing units (GPUs) and tensor processing units (TPUs). Recent advances have focused on optimizing ML/DL frameworks and libraries to take advantage of these accelerators. Frameworks such as TensorFlow and PyTorch have introduced support for GPU acceleration, enabling faster computations and improved training times. Additionally, specialized hardware architectures and accelerators designed specifically for ML/DL, such as Google's TPU and NVIDIA's A100 GPU, have further enhanced the performance of ML/DL in HPC.

AutoML and Neural Architecture Search (NAS): AutoML and NAS techniques have gained prominence in optimizing ML/DL models for HPC. These approaches leverage automated search algorithms and optimization techniques to discover optimal architectures and hyperparameters. AutoML methods streamline the model development process and enable researchers to efficiently explore the vast design space of ML/DL models, resulting in improved performance and efficiency in HPC settings.

Federated Learning: Federated learning has emerged as a promising approach for ML/DL in distributed and privacy-sensitive environments. This technique enables training ML/DL models directly on decentralized devices while preserving data privacy. Recent advancements in federated learning have focused on improving communication efficiency, scalability, and security. These advancements have led to the application of federated learning in various domains, including healthcare, the Internet of Things (IoT), and edge computing.

Explainability and Interpretability: As ML/DL models are increasingly being applied to critical domains, the need for model explainability and interpretability has grown. Recent research has focused on developing techniques to understand and interpret the decisions made by ML/DL models. This includes methods for feature importance analysis, model visualization, and generating explanations for ML/DL predictions. Improved explainability and interpretability contribute to better trust and adoption of ML/DL models in HPC applications.

9 Hardware and Software Architectures

The integration of machine learning (ML) and deep learning (DL) with high-performance computing (HPC) requires efficient hardware and software architectures to support the computational demands of ML/DL workloads.

Hardware Architectures:

Graphics Processing Units (GPUs): GPUs have emerged as a popular choice for accelerating ML/DL workloads in HPC. Their parallel processing capabilities and a large number of cores enable significant speedup compared to traditional central processing units (CPUs). GPUs are well-suited for executing highly parallel tasks such as matrix operations, which are fundamental to ML/DL computations.

Tensor Processing Units (TPUs): TPUs, developed by Google, are specialized hardware accelerators designed specifically for ML/DL workloads. TPUs offer high computational power with reduced energy consumption, making them suitable for both training and inference tasks. TPUs excel in executing large-scale matrix operations and are widely used in cloud-based ML/DL platforms.

Field-Programmable Gate Arrays (FPGAs): FPGAs provide hardware flexibility by allowing users to reconfigure the hardware architecture based on specific ML/DL requirements. FPGAs can be customized to accelerate specific computations and achieve higher performance and energy efficiency compared to general-purpose CPUs or GPUs. They offer fine-grained parallelism and can be utilized for both training and inference.

Software Architectures:

Frameworks and Libraries: ML/DL frameworks and libraries play a crucial role in enabling the efficient utilization of hardware resources in HPC. Popular frameworks such as TensorFlow, PyTorch, and Caffe provide abstractions for designing and deploying ML/DL models on various hardware architectures. These frameworks optimize computations and facilitate parallel execution across multiple devices, allowing seamless integration with HPC systems.

Distributed Computing: Distributed computing frameworks enable the execution of ML/DL workloads across multiple nodes in an HPC cluster. Systems like Apache Spark, Horovod, and TensorFlow's Distributed Training provide support for distributed ML/DL, allowing large-scale training and inference tasks to be distributed across multiple compute nodes. These frameworks optimize communication and synchronization to ensure efficient utilization of resources and scalability in HPC environments.

Compiler and Optimizations: ML/DL compilers and optimization techniques aim to improve the performance and efficiency of ML/DL workloads on HPC architectures. Compiler frameworks like XLA (Accelerated Linear Algebra) and TVM (Tensor Virtual Machine) optimize computations and generate efficient code for specific hardware targets. Additionally, techniques such as model quantization, pruning, and low-precision arithmetic help reduce memory usage and improve performance on resource-constrained HPC systems.

Co-design Approaches:

Hardware-Software Co-design: Hardware and software co-design approaches focus on jointly optimizing the design of ML/DL algorithms, models, and the underlying hardware architectures. By considering both aspects together, these approaches aim to maximize performance, energy efficiency, and scalability. Co-design techniques involve developing specialized hardware accelerators, designing algorithms tailored to the hardware architecture, and optimizing communication and memory access patterns to exploit the full potential of ML/DL in HPC.

Neural Architecture Search (NAS): NAS techniques automate the design of neural network architectures by leveraging machine learning algorithms to search for optimal architectures. NAS methods consider hardware constraints and performance metrics, enabling the discovery of network architectures that are well-suited for specific HPC platforms. NAS techniques facilitate the co-design of ML/DL models and hardware architectures, leading to improved performance and efficiency.

10 Conclusion

This chapter has explored the powerful synergy between Machine Learning (ML) and Deep Learning (DL) with High Performance Computing (HPC). It has been revealed how HPC empowers ML and DL by enabling the processing of massive datasets and training complex models in a timely manner. Conversely, ML and DL algorithms provide valuable tools for HPC, optimizing workflows and extracting deeper insights from the vast amount of data generated by advanced simulations and scientific research. Though deep learning gained popularity, both continue to evolve. Their combined potential to tackle ever-more challenging problems in scientific computing for science, engineering, and various other domains will undoubtedly continue to grow.

References

1. Luan, H., Tsai, C.-C.: A review of using machine learning approaches for precision education. Educ. Technol. Soc. **24**(1), 250–266 (2021)
2. Wang, X., Lin, X., Dang, X.: Supervised learning in spiking neural networks: a review of algorithms and evaluations. Neural Netw. **125**, 258–280 (2020)
3. Kadhim, A.I.: Survey on supervised machine learning techniques for automatic text classification. Artif. Intell. Rev. **52**(1), 273–292 (2019)
4. Li, N., Shepperd, M., Guo, Y.: A systematic review of unsupervised learning techniques for software defect prediction. Inf. Softw. Technol. **122**, 106287 (2020)
5. Dhal, P., Azad, C.: A comprehensive survey on feature selection in the various fields of machine learning. Appl. Intell. 1–39 (2022)
6. Li, Z., et al.: A survey of convolutional neural networks: analysis, applications, and prospects. IEEE Trans. Neural Netw. Learn. Syst. (2021)
7. Kiranyaz, S., et al.: 1D convolutional neural networks and applications: a survey. Mech. Syst. Signal Process. **151**, 107398 (2021)
8. Ajit, A., Acharya, K., Samanta, A.: A review of convolutional neural networks. In: 2020 International Conference on Emerging Trends in Information Technology and Engineering (IC-ETITE). IEEE (2020)
9. Pang, Z., Niu, F., O'Neill, Z.: Solar radiation prediction using recurrent neural network and artificial neural network: a case study with comparisons. Renew. Energy **156**, 279–289 (2020)
10. Yu, Y., et al.: A review of recurrent neural networks: LSTM cells and network architectures. Neural Comput. **31**(7), 1235–1270 (2019)
11. Fan, C., et al.: Assessment of deep recurrent neural network-based strategies for short-term building energy predictions. Appl. Energy **236**, 700–710 (2019)
12. Patel, P., Thakkar, A.: The upsurge of deep learning for computer vision applications. Int. J. Electr. Comput. Eng. **10**(1), 538 (2020)
13. Singh, B., et al.: A trade-off between ML and DL techniques in natural language processing. J. Phys. Conf. Ser. (IOP Publishing) **1831**(1) (2021)
14. O'Mahony, N., et al.: Deep learning vs. traditional computer vision. In: Advances in Computer Vision: Proceedings of the 2019 Computer Vision Conference (CVC), Volume 11. Springer International Publishing (2020)
15. Nagarhalli, T.P., et al.: The review of natural language processing applications with emphasis on machine learning implementations. In: 2022 International Conference on Electronics and Renewable Systems (ICEARS). IEEE (2022)

16. Qayyum, A., et al.: Secure and robust machine learning for healthcare: a survey. IEEE Rev. Biomed. Eng. **14**, 156–180 (2020)
17. Kumar, K., Chaudhury, K., Tripathi, S.L.: Future of machine learning (ml) and deep learning (dl) in healthcare monitoring system. In: Machine Learning Algorithms for Signal and Image Processing, pp. 293–313 (2022)
18. Ozbayoglu, A.M., Gudelek, M.U., Sezer, O.B.: Deep learning for financial applications: a survey. Appl. Soft Comput. **93**, 106384 (2020)
19. Ghoddusi, H., Creamer, G.G., Rafizadeh, N.: Machine learning in energy economics and finance: a review. Energy Econ. **81**, 709–727 (2019)
20. Hatcher, W.G., Yu, W.: A survey of deep learning: platforms, applications and emerging research trends. IEEE Access **6**, 24411–24432 (2018)
21. Dong, S., Wang, P., Abbas, K.: A survey on deep learning and its applications. Comput. Sci. Rev. **40**, 100379 (2021)
22. LeCun, Y., Bengio, Y., Hinton, G.: Deep learning. Nature **521**(7553), 436–444 (2015)
23. Sazli, M.H.: A brief review of feed-forward neural networks. In: Communications Faculty of Sciences University of Ankara Series A2-A3 Physical Sciences and Engineering 50.01 (2006)
24. Gu, J., et al.: Recent advances in convolutional neural networks. Pattern Recogn. **77**, 354–377 (2018)
25. Creswell, A., et al.: Generative adversarial networks: an overview. IEEE Signal Process. Mag. **35**(1), 53–65 (2018)
26. Weiss, K., Khoshgoftaar, T.M., Wang, DingDing: A survey of transfer learning. J. Big Data **3**(1), 1–40 (2016)
27. Zhuang, F., et al.: A comprehensive survey on transfer learning. Proc. IEEE **109**(1), 43–76 (2020)
28. He, K., Zhang, X., Ren, S., Sun, J.: Deep residual learning for image recognition. In: Proceedings of the IEEE Conference on Computer Vision and Pattern Recognition, pp. 770–778 (2016)
29. Zoughi, T., Homayounpour, M.M., Deypir, M.: Adaptive windows multiple deep residual networks for speech recognition. Expert Syst. Appl. **139**, 112840 (2020)
30. Shafiq, M., Gu, Z.: Deep residual learning for image recognition: a survey. Appl. Sci. **12**(18), 8972 (2022)
31. Wang, X., Liu, F., Ma, X.: Mixed distortion image enhancement method based on joint of deep residuals learning and reinforcement learning. SIViP **15**, 995–1002 (2021)
32. Njoroge Kahira, A.: Convergence of deep learning and high performance computing: challenges and solutions (2021)
33. Harlap, A., et al.: Pipedream: fast and efficient pipeline parallel DNN training. arXiv:1806.03377 (2018)
34. Chen, C.-C., Yang, C.-L., Cheng, H.-Y.: Efficient and robust parallel DNN training through model parallelism on multi-GPU platform. arXiv:1809.02839 (2018)
35. Jain, P., et al.: Parallelizing stochastic gradient descent for least squares regression: mini-batching, averaging, and model misspecification. J. Mach. Learn. Res. **18** (2018)
36. Bisong, E., Bisong, E.: Regularization for deep learning. In: Building Machine Learning and Deep Learning Models on Google Cloud Platform: A Comprehensive Guide for Beginners, pp. 415–421 (2019)
37. Zagoruyko, S., Komodakis, N.: Diracnets: training very deep neural networks without skip-connections. arXiv:1706.00388 (2017)
38. Cheng, Y., et al.: A survey of model compression and acceleration for deep neural networks. arXiv:1710.09282 (2017)

The Future of High Performance Computing in Biomimetics and Some Challenges

Lanston Pramith Fernandes, Palash Kharate, and Balbir Singh

Abstract The future of high-performance computing (HPC) in biomimetics holds immense promise for revolutionizing the way we draw inspiration from nature to advance technology and solve complex problems. Biomimetics, also known as bio-inspired design, involves emulating biological structures and processes to engineer innovative solutions in fields such as aerospace, materials science, robotics, and medicine. This chapter explores the pivotal role of HPC in the evolution of biomimetics while addressing some of the key challenges on the horizon. HPC provides the computational horsepower needed to simulate intricate biological systems and evaluate their potential applications in technology. From mimicking the aerodynamics of bird flight to designing self-healing materials inspired by biological regenerative processes, HPC enables researchers to model, optimize, and test these concepts efficiently. However, the integration of HPC in biomimetics faces several challenges. Ensuring the accuracy of simulations, handling vast datasets, and aligning computational methods with experimental data are among the complexities. Moreover, bridging the gap between the biological complexity of nature and the computational simplicity of models remains a significant challenge. As the future unfolds, the synergy between HPC and biomimetics promises groundbreaking innovations, but researchers must grapple with these challenges to fully unlock the potential of this interdisciplinary frontier. Addressing these obstacles will be critical for harnessing the transformative power of bio-inspired design in solving real-world problems and advancing technology.

Keywords Biomimetics · High performance Computing · Quantum computing · Computational performance · Computational future

---

L. P. Fernandes · P. Kharate · B. Singh (✉)
Department of Aeronautical and Automobile Engineering, Manipal Institute of Technology, Manipal Academy of Higher Education, Manipal, Karnataka 576104, India
e-mail: balbir.s@manipal.edu

B. Singh
Department of Aerospace Engineering, Faculty of Engineering, Universiti Putra Malaysia, 43400 Serdang, Selangor, Malaysia

K. A. Ahmad et al. (eds.), *High Performance Computing in Biomimetics*, Series in BioEngineering, https://doi.org/10.1007/978-981-97-1017-1_15

1 Introduction

High-performance computing (HPC) exerts a profound and far-reaching impact on various aspects of our daily lives. Its significance reverberates through our world, playing a pivotal role in scientific endeavors by enabling groundbreaking discoveries. Moreover, HPC functions as the linchpin propelling machine learning, especially the intricate domain of deep learning, into the mainstream. In an era awash with immense volumes of data, the absence of robust high-performance computing systems for the rapid and systematic analysis of massive datasets leaves these invaluable data reservoirs largely unexplored and, consequently, underutilized.

Throughout its evolutionary trajectory, high-performance computing has been inextricably tied to the momentum of Moore's law, which has consistently delivered more advanced computing cores. This progression has featured notable technological leaps such as pipelining, speculative execution, out-of-order execution, wider superscalar capabilities, vector engines, and simultaneous multithreading [1]. These technological milestones have driven the HPC field forward.

Subsequently, there was a shift towards multicore processors, with an increasing number of cores integrated into chips [2]. Scaling high-performance computing to larger levels was accomplished through the connection of multiple chips in an intricate hierarchy, including multi-sockets, nodes, racks, and other components, all equipped with distributed memory. This architecture capitalizes on different tiers of parallelism. Initially, a single core exploited instruction-level parallelism [3], where programmers anticipated their code would naturally gain speed with each successive generation of cores without requiring additional effort. However, this era of effortless advancement concluded as we transitioned to multicore processors [4]. Progressing beyond multicore processors to multi-socket nodes and extensive multi-node distributed memory systems necessitates the development of parallel code. When working within a single node where all cores share memory, programmers are required to craft multithreaded applications using tools such as pthreads and OpenMP [5], among others. On the other hand, when operating across nodes with distributed memory, programmers must employ multiple processes by leveraging libraries like MPI [6] and Unified Parallel C (UPC) [7].

Over the past couple of decades, a notable transformation has been underway in the realm of high-performance computing (HPC). A significant development in this space has been the pervasive integration of accelerators into HPC systems. These accelerators have rapidly become instrumental components of the HPC ecosystem, altering the landscape of computation.

Exemplifying this transformation, numerous supercomputing systems featured on the prestigious Top500 list [8] have adopted Graphics Processing Units (GPUs) as integral accelerators. These GPUs, typically known for rendering graphics, have been harnessed to perform an array of computational tasks within HPC environments, significantly amplifying their processing capabilities.

Moreover, the evolution of specialized accelerators designed to cater to specific application types has been striking. For instance, Google's Tensor Processing Units

(TPUs) [9] have been meticulously engineered to cater to machine learning workloads, with a particular emphasis on optimizing the training phase. The advent of these purpose-built accelerators has ushered in a new era of efficiency and performance in machine learning.

Another noteworthy renaissance in accelerator technology has been the resurgence of Field Programmable Gate Arrays (FPGAs). In recent years, FPGAs have witnessed a resurgence, particularly for their prowess in accelerating specialized tasks, such as inference in machine learning. Microsoft's server infrastructure has prominently incorporated FPGAs to bolster acceleration efforts, attesting to the versatility and impact of these devices in HPC [10].

However, it is essential to recognize that to harness the full potential of these accelerators, programmers are compelled to work with specialized programming languages and frameworks. These tools come in varying shades, balancing user-friendliness with the degree of control they grant to the programmer. In the user-friendly category, OpenACC [11] serves as a salient example. It offers a more approachable avenue for leveraging accelerators without necessitating an extensive background in low-level programming. This contrast illustrates the dynamic interplay between accessibility and fine-grained control that characterizes the world of accelerator programming.

In the second category, we encounter a spectrum of programming options tailored for high-performance computing (HPC) that provide varying degrees of user-friendliness and control. Among these, you'll find CUDA [12], OpenCL [13], the latest iterations of OpenMP optimized for GPUs, and hardware description languages like VHDL and Verilog designed specifically for Field Programmable Gate Arrays (FPGAs). As of now, this represents the prevailing paradigm in the field of high-performance computing. However, the landscape is undergoing significant transformations.

Moore's law, which historically underpinned the relentless advance of computational power, is no longer economically sustainable and is poised to cease its progress in the very near future. This impels the HPC community to seek alternative avenues for achieving enhanced performance. The present-day objective within this community is to attain exascale performance, defined as the capacity to execute 10^18 floating-point operations per second (FLOPS) while maintaining power consumption within the range of 20–30 megawatts (MW) per exascale system. The pressing question that arises is how to attain this level of performance, and even surpass it, without the continued support of Moore's law.

Achieving such remarkable performance hinges on three primary factors that impact the operation of HPC systems: computational capabilities, communication infrastructure, and memory management. Notably, among these factors, computational capacity is the most cost-effective to enhance. Consequently, the primary bottlenecks obstructing performance within contemporary systems are the effectiveness of communication channels among various components of the system (including processors, memory, and disk storage) and the efficiency of memory access. To provide a visual reference, Fig. 1 illustrates a generic HPC machine. In the ensuing sections, we will now look into each of these components in greater detail.

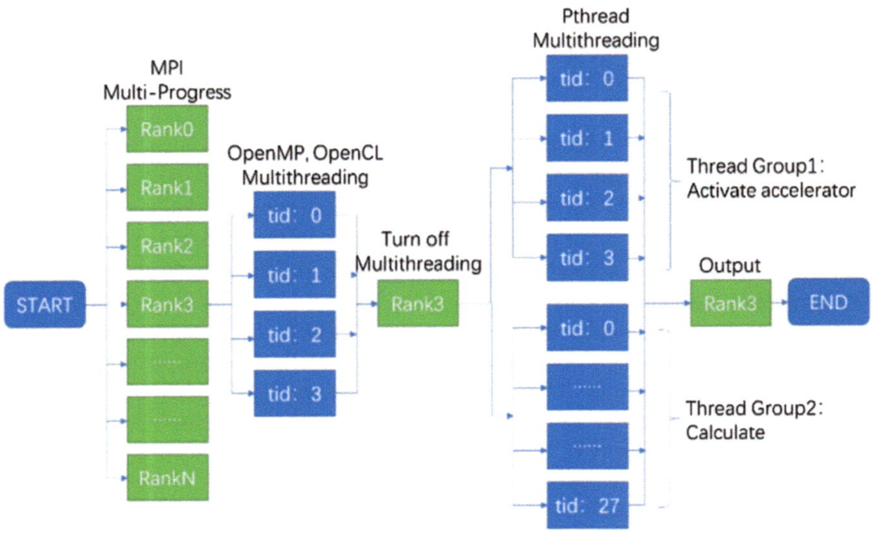

Fig. 1 Design blueprint for a heterogeneous parallel computing system [14]

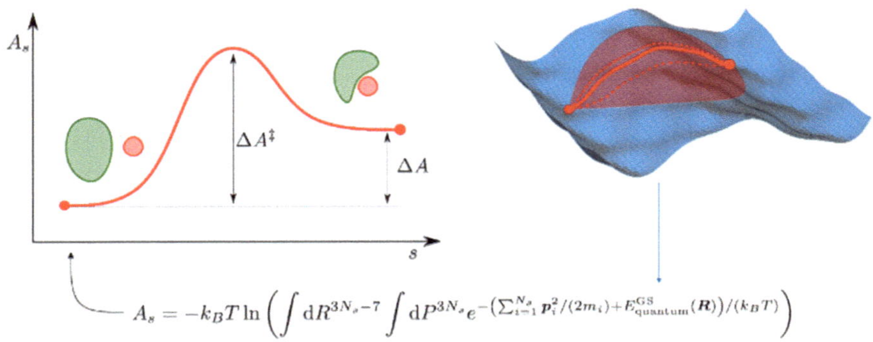

Fig. 2 A visual depiction illustrating the sequence of operations required to compute the potential of mean force for a biochemical process [15]

Furthermore, our discussion extends to encompass the future of software within the realm of HPC systems and explores the evolving role of artificial intelligence (AI) in shaping the future of high-performance computing.

2 Computational Strength

The execution of programs fundamentally revolves around computational operations, constituting the bedrock of computing. However, the less glamorous yet indispensable aspects of computing entail communication and memory access. These facets are the necessary but often less-heralded components that orchestrate the seamless functioning of digital systems.

In the quest for computational supremacy, a pivotal decision that architects and engineers must grapple with is whether to opt for general-purpose processors or tailor-made, application-specific ones. The balance between versatility and specialization becomes a focal point of the design process, setting the trajectory for the system's capabilities.

In the contemporary landscape, the prevailing paradigm involves an intricate tapestry of heterogeneous processing cores within a single computing machine. This diversity within the processing cores reflects the versatility and adaptability of modern computing systems. To provide further insight into this multifaceted domain, here are a few illustrative examples:

- In contemporary computing, general-purpose architectures have embraced the concept of heterogeneity, marking a significant departure from the traditional homogenous core models. A case in point is the introduction of heterogeneous architectures within prominent processors, exemplified by Intel's Alder Lake and Apple's M1 Max. These processors incorporate two distinct types of cores, catering to different computing needs: performance cores for high-speed tasks and efficiency cores for power-conscious operations. This innovative approach aligns with the fundamental principle that not all segments of a program demand rapid execution. By employing this heterogeneous approach, processors can expertly strike a balance between performance and power efficiency, allocating computing resources where they are most needed. This adaptability is emblematic of the contemporary drive for computing systems that can dynamically optimize their resources in response to diverse workloads and energy considerations.
- Within the realm of traditional multicore processors, there is a subtle yet intrinsic heterogeneity even in what may outwardly appear to be identical cores. This subtlety lies in the realm of voltage and frequency scaling, often referred to as Dynamic Voltage and Frequency Scaling (DVFS). The essence of this variation is that each core possesses its own capacity to modulate voltage and frequency based on its specific operational requirements.

As a result, depending on the nature of the software being executed, a given core within a multicore processor can exhibit variations in its performance. In this context, it's essential to recognize that not all cores within the same chip operate at a uniform pace. Instead, the performance of each core dynamically fluctuates in response to the software's demands and the workload it is managing. This inherent heterogeneity, driven by DVFS mechanisms, introduces a layer of adaptability that ensures that each core optimizes its operation to efficiently address the specific tasks assigned to it. Thus, even in what might initially appear to be a uniform array of cores, there exists a nuanced form of heterogeneity driven by the core's autonomous ability to fine-tune its performance characteristics.

- The prevailing trend in contemporary computing is to augment general-purpose chips with application-specific counterparts, effectively harnessing the synergy of both to optimize performance for specific workloads. This strategic amalgamation of general-purpose and specialized processing units aligns seamlessly with the core objectives of high-performance computing (HPC), where tailored solutions can deliver exceptional efficiency for the types of applications that HPC systems are purpose-built to address.

This trend is vividly evident in the computing landscape. Prominent examples include the integration of Graphics Processing Units (GPUs) into the infrastructure of many supercomputers listed in the prestigious Top500 ranking. These GPUs serve as potent accelerators, proficiently handling a wide array of computational tasks and dramatically enhancing overall system performance.

In the domain of artificial intelligence and deep learning, a domain closely intertwined with HPC, companies like Google have adopted Tensor Processing Units (TPUs) within their server architecture to specifically expedite the training phase of deep learning models. Similarly, Microsoft has embarked on an innovative path, incorporating Field Programmable Gate Arrays (FPGAs) into a substantial portion of its server infrastructure. These FPGAs are instrumental in accelerating various computational operations, contributing to enhanced computational throughput and efficiency.

Beyond the realm of supercomputers and enterprise data centers, the reach of application-specific processors extends to portable devices as well. For instance, processors designed for mobile devices, exemplified by the Apple M1 Max, are equipped with embedded GPUs and specialized neural engines. These elements bolster the processing capabilities of these devices, facilitating a range of applications, from gaming to artificial intelligence tasks, and enhancing the overall user experience.

In essence, this evolving trend underscores the importance of a well-balanced computing ecosystem that capitalizes on the unique strengths of both general-purpose and application-specific chips. It provides a dynamic framework for high-performance computing, tailored to address a wide spectrum of computational challenges, from scientific simulations to machine learning and beyond.

The prevailing trajectory in the field of high-performance computing (HPC) is unmistakably characterized by the integration of ever-increasing levels of heterogeneity in computational engines [16]. This strategic shift is not merely a matter of preference but a necessity, particularly as Moore's law appears to have exhausted its traditional route to boosting performance, at least for the foreseeable future. To counter this limitation, the path forward revolves around embracing greater parallelism and heterogeneity, as aptly portrayed in Fig. 1 [17].

The burgeoning significance of specialized chips stems from their intrinsic ability to deliver peak performance with a remarkable economy of power consumption, chiefly tailored to specific families of applications. The essence of their design is precision—to excel in their dedicated tasks, while economizing on power resources. This specialization, however, comes with a trade-off—a pronounced lack of versatility. These chips are inherently constrained when it comes to accommodating a diverse range of applications, often delivering suboptimal performance and consuming excess power in scenarios beyond their specialized domain. For instance, deploying non-data-parallel applications on Graphics Processing Units (GPUs) is counterproductive, as GPUs are not engineered for such tasks, leading to subpar performance and heightened power consumption. This limitation has given rise to a fundamental shift in the design philosophy of HPC systems.

In response to the constraints posed by specialized chips, HPC machines are now meticulously conceived with specific application families, or even multiple families, in mind. This shift enables the tailoring of hardware configurations to align perfectly with the unique demands of a particular category of applications, optimizing performance and power efficiency within that niche.

In practical terms, the prevailing programming model, at least until now, revolves around general-purpose processors initiating the execution of the application and subsequently delegating specific tasks to the specialized chips. This approach underscores the indispensable role of heterogeneity in contemporary HPC systems. However, it is essential to recognize that this arrangement necessitates the movement of substantial volumes of data from storage to memory and subsequently to the various specialized chips. This data transfer inherently underscores the persistent significance of memory, which has remained a critical bottleneck for performance, even during the era of single-core processors [18]. The challenge of efficiently managing this data flow within heterogeneous architectures further underscores the dynamic landscape of high-performance computing, as the demand for increasingly sophisticated memory and data management solutions continues to evolve.

3 Memory and Storage

Memory has long been a persistent bottleneck in computer performance due to the inherent time it takes to access data stored in memory. This delay, referred to as latency, encompasses the time it takes to locate the required data within memory and then transmit it over the memory bus to the last-level cache for immediate processing. Historically, memory hierarchies were constructed using two primary technologies: Dynamic Random-Access Memory (DRAM) for main memory and Static Random-Access Memory (SRAM) for caches. DRAM often served as the last-level cache, sometimes referred to as embedded DRAM or eDRAM, in select design configurations [19]. Notably, DRAM offered higher memory density but was associated with higher latency, while SRAM exhibited precisely the opposite characteristics.

A substantial body of research has been dedicated to optimizing memory hierarchies. This research spans various aspects, ranging from cache management strategies to enhancing cache reliability and achieving more robust cache performance, among others [20]. The efficacy of memory hierarchy access latency is influenced not only by hardware design but also by access patterns. For instance, programming languages like C/C++ store matrices in a row-major fashion. Therefore, if a thread accesses a matrix row by row, this access pattern aligns with cache-friendly behavior, resulting in significantly lower memory access latency compared to accessing the same matrix in a column-major format. Notably, languages like Fortran adopt a column-major array storage scheme. Consequently, programs developed in Fortran must access matrices in a column-major way to optimize memory access latency.

Furthermore, parallel program performance is significantly influenced by the coherence protocol implemented among caches [21]. The introduction of coherence incurs additional bandwidth consumption, heightened latencies in data access, and an increased incidence of cache misses. Programmers can mitigate some of these challenges by adopting strategies that minimize false sharing, a phenomenon related to how threads access memory, including the cache hierarchy bridging the processor and main memory.

However, it's important to note that main memory does not represent the lowest level within the memory hierarchy. Data also traverses between memory and storage (such as hard-disk-drives or HDDs) through comparatively slower interconnections. The transition from traditional HDDs to solid-state-drives (SSDs) has delivered a performance boost. Nevertheless, it's essential to recognize that SSDs still lag behind traditional memory by at least two orders of magnitude in terms of access speed.

These aspects described above represent the current state of affairs. Nevertheless, three pivotal advancements have emerged to enhance memory hierarchy performance. Each of these developments will be explored in greater detail in the subsequent subsections.

A. Non-Volatile Memory

Both DRAM and SRAM are volatile memory technologies. In the past decade, substantial efforts, spanning academia and industry, have been dedicated to exploring alternative technologies that give rise to non-volatile memories (NVMs). These NVMs promise to combine the density of DRAM with the latency attributes of SRAM, albeit with the caveat of additional power consumption for reading and/or writing. Various non-volatile memory technologies have been investigated, including Phase-Change Memory (PCM), Spin-Transfer Torque RAM (STT-RAM), Magnetic RAM (MRAM) [22], and more. NVMs present an enticing proposition for serving as faster main memory, provided that their heightened power requirements, reduced endurance, and the potential for partial memory updates during power fluctuations are managed effectively. In practice, NVMs are often employed in a hybrid memory configuration alongside DRAM to leverage the strengths of both memory types.

B. 3D Stacked Memory

Memory chips have evolved from traditional 2D designs to 2.5D configurations and, most recently, to 3D stacked memory architectures. 3D stacked memory involves multiple memory layers interconnected vertically using through-silicon vias (TSVs). Examples of 3D memory technologies include High-Bandwidth Memory (HBM), which has become a standard interface in contemporary GPUs, and the Hybrid Memory Cube (HMC), a collaborative innovation by Samsung and Micron [23]. 3D-stacked memories provide two key advantages: higher memory density and enhanced bandwidth. These attributes contribute to a reduction in data access latency. Beyond their bandwidth and capacity benefits, 3D-stacked memories open the door to a novel data-centric architectural paradigm.

C. Near Storage Processing

The chief reason for delays experienced by programs when accessing data is the inefficiency of moving data to where processing occurs, particularly in the era of big data. A transformative trend is to invert this conventional approach by moving computation to the data itself. This concept of embedding processing power within memory is not entirely novel; it was proposed as far back as the 1970s [24]. However, at that time, Moore's law offered a more economically viable path to performance improvement, and the necessary technology for integrating processing within memory or in close proximity was lacking.

Today, the conditions are ripe for embracing near-data processing, where computational functions are integrated into one layer of a 3D-stacked memory. Other research endeavors explore embedding rudimentary data-level processing directly within DRAM, capitalizing on the inherent parallelism within the DRAM matrix [24]. Numerous high-performance applications heavily rely on the manipulation of multi-dimensional arrays, a task that can often be accomplished in situ without the need to transport these massive arrays up the memory hierarchy to the primary processor. This heralds an era of data-centric machines, where a substantial portion of processing is executed within memory or in close proximity to storage.

Research initiatives are also exploring the integration of processing power within caches and SSDs [25, 26], further expanding the paradigm of data-centric computing. As Fig. 1 illustrates, memory and storage equipped with embedded processing power are characterized as "smart," in contrast to the "dumb" traditional memory and storage configurations. These advancements collectively represent a dynamic shift toward a more efficient, data-centric approach to computation and memory access.

4 The Function of Artificial Intelligence

Starting around 2012, the term "machine learning" (ML) began its widespread emergence across various domains, becoming a pervasive buzzword that continues to hold sway in contemporary discourse. More recently, "deep learning" (DL), a specialized branch of ML, has made profound inroads into numerous domains, demonstrating remarkable success. This surge in interest and application prompted the architecture community to embark on a journey of intense engagement in providing hardware support for the broader field of artificial intelligence (AI), which primarily revolves around ML. The demand for architecture support for AI has remained unwavering and dynamic, evident in the escalating number of research papers presented at esteemed conferences.

Within this landscape, researchers have been fervently devising and proposing a plethora of hardware designs to enhance the efficiency of ML, spanning both the training and inference phases. These designs manifest in various forms, from neuromorphic chips that painstakingly emulate neurons and their synaptic connections in hardware [27] to dedicated tensor cores seamlessly integrated into Graphics Processing Units (GPUs), custom-designed AI-accelerating chips, innovative microarchitecture techniques engineered to expedite inference processes, and a burgeoning interest that extends down the computing stack to the very circuitry of computing systems [28].

In a parallel vein, another facet of AI-related research seeks to establish a reciprocal relationship between AI and high-performance computing (HPC) architecture. This endeavor aims to harness advances in AI to orchestrate more efficient hardware execution of programs. Here, efficiency is a multifaceted term encompassing diverse objectives like performance optimization, power efficiency, reliability, and more. An enduring dimension of AI support for architecture pertains to the use of optimization techniques for computer-aided design. These techniques range from behavioral analysis to placement and routing, enhancing the adaptability of hardware systems when confronted with new programs based on their performance with other software applications [29].

This multifaceted convergence of ML, AI, and HPC architecture underscores the dynamic and transformative nature of modern computing. The interplay between hardware innovation and AI advancements is a testament to the ever-evolving landscape of computational technology. The fruitful cross-pollination of these fields promises to redefine the future of computing, amplifying its efficiency, capabilities, and adaptability.

5 The Internet of Things IoT and Smart Cities

The Internet of Things (IoT) serves as the interconnection of physical devices, including computers, sensors, and electronics, with the Internet and network infrastructure. This connectivity enables these devices to communicate and share data. A unified IoT platform acts as the converging point for diverse information sources, allowing seamless communication by establishing a common language and framework. According to Gartner's projections [30], the installed base of IoT units is expected to reach a staggering 20.8 billion by 2020. This proliferation of IoT devices generates vast amounts of data, amplifying the significance of addressing critical challenges such as security, customer privacy, efficient data storage management, and the development of data-centric networks.

The concept of smart cities is at the forefront of leveraging IoT technology to enhance urban living by providing innovative services that ensure the seamless functioning of the entire city infrastructure. This approach seeks to improve people's quality of life through the intelligent utilization of data. Smart cities and IoT applications represent two of the burgeoning areas where High-Performance Data Analytics (HPDA) is making a substantial impact. HPC has long been involved in managing critical aspects like power grids, transportation systems, vehicle design, and urban traffic management. Its application continues to evolve and expand, particularly in markets related to cognitive computing, artificial intelligence (AI), autonomous vehicles, and healthcare organizations.

Baz's exploration explains intricate relationship between IoT and HPC, highlighting the challenges and opportunities within smart applications that are integral to a rapidly advancing world. These applications span various domains, including smart building management, smart logistics, and smart manufacturing. HPC-driven solutions hold the potential to address these challenges and unlock opportunities for transformative advancements.

Furthermore, some countries and their ambitious HPC-IoT plan for 2030 underscores the strategic utilization of High-Performance Computing in the management and security of IoT networks. This initiative reflects the pivotal role of HPC in ensuring the wellness of IoT networks, an endeavor with far-reaching implications for the future of connected systems [31]. With the advent of quantum computing and the speed at which it is progressing, the concept of computational speed will completely change in future. Look at the example below in Sect. 6.

6 Role of Quantum Computing for Structural Biology

This case study is driven from ref [15] to understand the impact and application of quantum computing. In future it is possible that there will be an integration of near-term quantum computers with high performance computing systems. So performance wise, this case study shows the quantum simulations for molecular biology. Molecular biology is primarily governed by reversible, non-covalent interactions among biomolecules [15]. While these interactions can be theoretically described using quantum approaches, such an approach is both impractical and unnecessary. It is impractical because the curse of dimensionality makes full atomistic quantum mechanical simulations of large biomolecules unfeasible for any simulation method, whether classical or quantum computing. Furthermore, it is unnecessary because weak interactions can be effectively approximated by classical surrogate potentials, known as force fields [15]. A force field models a molecule's Potential Energy Surface (PES) using classical interactions among its atoms. Covalent bond energies are described by a series involving bond lengths and angles, while non-covalent interactions are expressed as sums of Lennard–Jones potentials. Force field parameters, like bond constants and interaction terms, are often determined by fitting to quantum calculations or experimental data [32–35].

It's crucial to understand that biological processes are intertwined with their environment, operating at room temperature with continuous energy exchange. Within the cellular context, biomolecules, like proteins, exist not as fixed structures but as dynamic ensembles of configurations, each influenced by thermodynamic factors. Hence, rather than focusing on the energy of individual configurations, the free energy derived from the partition function, accounting for all possible states, is more relevant. This approach grants access to various thermodynamic properties. Consequently, in analyzing reactivity within a thermodynamic ensemble, the Potential Energy Surface (PES) isn't ideal; the driving force of a reaction lies in the free energy [15, 36, 37] as shown in Fig. 2.

The potential of mean force, a generalization of the Potential Energy Surface (PES) for thermal systems, typically incorporates electronic energies. These electronic energies play a significant role in calculating the partition function and various thermodynamic properties. While quantum algorithms have been proposed to compute partition functions on quantum hardware, extending these techniques to classical thermal averages remains underexplored. Some implementations of quantum Monte Carlo algorithms exist, suggesting potential for computing thermal averages via sampling techniques. However, these algorithms face challenges in mapping classical probability distributions onto qubit states [15, 38–40].

To tackle the protein folding problem, simulating a protein's dynamic behavior under thermal equilibrium conditions is effective. Over time, a protein naturally adopts its most stable conformation, solving the folding problem. Classical simulations are adept at this task, as they aren't limited by dimensionality constraints and can explore larger biomolecules. Yet, the potential contribution of quantum computing to this field remains an open question [15, 32, 41, 42].

An approach to merge quantum and classical methods involves calibrating a force field using energies from quantum computations. For example, refining a force field for protein molecular interactions based on precise quantum simulations of smaller polypeptides is a viable strategy [15]. As quantum computing progresses, its capacity to handle larger molecules may enhance the precision of force field parametrization. Furthermore, fault-tolerant quantum computers could offer new avenues for exploring these challenges in the future.

Additionally, quantum computing could be applied to classical simulations that rely on a given force field. For instance, problems like protein folding, which can be reformulated as an optimization problem, could benefit from quantum computing. The stable structure of a protein corresponds to the structure with the lowest free energy, making it an optimization challenge. Classical solutions to this problem are known to be complex and computationally demanding. Quantum computing has the potential to offer more efficient solutions to these optimization problems [43, 44].

The intersection of quantum computing and classical simulations represents a rich area for exploration, offering the possibility of significant advancements in molecular biology and computational chemistry.

7 The Future of HPC in Biomimetics

The future of HPC in biomimetics is very promising. HPC can be used to accelerate and improve the development of new biomimetic materials, devices, and systems, as well as to better understand and optimize existing biomimetic designs.

One of the key trends in the future of HPC in biomimetics is the use of machine learning. Machine learning can be used to analyse large datasets of biomimetic data, such as images of plant and animal surfaces. This can help identify new biomimetic patterns and designs, and develop new algorithms for biomimetic design.

Another key trend in the future of HPC in biomimetics is the development of new biomimetic algorithms for solving complex problems in various fields, such as engineering, medicine, and finance. For example, researchers are using biomimetic algorithms to design more efficient algorithms for routing traffic, scheduling jobs, and optimizing financial portfolios.

The integration of HPC and biomimetics holds great promise for addressing complex challenges and driving innovation across a wide range of applications. This multidisciplinary approach, drawing inspiration from nature and leveraging the power of high-performance computing, is poised to revolutionize the way we approach problem-solving and design in the coming years. However, addressing challenges related to computational resources, interdisciplinary collaboration, data integration, ethics, and education will be crucial for fully realizing this potential.

The future of high-performance computing (HPC) in biomimetics holds significant promise, with several key trends and developments on the horizon:

- Machine Learning and AI Integration: HPC will increasingly integrate machine learning and artificial intelligence (AI) to improve biomimetic research. These technologies will assist in analysing complex biological data, recognizing patterns, and optimizing biomimetic designs more effectively. Machine learning algorithms can help identify new biomimetic solutions and refine existing ones.
- Advanced Simulation Techniques: HPC will continue to evolve to support more advanced and detailed simulations of biological processes and natural systems. This will enable researchers to gain deeper insights into the behaviour of biological materials, tissues, and ecosystems, leading to more accurate biomimetic designs.
- Cross-Disciplinary Collaboration: The future of HPC in biomimetics will emphasize even greater collaboration across diverse fields, including biology, engineering, materials science, computer science, and environmental science. This multidisciplinary approach will foster more creative problem-solving and drive innovation.
- Quantum Computing in Biomimetics: As quantum computing technology matures, it holds the potential to revolutionize biomimetics. Quantum computers can handle complex simulations and data analysis at speeds unimaginable for classical computers. This could open up new frontiers in biomimetic research and problem-solving.
- Sustainability and Environmental Applications: The future of biomimetics will increasingly focus on sustainability and environmental conservation. HPC will

be used to model and simulate natural systems to develop eco-friendly technologies and practices, addressing pressing environmental challenges such as climate change and resource depletion.
- Healthcare and Medicine: HPC will continue to advance biomimetic solutions in healthcare, with applications ranging from the development of patient-specific medical treatments to the creation of highly realistic organ and tissue simulations. This will improve patient care and enable personalized medicine.
- Optimization Algorithms: Biomimetic algorithms will be further developed and optimized for various fields, including engineering, transportation, and finance. These algorithms will draw inspiration from natural processes to solve complex problems efficiently.
- Education and Workforce Development: The future of HPC in biomimetics will require a skilled workforce with a deep understanding of both computational techniques and biological systems. Educational programs will adapt to provide specialized training, fostering the next generation of biomimetic researchers and engineers.
- Ethical Considerations: As biomimetic technologies advance, ethical considerations surrounding the use of biological materials and technologies will become increasingly important. Researchers and policymakers will need to address these ethical issues to ensure responsible development and application of biomimetics.
- International Collaboration: Global collaboration and the sharing of resources and knowledge will become more critical in advancing biomimetics through HPC. International partnerships will accelerate progress and ensure the broadest possible impact.

8 Challenges in Harnessing HPC for Biomimetics

Harnessing high-performance computing (HPC) for biomimetics comes with several challenges, despite its significant potential. These challenges must be addressed to fully realize the benefits of using HPC in this multidisciplinary field:

- Computational Resources: Access to powerful HPC resources can be expensive and limited, especially for smaller research teams and organizations. The cost of acquiring and maintaining HPC infrastructure, including supercomputers and specialized software, can be a significant barrier.
- Interdisciplinary Collaboration: Biomimetics inherently requires collaboration between researchers from diverse fields, such as biology, engineering, materials science, and computer science. Bridging the gap between these disciplines and effectively integrating their knowledge is often challenging, as each discipline has its own jargon and methodologies.
- Data Complexity and Integration: Biological data is often complex and diverse, coming from various sources and in different formats. Integrating and analysing this data coherently can be a significant challenge, as it often requires specialized tools and expertise for data integration and preprocessing.

- Ethical and Regulatory Considerations: As biomimetic research advances, ethical considerations regarding the use of biological materials and technologies may become more complex. Regulatory frameworks may need to adapt to address these emerging challenges, especially when biological materials or animal models are involved.
- Education and Training: Developing a skilled workforce capable of leveraging HPC for biomimetic research is essential. Traditional education and training programs may not provide researchers with the necessary computational skills required for this multidisciplinary field. This gap needs to be filled through specialized training programs and interdisciplinary educational initiatives.
- Scaling and Optimization: Biomimetic simulations often require significant computational power and parallel processing capabilities. Scaling algorithms and simulations to leverage the full potential of HPC systems while ensuring efficient resource utilization can be a complex task.
- Integration of Machine Learning: Integrating machine learning and artificial intelligence into biomimetic research introduces new challenges related to data labelling, model training, and the interpretability of results. Additionally, it may require substantial computational resources to train and deploy machine learning models effectively.
- Data Security and Privacy: Biomimetic research may involve the collection and analysis of sensitive biological data. Researchers must address issues related to data security, privacy, and compliance with data protection regulations.
- Cross-Domain Knowledge Transfer: Translating biological knowledge into actionable design principles often requires cross-domain knowledge transfer, which can be challenging. It involves understanding the relevant aspects of biology and engineering to create biomimetic solutions effectively.
- Validation and Real-World Implementation: Validating biomimetic designs and ensuring their real-world implementation can be complex. Bridging the gap between simulated models and practical applications requires extensive testing, validation, and iterative refinement.

Addressing these challenges will be crucial for harnessing HPC for biomimetics effectively. Collaboration, education, technological advancements, and a clear ethical framework will play key roles in overcoming these obstacles and advancing biomimetic research and innovation.

References

1. Tullsen, D.M., Eggers, S., Levy, H.M.: Simultaneous multithreading: maximizing on-chip parallelism. In: Proceedings of the 22th International Symposium on Computer Architecture (1995)
2. Nayfeh, B.A.: The Case for a Single-Chip Multiprocessor. PhD thesis, Stanford University (1998)

3. Zahran, M., Franklin, M.: Hierarchical multi-threading for exploiting parallelism at multiple granularities. In: Proceedings of the 5th Workshop on Multithreaded Execution, Architecture and Compilation (MTEAC-5), pp 35–42 (2001)
4. Sutter, H.: The free lunch is over: A fundamental turn toward concurrency in software. Dr. Dobb's J. **30** (2005)
5. der Pas, R.V., Stotzer, E., Terboven, C.: Using OpenMP—The Next Step. The MIT Press (2017)
6. Pacheco, P.: An Introduction to Parallel Programming. Elsevier Morgan Kaufmann (2011)
7. Cantonnet, F., Yao, Y., Zahran, M., El-Ghazawi, T.: Productivity analysis of the upc language. In: 3rd International Workshop on Performance Modeling, Evaluation, and Optimization of Parallel and Distributed Systems (PMEO-PDS), pp. 254– (2004)
8. Khan, A., Sim, H., Vazhkudai, S.S., Butt, A.R., Kim, Y.: An analysis of system balance and architectural trends based on top500 supercomputers. In: The International Conference on High Performance Computing in Asia-Pacific Region, HPC Asia 2021 (New York, NY, USA), pp. 11–22. Association for Computing Machinery (2021)
9. Jouppi, N.P., Yoon, N.P., Ashcraft, M., Gottscho, M., Jablin, T.B., Kurian, G., Laudon, J., Li, S., Ma, P., Ma, X., Norrie, T., Patil, N., Prasad, S., Young, C., Zhou, Z., Patterson, D.: Ten Lessons from Three Generations Shaped Google's TPUv4i, pp. 1–14. IEEE Press (2021)
10. Damiani, A., Fiscaletti, G., Bacis, M., Brondolin, R., Santambrogio, M.D.: Blastfunction: a full-stack framework bringing fpga hardware acceleration to cloud-native applications. ACM Trans. Reconfigurable Technol. Syst. **15** (2022)
11. Chandrasekaran, S., Juckeland, G.: OpenACC for Programmers: Concepts and Strategies, 1st edn. Addison-Wesley Professional (2017)
12. Sanders, J., Kandrot, E.: CUDA by Example: An Introduction to General-Purpose GPU Programming, 1st edn. Addison-Wesley Professional (2010)
13. Banger, R., Bhattacharyya, K.: OpenCL Programming by Example. Packt Publishing (2013)
14. Qi, Y., Li, Q., Zhao, Z., Zhang, J., Gao, L., Yuan, W., Lu, Z., Nie, N., Shang, X. and Tao, S.: Heterogeneous parallel implementation of large-scale numerical simulation of saint-Venant equations. Appl. Sci. **12**(11), 5671 (2022). https://doi.org/10.3390/app12115671
15. Baiardi, A., Christandl, M., Reiher, M.: Quantum computing for molecular biology. ChemBioChem **24**(13), e202300120 (2023). https://doi.org/10.1002/cbic.202300120
16. Zahran, M.: Heterogeneous Computing: Hardware and Software Perspectives. Association for Computing Machinery, New York, NY, USA (2019)
17. Leiserson, C.E., Thompson, N.C., Emer, J.S., Kuszmaul, B.C., Lampson, B., Sanchez, D., Schardl, T.B.: There is plenty of room at the top. Science **368**, 1–7 (2020)
18. McKee, S.A.: Reflections on the memory wall. In: Proceedings of the 1st Conference on Computing Frontiers, CF'04, (New York, NY, USA), p. 162. Association for Computing Machinery (2004)
19. Park, Y.-H., Yoo, H.-J., Kook, J.: Embedded dram (edram) powerenergy estimation for system-on-a-chip (soc) applications. In: Proceedings of the 2002 Asia and South Pacific Design Automation (2002)
20. Yang, H., Govindarajan, R., Gao, G.R., Hu, Z.: Improving power efficiency with compiler-assisted cache replacement. J. Embedded Comput. **1**, 487–499 (2005)
21. Archibald, J., Baer, J.-L.: Cache coherence protocols: evaluation using a multiprocessor simulation model. ACM Trans. Comput. Syst. (1986)
22. Das, J., Scott, K., Bhanja, S.: Mram puf: using geometric and resistive variations in mram cells. J. Emerg. Technol. Comput. Syst. **13** (2016)
23. Ibrahim, K.Z., Fatollahi-Fard, F., Donofrio, D., Shalf, J.: Characterizing the performance of hybrid memory cube using apexmap application probes. In: Proceedings of the Second International Symposium on Memory Systems, MEMSYS'16, (New York, NY, USA), pp. 429–436. Association for Computing Machinery (2016)
24. Peskin, A.M.: A logic-in-memory architecture for large-scaleintegration technologies. In: Proceedings of the ACM Annual Conference—Volume 1, ACM'72, (New York, NY, USA), pp. 12–25. Association for Computing Machinery (1972)

25. Fujiki, D., Wang, X., Subramaniyan, A., Das, R.: In-/Near-Memory Computing. Synthesis Lectures on Computer Architecture. Morgan & Claypool Publishers (2021)
26. Wang, J., Park, D., Kee, Y.-S., Papakonstantinou, Y., Swanson, S.: Ssd in-storage computing for list intersection. In: Proceedings of the 12th International Workshop on Data Management on New Hardware, DaMoN'16, (New York, NY, USA). Association for Computing Machinery (2016)
27. Modha, D.S., Ananthanarayanan, R., Esser, S.K., Ndirango, A., Sherbondy, A.J., Singh, R.: Cognitive computing. Commun. ACM **54**, 62–71 (2011)
28. Sayyaparaju, S., Chakma, G., Amer, S., Rose, G.S.: Circuit techniques for online learning of memristive synapses in cmos-memristor neuromorphic systems. In: Proceedings of the on Great Lakes Symposium on VLSI 2017, GLSVLSI'17, (New York, NY, USA), pp. 479–482. ACM (2017)
29. Quackenbush, C., Zahran, M.: Beyond profiling: Scaling profiling data usage to multiple applications. CoRR, vol. abs/1711.01654 (2017)
30. Egham: Gartner Says 8.4 Billion Connected "Things"; Will Be in Use in 2017, Up 31 Percent From 2016. http://www.gartner.com/newsroom/id/3598917
31. Conway, S.: High Performance Data Analysis (HPDA): HPC—Big Data Convergence—insideHPC (2017)
32. Kirkwood, J.G.: J. Chem. Phys. **3**, 300 (1935)
33. Lewis, A.M., Fay, T.P., Manolopoulos, D.E.: J. Chem. Phys. **145**, 244101 (2016)
34. Lindoy, L.P., Fay, T.P., Manolopoulos, D.E.: J. Chem. Phys. **152**, 164107 (2020)
35. Wang, B.-X., Tao, M.-J., Ai, Q., Xin, T., Lambert, N., Ruan, D., Cheng, Y.-C., Nori, F., Deng, F.-G., Long, G.-L.: NPJ Quant. Inf. **4**, 52 (2018)
36. Potočnik, A., Bargerbos, A., Schröder, F.A.Y.N., Khan, S.A., Collodo, M.C., Gasparinetti, S., Salathé, Y., Creatore, C., Eichler, C., Türeci, H.E., Chin, A.W., Wallraff, A.: Nat. Commun. **9**, 904 (2018)
37. Scott, W.R.P., Hünenberger, P.H., Tironi, I.G., Mark, A.E., Billeter, S.R., Fennen, J., Torda, A.E., Huber, T., Krüger, P., van Gunsteren, W.F.: J. Phys. Chem. A **103**, 3596 (1999)
38. Wang, J., Cieplak, P., Kollman, P.A.: J. Comput. Chem. **21**, 1049 (2000)
39. Ponder, J.W., Case, D.A.: Force fields for protein simulations. Adv. Protein Chem. **66**, 27–85 (2003)
40. Nerenberg, P.S., Head-Gordon, T.: Curr. Opin. Struct. Biol. **49**, 129 (2018)
41. Motta, M., Sun, C., Tan, A.T.K., O'Rourke, M.J., Ye, E., Minnich, A.J., Brandão, F.G.S.L., Chan, G.K.-L.: Nat. Phys. **16**, 205 (2020)
42. Temme, K., Osborne, T.J., Vollbrecht, K.G., Poulin, D., Verstraete, F.: Nature **471**, 87 (2011)
43. Frenkel, D., Smit, B.: Statistical mechanics. In: Understanding Molecular Simulation, pp. 9–22. Elsevier (2002)
44. Micheletti, C., Hauke, P., Faccioli, P.: Phys. Rev. Lett. **127**, 080501 (2021)

MIX
Papier aus verantwortungsvollen Quellen
Paper from responsible sources
FSC® C105338

If you have any concerns about our products,
you can contact us on
ProductSafety@springernature.com

In case Publisher is established outside the EU,
the EU authorized representative is:
Springer Nature Customer Service Center GmbH
Europaplatz 3, 69115 Heidelberg, Germany

Printed by Libri Plureos GmbH
in Hamburg, Germany